A DISTANT LIGHT
SCIENTISTS AND PUBLIC POLICY

Masters of Modern Physics

Advisory Board

Dale Corson, Cornell University
Samuel Devons, Columbia University
Sidney Drell, Stanford Linear Accelerator Center
Herman Feshbach, Massachusetts Institute of Technology
Marvin Goldberger, University of California, Los Angeles
Wolfgang Panofsky, Stanford Linear Accelerator Center
William Press, Harvard University

Published Volumes

The Road from Los Alamos by Hans A. Bethe
The Charm of Physics by Sheldon L. Glashow
Citizen Scientist by Frank von Hippel
Visit to a Small Universe by Virginia Trimble
Nuclear Reactions: Science and Trans-Science by Alvin M. Weinberg
In the Shadow of the Bomb: Physics and Arms Control
    by Sydney D. Drell
The Eye of Heaven: Ptolemy, Copernicus, and Kepler
    by Owen Gingerich
Particles and Policy by Wolfgang K.H. Panofsky
At Home in the Universe by John A. Wheeler
Cosmic Enigmas by Joseph Silk
Nothing Is Too Wonderful to Be True by Philip Morrison
Arms and the Physicist by Herbert F. York
Confessions of a Technophile by Lewis M. Branscomb
Making Waves by Charles H. Townes
Of One Mind: The Collectivization of Science by John Ziman
Einstein, History, and Other Passions by Gerald Holton
Bright Galaxies, Dark Matters by Vera Rubin
Atomic Histories by Rudolf E. Peierls
Essays of a Soviet Scientist by Vitaliĭ I. Gol'danskiĭ
This Gifted Age: Science and Technology at the Millenium
    by John H. Gibbons
A Distant Light: Scientists and Public Policy
    by Henry W. Kendall

A DISTANT LIGHT
SCIENTISTS AND PUBLIC POLICY

Henry W. Kendall

Foreword by
Howard Ris

Springer

Henry W. Kendall (deceased)
Department of Physics
Massachusetts Institute of Technology
Cambridge, MA 02139-4307
USA

Library of Congress Cataloging-in-Publication Data
Kendall, Henry Way, 1926–
   A distant light : scientists and public policy / Henry W. Kendall.
     p.   cm.—(Masters of modern physics)
   Includes bibliographical references and index.
   ISBN 978-1-4612-6423-1   ISBN 978-1-4419-8507-1 (eBook)
   DOI 10.1007/978-1-4419-8507-1
   1. Science and state—United States. 2. Union of Concerned
Scientists. I. Title. II. Series.
Q127.U6K46   1999   99-20533
338.97306—dc21

Printed on acid-free paper.

© 2000 Springer Science+Business Media New York
Originally published by Springer-Verlag New York, Inc. in 2000
Softcover reprint of the hardcover 1st edition 2000

All rights reserved. This work may not be translated or copied in whole or in part without the written permission of the publisher Springer Science+Business Media, LLC, except for brief excerpts in connection with reviews or scholarly analysis. Use in connection with any form of information storage and retrieval, electronic adaptation, computer software, or by similar or dissimilar methodology now known or hereafter developed is forbidden. The use of general descriptive names, trade names, trademarks, etc., in this publication, even if the former are not especially identified, is not to be taken as a sign that such names, as understood by the Trade Marks and Merchandise Marks Act, may accordingly be used freely by anyone.

Production coordinated by Matrix Publishing Services, Inc., and managed by Timothy Taylor; manufacturing supervised by Joe Quatela.
Typeset by Matrix Publishing Services, Inc., York, PA.

9 8 7 6 5 4 3 2 1

ISBN 978-1-4612-6423-1   SPIN 10721030

*He was born with the gift of laughter
and a sense that the world was mad.*

Opening line, *Scaramouche*, Raphael Sabatini

The West Peak of Nevado Huandoy in the Cordillers Blanca, Peru. Photograph by the author.

FOREWORD

Henry Kendall reached the end of his life unexpectedly in February 1999. He died while scuba diving in the company of a National Geographic Society expedition in Wakulla Springs, Florida, doing what he loved best: exploring the world's great natural treasures.

At the time of Henry's death, he had not yet completed the task of assembling and proofing the final manuscript for this book. I volunteered for that task, having worked closely with Henry at the Union of Concerned Scientists for the last eighteen years and having provided some help in reviewing early drafts of the introductory material that he prepared for each section of the book. Working on this book reminded me yet again of the staggering breadth and depth of Henry's contributions to a great many fields, ranging from physics to the environment to national security. He was a truly remarkable human being; I was privileged to be his friend and colleague.

I know that Henry very much wanted this book to serve not just as a summary or memoir of his work, but as a guide and inspiration for other scientists whose participation in the public policy process is badly needed. I hope any changes or corrections I have made will serve that purpose well, and I am grateful to the staff at Springer-Verlag and WordCrafters Editorial Services for remaining steadfast in their desire to see this book published.

Howard Ris
Executive Director
Union of Concerned Scientists
Cambridge, Massachusetts

ABOUT THE SERIES

Masters of Modern Physics introduces the work and thought of some of the most celebrated physicists of our day. These collected essays offer a panoramic tour of the way science works, how it affects our lives, and what it means to those who practice it. Authors report from the horizons of modern research, provide engaging sketches of friends and colleagues, and reflect on the social, economic, and political consequences of the scientific and technical enterprise.

Authors have been selected for their contributions to science and for their keen ability to communicate to the general reader—often with wit, frequently in fine literary style. All have been honored by their peers and most have been prominent in shaping debates in science, technology, and public policy. Some have achieved distinction in social and cultural spheres outside the laboratory.

Many essays are drawn from popular and scientific magazines, newspapers, and journals. Still others—written for the series or drawn from notes for other occasions—appear for the first time. Authors have provided introductions and, where appropriate, annotations. Once selected for inclusion, the essays are carefully edited and updated so that each volume emerges as a finely shaped work.

Masters of Modern Physics is overseen by an advisory panel of distinguished physicists. Sponsored by the American Institute of Physics, a consortium of major physics societies, the series serves as an authoritative survey of the people and ideas that have shaped twentieth-century science and society.

PREFACE

Like many scientists, I have always had interests far different from those of my professional life. Among them has been the challenges of dealing with the array of problems humanity faces, including how to exploit the earth's bounty without damaging it and how to deal with the conflicts between nations, for which extraordinarily destructive weapons stand at the ready. At issue are not just the unwanted side effects of a few technologies but the cumulative impact of a host of technologies and human activities that appear to threaten humanity and the global environment in a deep way.

For many years such matters—what I call survival issues—have claimed much of my time and attention, providing extensive experience in a variety of public controversies, most involving substantial scientific or technical matters. This volunteer work in the quasipolitical world of public policy debate has proved vastly different from my professional work as an experimental particle physicist. It was my good fortune to find a second home base—the Union of Concerned Scientists—whose course I helped guide and from which I could research issues, publish my findings, and involve myself and others in the effort to shape public policy.

In 1992 the American Institute of Physics' publishing division approached me with an invitation to prepare a volume for its series *Masters of Physics*. A typical volume in this series is based mainly on the author's principal scientific publications. In the course of preparing some material for the World Bank on public policy problems, a different approach seemed appropriate in my case: to devote the proposed volume primarily to material dealing with science and public policy problems. The book could serve as a stimulus and a resource for others in the scientific community who might also wish to turn their attention to such concerns. The scientific community has many fine minds that can advance understanding of public problems and nudge society toward sensible policies. The key is getting enough scientists involved in sustained and constructive ways.

This focus would require a very different mix of material than would a collection of physics reprints. It would also include written matter that customarily does not appear on publication lists, such as news releases, public statements, and declarations, for example, that can be of great importance to those engaged in debates over public policy; they frequently require very careful writing. Based on this approach, the form and selection of material for the volume took its present form: a volume intended to communicate the nature of the controversies that swirl about some of humanity's great problems. My challenge was to illuminate the behavior both of allies and opponents, to set out the lessons that I, with numerous colleagues, have learned, and, in some small way, to make a contribution that could provide useful guidance for other scientists.

The arrangement of the book's material is very roughly chronological. The Introduction concerns public controversies and includes material regarding the Union of Concerned Scientists (UCS), a public interest organization to which I have devoted much time and effort. Nuclear power and the major controversy that was associated with reactor emergency systems is next, followed by a section on the only physics topic included, a technical article and a popular article on experimental studies of the internal structures of the proton and neutron that I helped to guide. This is followed, in turn, by sections on national security issues, global environmental problems, radioactive contamination of the US nuclear weapons production facilities and, before the Epilogue, articles concerned with burning oil wells and antipersonnel mine clearing.

The illustrations, aside from those that accompany the reprinted articles, are drawn, for the most part, from an exhibit of my pictures, *Arctic and Expeditionary Photographs*, at the MIT Museum in 1992.

Henry W. Kendall
Cambridge, Massachusetts

CONTENTS

Foreword   vii
About the Series   ix
Preface   xi

INTRODUCTION   1

BEYOND MARCH 4   9

PART ONE   NUCLEAR POWER   13

    Press Release: Nuclear Power   19

    Nuclear Reactor Safety: An Evaluation of New Evidence   21

    Cooling Water   37

    Nuclear Safety   49

    The Failure of Nuclear Power   64

PART TWO   SCIENCE   69

    Structure of the Proton and the Neutron   71

    Deep Inelastic Scattering: Experiments on the Proton and the Observation of Scaling   94

PART THREE   NATIONAL SECURITY AND NUCLEAR WEAPONS   127

    News Release: Announcing Declaration on the Nuclear Arms Race   133

Declaration on the Nuclear Arms Race	135
Spaced-Based Ballistic Missile Defense	139
The Fallacy of Star Wars	159
Accidental Nuclear War	165
Dismantle Armageddon	177

PART FOUR GLOBAL ENVIRONMENTAL AND RESOURCE ISSUES — 179

Appeal by American Scientists to Prevent Global Warming	183
Speech to the United Nations Intergovernmental Negotiating Committee	189
Press Release: Announcing World Scientists' Warning to Humanity	193
World Scientists' Warning to Humanity	198
Constraints on the Expansion of the Global Food Supply	202
Meeting Population, Environment and Resource Challenges: The Costs of Inaction	224
Bioengineering of Crops	236

PART FIVE CLEANUP OF RADIOACTIVE CONTAMINATION — 269

Alternative Futures for the Department of Energy National Laboratories	271

PART SIX CURRENT ISSUES — 283

Quenching the Wild Wells of Kuwait	285
Post-Conflict Mine Clearing	296

EPILOGUE — 303

INDEX — 305

INTRODUCTION

The world's problems are pressing in on us all. The scale and impact of human activities now affects a great portion of the global resources important to human welfare. These activities are putting growing, often destructive pressure on the global environment, pressure that appears likely to increase as human numbers swell toward the doubling of the world's population that evidently lies ahead. These pressures can spawn or aggravate conflict that, in a world with so much destructive weaponry, generates important national security problems. Great changes are necessary to help ensure a humane future for the generations to come.

Most of the world's scientific community and many people in the environmental movement are aware of the gravity of the problems. Yet despite sober warnings from these and other groups, in both the industrial and the developing world, remedial efforts frequently appear powerless and ineffectual. The scientific community has not taken a sustained, powerful role in the public arena where this great cluster of issues is debated and where the problems must be resolved. There is much more that our community can contribute to assessment, warning, and proposals for new patterns of behavior.

This volume is mainly directed to those scientists who do or may wish to engage these or similar challenges. The scientific community has much to offer aside from its purely technical skills. The public generally holds it in high regard and, given the right circumstances, listens to its voice on controversial matters. But if scientists do not speak out, then significant opportunities are lost.

Some in the scientific community have always had an interest in major public problems and challenges and, from time to time, have devoted quite substantial effort to them, sometimes joining in public debate or working quietly for the government or other organization. Occasionally, some striking need stimulates a great response. The enormous participation of scientists in the U.S. and British effort during World War II is perhaps the most outstanding example. From this wartime involvement came several postwar groups: Pugwash, the Federation of American Scientists, and the *Bulletin of the Atomic Scientists* are organizations that have provided support for sustained activities of scientists searching for control of nuclear weapons. More recently, the risk of climatic disruption from the greenhouse effect has become a central issue. A global consortium of scientists, the Intergovernmental Panel on Climate Change, with over 2000 members, has been actively assessing the prospects. It has been a major player in an occasionally unpleasant public debate.

Much greater participation is still needed, however, to deal with a host of unsustainable environmental, resource, and population activities as well as conflict. As with any environmentally destructive activity, they must, at some point, end. The scientific community's help is required to ensure a smooth and well-designed transition to less damaging behavior and to generate efficient repair of the damage already inflicted. There is special need for more specialists from the life sciences, as so many of the adverse pressures affect the plant and animal world, and for scientists who can understand and illuminate the links between the great problems we face.

Union of Concerned Scientists

The major portion of my activities in the public domain have been spent in collaboration with colleagues in the Union of Concerned Scientists (UCS). UCS began as an informal organization of faculty and staff at the Massachusetts Institute of Technology (MIT) in 1969, spawned by the unrest on college campuses mainly arising from the war in Southeast Asia but driven also by the nuclear arms race, growing environmental concerns, the women's movement, and the pressure for civil rights.[1] A strike against MIT had been proposed by a radical student group but there was so much sympathy among the faculty for public discussion of the concerns in these areas that no strike was necessary: the institution closed down for several days, starting on March 4, 1969, for reappraisal and discussion. UCS was the faculty group's vehicle, and through it many colleagues organized and executed the week's activities. Kurt Gottfried, a theoretical physicist and close friend since graduate school days in the early 1950s, was one of the central group that organized and launched UCS. Other theoretical physicists from MIT's department included Herman Feshbach, Francis Low, and Victor F. Weisskopf.

The principal issues that the Union of Concerned Scientists focused on at its start in 1969 were military: the antiballistic missile debate, chemical and biological weapons, and the war in Vietnam. I had been a consultant to the Department of Defense, through membership in the Jason Panel of the Institute for Defense Analyses. The panel's membership was drawn from university scientists who were willing to devote substantial time and effort to a great range of national security matters, including nuclear weapons, ballistic missiles and missile defense, radar problems, undersea warfare, and the like. Owing to this work, I chose to help organize activities related to environmental and resource problems. It was not for lack of interest in the national security challenges but rather from the complications that ensue from an attempt to operate in the public domain while holding a security clearance. This is possible to accomplish, and it can be very valuable, but it was not until many years after this period that I felt experienced enough to be confident that no breach of secrecy would occur.

It was the environmental arm of UCS that persisted and ultimately flourished over the years; other activities dwindled and, within about two years of UCS's start, disappeared, as volunteers returned to their responsibilities on campus and elsewhere.

In the early 1970s UCS was transformed into an independent public interest organization, no longer associated with MIT.[2] At this writing, it is an organization with a staff of about 50 persons, offices in three locations, an annual budget exceeding $6 million, and close to 60,000 supporters drawn from the public and the scientific community. Kurt is a member, and I am chairman of its Board of Directors. We are, along with Victor Weisskopf and Jay Fay, the last people remaining from the organization's formative days.

The brief piece that follows is a statement written by Kurt and me for the activities of the Union of Concerned Scientists on March 4, the day in 1969 when UCS held its first public gathering. The organization's present array of programs and its modes of operation have a thrust that is remarkably close to that outlined in this statement. It remains a good mission statement.

Public Issues

The nature of the issues that are important to public well-being can vary over a great range, giving rise to great variation in the strategies for engaging in public debate and for reaching settlement or solution.

Global environmental and resource challenges fall broadly into two categories. The first consists of the activities that produce prompt and palpable harm or serious injury to large segments of the public. These generate political forces that have the strength to bring the problems under control. Contaminated water supplies, fouled air, Love canals, *Silent Spring*—these are what spawned the powerful environmental movement here and abroad and enabled the passage of corrective legislation.

Thus, in many important respects, the environmental movement is a great success. Environmental concerns are now deeply embedded in the consciousness of our society: they are in our bones, so to speak, and are irreversible. But the movement's limited focus on activities that have already hurt us is a serious defect.

A second set of activities consists of those whose damaging consequences may not be apparent until well into the future; because they have not yet captured society's attention, they do not induce society to correct them. Some of these problems are rather well known, such as global warming and consequent possible climatic disruption and population growth, the near-doubling of human numbers that will probably occur by about the middle of the twenty-first century. Others are nearly invisible, except to experts, including gravely unsustainable practices in agriculture and numerous energy-related activities. It also includes species loss, deforestation, and reckless use of fresh water resources. The nuclear arms race for many years fell in this category.

All these activities, visible and invisible alike, share the feature that there are no powerful forces driving their remediation. Some are examples of "the tragedy of the commons": they have narrow, short-term benefits to individuals or selected groups but impose broad long-term damage to the community, the nation, or the globe. Others can involve benefit to the public or to a vested interest from continuing hurtful activity but with future costs inadequately discounted. Overfishing, depletion of water supplies, deforestation, and fossil fuel use are current examples. Swelling populations, including that of the United States, are in this group. Solutions can be disruptive and costly. An irate, injured, or scared community with sufficient political clout may not exist. This makes solutions very hard to achieve.

The global problems comprise a disparate array; they have a thousand faces. Approaches to dealing with them also have many faces, so the ways the controversies proceed and how they are best managed vary greatly. There are, however, some common elements and common behavior patterns.

Common Threads

Many issues have in common a particularly vexing element stemming from the broad ignorance and misunderstanding of science and technology of the general public and many in public life. While science and technology play critical roles in sustaining modern civilization, they are not part of our culture in the sense that they are not commonly studied or well comprehended. Neither the potential nor the limitations of science are understood so that what can be achieved and what is beyond reach are not comprehended. The line between science and magic becomes blurred so that public judgments on technical issues can be erratic or badly flawed.[3] It frequently appears that some people will believe almost anything.[4] Thus judgments can be manipulated or warped by unscrupulous groups. Distortions or outright falsehoods can come to be accepted as fact.

Nowhere is misunderstanding and misuse of science more important than in controversies with important scientific or technical elements. They have plagued the nuclear power debate, the Star Wars controversy, and numerous environmental issues. The embrace of bad science, what some call "junk science," is not confined to purveyors of miscast projects or to the general public, either. Many in the environmental and peace movements, poorly equipped for technical debate, succumb to illusory ideas and unrealistic solutions. This can badly cloud the search for practical solutions, for junk science can be very difficult to dislodge.

Distortion and false statements have a sturdy history in public discourse. Neither the government nor large organizations can be depended on to support their objectives honestly and with integrity. Replying in kind turns out not to be an option, not just to retain scientific integrity but for practical reasons. Critics, whether individuals or public interest groups, cannot afford to slant the truth, ever. Scientists are far more vulnerable to the consequences of their own ill-considered words than are laypeople, owing to the care and integrity that is believed to characterize the scientific approach to problems. Intentional distortions are almost always uncovered and the purveyors pilloried without mercy. It may not be forgotten for years and surfaces over and over again. So too will honest mistakes which, along with even minor exaggerations, are seized on and exploited mercilessly. Not a bad rule—one that I and some colleagues observe—is to pull back a bit in most argument. Not only should one never distort nor exaggerate, it is best, I believe, to understate. If the lethal range of some toxic effect is arguably 150 miles, one can set it out as 120 miles, for there is then recourse if challenged. And so forth.

A related useful approach to public argument, especially helpful for scientists, is to focus on the factual aspects of a case, not assessments and opinion. An audience that reaches your conclusions on their own, based on the facts presented, is more supportive than when asked to agree with you based on your say-so. Present "facts, not adjectives," as I was told by William Shurcliff, a physicist whose public skills were deployed with great success in opposition to the supersonic transport proposal of the early 1970s.

Another common thread is failure. Almost no controversy is resolved favorably either as quickly or as completely as one might wish. Indeed, many drag on indefinitely or worsen as one works on them. It can be frustrating, demoralizing, and, not infrequently, destructive. Failure is at the root of the paranoia that is endemic in parts of the environmental and peace communities and that can lead to unfortunate consequences. As with many other aspects of public controversies, there is no certain cure. But recognizing and dealing with failure in stable, thoughtful ways is an important part of participation in these activities, for failure is far more common than success.

Finally, there is the role chance plays in public matters. One can never know whether some wholly unexpected event will occur that proves to be either a major obstacle or a heaven-sent opportunity. The public arena is full of surprises, and an issue can become ripe with astonishing speed or can disappear with little warning. No one was expecting the Three Mile Island accident, Chernobyl, the ozone hole in the stratosphere, or the public reaction to "winnable nuclear war." A nimble, flexible mode of operation, ready to exploit an opening or change course to deal with a new difficulty, is essential. There are several examples in the material that follows.

Working with the Media

Keeping an issue in the media spotlight is essential to keeping it on the policy-making agenda. This is particularly true of issues that do not have high salience with voters. Because our opponents, whether the fossil fuel industry or defense contractors, have an advantage from political contributions and high-level lobbying, it is necessary to amplify a message through the press. Understanding how to work with the media is therefore a must.

Finding a reporter in search of news linked to a story generated by an environmental, resource, or or national defense issue that hinges on technical or scientific elements can be difficult. Reporters are rarely scientists, and it is often difficult for them to learn a subject well enough to tell good science from bad and, sometimes, good sources from bad. According to a report from a Freedom Forum Environmental Journalism Summit, "the beat puts reporters at odds with scientists who mine arcane detail, public interest groups that push political agendas, and businesses that push economic ones."

Complicating the picture is the fact that journalists are trained to be "balanced" in their coverage. But achieving balance is often more difficult for reporters covering science matters than it is for those who cover, say, politics. In politics, news sources express their own opinions or values, few of which have more authority than those of others. In science, however, some views are more valid than others because they represent the scientific community's consensus on what is known. One should not "balance" an argument based on Newton's laws of motion against those that contradict them. This can sometimes put "balance" at odds with another important journalists goal: accuracy.

Science and environmental journalists feel obliged to provide equal weight to opposing points of view. Most are sincerely struggling with the challenge of how to accurately describe a scientific debate and give correct weight to different scientific opinions.

Another factor that complicates such reporting is the nature of scientific "truth." Rarely is certainty perfect; residual uncertainty at some level accompanies most scientific matters of public importance. Given this uncertainty, many scientists are understandably reluctant to make the kinds of unequivocal statements demanded by the media—the "sound bite"—or they are uncomfortable drawing conclusions about policy from their data. Without the time or training to appreciate the subtleties of the debate or to track closely the dialogue among the scientists themselves, reporters can end up feeling confused and, possibly, misled. Creating good, crisp sound bites that do not distort the truth is a knack that is important to develop, for the media lives off them. It is necessary to exert great control over the scientist's tendency to "explain it all," for a garrulous source of news is an ineffective source. As an environmental reporter from the *Washington Post* would say to her potential sources, "Speak English and be honest."

Nothing helps more than building a relationship with a reporter based on mutual respect, mutual understanding, and common interests. As a rule, reporters work hard not to take sides in policy debates. However, scientists and science journalists do share a common interest in fostering public understanding of science. The two groups also share the need to continually verify and confirm. These common interests can form the basis for building a relationship with a reporter who faces the challenge of communicating complicated science questions to his or her audience.

Many scientists I know are reluctant to talk to the press or other media because "they twist the story" and it comes out slanted and wrong. It is certainly true that much scientific reporting is not entirely accurate but, in close to 30 years of dealing with the press, I have never known a reporter to deliberately twist what I had to say. Getting it wrong is hardly unknown, but this nearly always arises from innocent causes, especially poor communication. Editorial boards are another matter. I have seen numerous examples of judgments and editorial columns made from distorted rendering of material that I or colleagues had set out, and in many cases the distortion was quite clearly deliberate.

Dealing with persistently adverse editors can be a tricky matter. First, there is the old dictum that one should not offend anyone who buys ink by the gallon. The best approach appears to be the low-key, factual response with the hope that it will prove effective. Not infrequently, nothing can be done directly and one can win only by placing material elsewhere.

References and Notes

1. Jonathan Allen, editor, *March 4—Scientists, Students and Society* (MIT Press, Cambridge, 1970). This volume recounts the issues and discussions during the activities sponsored by UCS in its first public presentations.
2. For an account of the early years of UCS, see Brian Balogh, *Chain Reaction: Expert Debate and Public Participation in American Commercial Nuclear Power, 1945–1975* (Cambridge University Press, Cambridge and New York, 1991).
3. In this regard, see especially the two chapters "What Place for Science in Our Culture at the 'End of the Modern Era'?" and "The Public Image of Science" in Gerald Holton, *Einstein, History, and Other Passions* (American Institute of Physics Press, Woodbury, NY, 1995).
4. A 1997 Gallup poll found that 42% of American college graduates believe that flying saucers have visited Earth in some form, whereas 20 years ago, a Roper Center survey found that 30% of college graduates believed in unidentified flying objects (UFOs) (reported in the *San Francisco Chronicle*, June 14, 1997). This example is only one of numerous others that have developed in astrology, creationism and other areas.

Beyond March 4

> ... the Stone Age may return on the gleaming wings of Science, and what might now shower immeasurable material blessings upon mankind, may even bring about its total destruction. Beware, I say; time may be short.
> —Winston Churchill

We are immersed in one of the most significant revolutions in man's history. The force that drives this revolution is not social dissension or political ideology, but relentless exploitations of scientific knowledge. There is no prospect that this revolution will subside; on the contrary, it will continue to transform profoundly our modes of living and dying. That many of these transformations have been immeasurably beneficial goes without saying. But, as with all revolutions, the technological revolution has released destructive forces, and our society has failed to cope with them. Thus we have become addicted to an irrational and perilous arms race, and we are unable to protect our natural environment from destruction.

Can the American scientific community—all those who study, teach, apply or create scientific knowledge—help to develop effective political control of the technological revolution? Our proposal is based on the conservative working hypothesis that the existing political system and legal tradition have sufficient powers of adaptation to eventually contain and undo the destruction and peril generated by technological change.

At first sight recent history appears to indicate that our democratic system of government will fail the test just described. Let us recognize, however, that an intricate pattern of political and technological developments have conspired to subvert profoundly the democratic process. The government that we see today is in many respects no longer democratic because the vast bulk of its constituency cannot begin to scrutinize some of the gravest issues. One can claim that this has always been

---

Union of Concerned Scientists, 1968.

the case, but in so doing one tacitly ignores the radically altered relationship between the government and the citizenry. For we must remember that before the advent of nuclear weapons, government decision could only on rare occasion threaten the existence of any large portion of mankind. Now such decisions are a common occurrence, and they pass virtually unnoticed. This is not simply due to the habitual apathy of the electorate. A more important cause is surely the shroud of secrecy that enfolds so much of the government's operations. Today many tens of billions of dollars pass through the hands of classified government programs and agencies, and hundreds of thousands of Americans hold security clearances. Both of these figures have grown enormously in recent decades.

The technological revolution tends to erode democracy even in the absence of secrecy. The vastly increased importance and complexity of technology has, in effect, increased the ignorance of the public and its elected representatives, and thereby concentrated power in the administration and the military. This trend has been greatly amplified by external threats, both real and imagined. In the face of these developments the Congress has largely surrendered its constitutional duties.

The scientific community has various responsibilities, most of which it has ably discharged. Thus it has created the basic knowledge and developed the applications that make the continuing technological revolution possible; it has trained an ample supply of technical manpower, and it has advised the administrative and military branches of the government.

Our community has the additional responsibility to educate the public, to evaluate the long term social consequences of its endeavor, and to provide guidance in the formation of relevant public policy. This is a role it has largely failed to fulfill and it can only do so if it enters the political arena.

Only the scientific community can provide a comprehensive and searching evaluation of the capabilities and implications of advanced military technologies. Only the scientific community can estimate the long-term global impact of an industrialized society on our environment. Only the scientific community can attempt to forecast the technology that will surely emerge from the current revolution in the fundamentals of biology.

The scientific community must meet the great challenges implied by its unique capacity to provide these insights and predictions. It must engage effectively in planning for the future of mankind, a future free of deprivation and fear. This important endeavor, in which we seek your active participation, stems from our conviction that even though the technological revolution has greatly benefited mankind, it has also released destructive forces that our society has failed to control.

Far-reaching political decisions involving substantial applications of technology are made with virtually no popular participation. It is our belief that a strengthening of the democratic process would lead to a more humane exploitation of scientific and technical knowledge, and to a reduction of the very real threats to the survival of mankind.

We ask you, as a member of the scientific community, to join us in a concerted and continuing effort to influence public policy in areas where your own scientific knowledge and skill can play a significant role. The issues which are of primary concern to us are "survival problems"—where misapplication of technology literally threatens our continued existence. These fall into three categories: those that are upon us, as is the nuclear arms race; those that are imminent, as are pollution-induced climatic and ecological changes; and those that lie beyond the horizon, as, for example, genetic manipulation.

In order to affect public policy in the domain of concern to us we propose to:

1. Initiate a continuing critique of governmental policy—based, if necessary, on our own research—which hopefully would lead to new alternatives where appropriate;
2. Institute a broad program of public education wherein the membership would disseminate information to the news media and a variety of institution and organizations: schools, church groups, community clubs, unions, and so forth;
3. Establish contacts with and support appropriate political figures, locally and nationally;
4. Encourage universities to establish interdisciplinary educational programs on the frontiers between science and public affairs;
5. Maintain liaison and establish coordinated programs with politically active engineers and scientists throughout the nation and world;
6. Investigate innovative forms of political action.

PART ONE
NUCLEAR POWER

The controversy concerning the safety of nuclear power reactors has required more of my effort and time and has lasted longer than any other I have been involved in. It started in 1971, when UCS was two years old, with a membership that had declined from the several hundred of its formative year to only about a dozen, working as volunteers on local and state-level environmental and health issues. The technical issues are the subjects of the papers reprinted in this part, but some additional background is helpful to understand how the controversy unfolded.[1]

Daniel F. Ford, a recent Harvard graduate in economics, had become interested in the economics of nuclear power and, in the course of his studies, became aware of some troubling safety issues that environmental groups had raised. Boston Edison was organizing to build two nuclear plants in Plymouth, south of Boston, and he learned that no one had moved to intervene in the hearings preliminary to the Atomic Energy Commission's (AEC) issuance of a construction permit to the electric utility, despite the existence of a host of unanswered safety questions. He came to UCS with a request for technical assistance in the preparation of a petition to raise and get answers to these questions. We moved to help him and, as the safety issues became clearer, joined with him as an intervener group.

Several of us in UCS at the time became interested in the project and started to gather other, more knowledgeable people into it. Ian Forbes, a graduate student in nuclear engineering at MIT, was one of the earliest. Another was David Comey, head of an organization in Chicago, Businessmen in the Public Interest. He had been involved in nuclear safety issues for several years and was able to outline for us hints of very serious weaknesses in the claims for reactor safety, especially government safety tests that had failed unexpectedly and whose results had been suppressed by the AEC's Reactor Division. He also led us to a study of accident consequences, also suppressed. Even these partial glimpses of the difficulties were startling and led several of us to join the effort.

Once started, we were soon able to acquire far more information through colleagues in the physics community than had been available to environmental groups or to David Comey. We discovered great vulnerabilities in the emergency systems required in all nuclear power plants. In a July 1971 news conference, we released our first study on the consequences of a large reactor accident and the weaknesses of reactor safety systems. A second report followed in October, criticizing the AEC's safety criteria, as, at the same time, we were preparing to introduce the issues in the Pilgrim construction permit hearing.

Because our material was new to the then-small public debate on reactor safety, it was widely reported in the press and so gained considerable attention. The safety issues we were documenting were quickly raised by intervenors in nuclear power plant construction and licensing hearings at a number of sites in the United States. So that the same safety matters would not be contested in numerous duplicate hearings, we and the AEC agreed that the issues would be pulled out of all local hearings and consolidated in a single rule-making hearing in Washington. Its principal subject was the claimed effectiveness of the emergency core cooling systems (ECCS) that were required on all nuclear power reactors to prevent catastrophic consequences should certain plant mishaps occur.

The ECCS hearings started in the spring of 1972, planned and expected by the AEC to occupy no more than six weeks or so. In preparing for the hearing, we very quickly discovered that we had uncovered a hornet's nest. The AEC had engaged in a far more extensive program of suppression of disconcerting safety information than anyone had ever imagined, had censored safety-related information, had pressured their own researchers to keep quiet on key issues, and was sitting on a mass of disquieting research results. In some cases, commission officials made public statements that were contrary to statements they had made in their internal records or reports. It appeared that the most serious deficiencies were in the assertions that accident risks were satisfactorily low.

In the hearing, Ford, who had the instincts and the skills of a fine lawyer, although without formal legal training, carried out extensive cross examination both of friendly and of hostile nuclear safety experts and managed, indeed stimulated, a flow of safety documents from whistle-blowers in the AEC laboratories that had never been destined to see the light of day. These activities attracted great press interest and served to spread knowledge of the safety deficiencies throughout the environmental community.

My part was to digest the intricacies of the safety debate, help prepare our technical testimony, and defend it against attack by the 17 lawyers representing the electric utilities, reactor manufacturers, and the AEC who were participants in the hearing as well as having chosen the board who conducted the hearings and sat in judgment. The hearings failed to end in six weeks; instead they ran on for nearly two years, producing 20,000 pages of transcript of oral testimony and another 20,000 pages of documents of record. I was on the witness stand five days a week for nearly a month, which must be close to a record for this sort of thing. With support from nuclear experts, some known only to us or wholly anonymous, we were able to withstand the numerous attempts to discredit us and our case. Nevertheless, the length and tension involved proved to be very wearing. A sadder consequence of the hearings was that the careers of a number of whistle blowers from the National Laboratories, who were identified during the hearings, were ruined by the AEC.

While the direct result of the ECCS hearings was at best a minor improvement in reactor safety, Ford's work, combined with our written testimony, proved a major embarrassment for nuclear power. The testimony he elicited and the safety documents that were released were extraordinarily damaging to the AEC and contributed to the breakup of that agency by the Congress in January 1975.

In an attempt to recover credibility in the wake of the broad disclosures of safety lapses in the ECCS hearings, the AEC launched a major reactor safety study in 1972, intended to demonstrate that the safety risks were, in fact, infinitesimally small. A major enterprise, it was released in draft form in 1974 and in final form, designated as WASH1400, in 1975.[2] This got the AEC into further difficulty at the hands of UCS and others, for it proved to be unacceptably slanted. I organized a UCS study group that prepared a review of its weaknesses.[3] The consequence was to force the Nuclear Regulatory Commission (NRC), the successor agency to the AEC, to withdraw the report's principal conclusions, a further embarrassment.

The mandate given the AEC in its enabling legislation—both to regulate and to expand nuclear power—was a key to the extensive failures that followed. Bureaucracies appear to be fundamentally incapable of setting and maintaining a balance between opposing mandates, as these clearly were.

In its early years, the AEC performed its regulatory function with little public visibility, yet it was sowing the seeds for later trouble; it started its reliance on industry self-regulation. By the middle of the 1960s, the agency had come to focus its energies on promotion, having become a captive of the industry it was intended to regulate. Regulation was given a lower priority and safety research was constrained, just as serious safety matters started to surface. The agency and the nuclear industry had a lot of bad luck as well. Neglect, incompetence, narrow vision, and unexpected technical troubles interacted synergistically. Pressures on their own research personnel and the unresolved safety problems became a minefield, exploited by UCS as well-hidden matters slowly came to light.

In 1975 an interesting series of events began that made UCS the lead nuclear power critic and watchdog in the nation, a role it has not abandoned. Robert Pollard, an NRC staff safety engineer, approached Dan Ford with the wish to leave the Commission and come to work for UCS. He had been in the nuclear navy, with assignments on nuclear submarines, had later earned a nuclear engineering degree, and, among other responsibilities with the NRC, was chief safety reviewer of the Indian Point III nuclear power plant. He had come to believe that regulation of the industry was inadequate to protect the public and wished to make his views, and the supporting evidence, widely known.

Dan Ford had earlier asked the widely viewed TV show *60 Minutes* to cover safety deficiencies but found that the producers required a focus before tackling the issues. He concluded that the resignation of a government safety officer on the air would be the "peg" needed for the producers to move. That proved correct, and a powerful show aired in February 1976. Quite accidentally, the week before Dan had gone to the West Coast at my urging to stage manage the resignation of three nuclear engineers from General Electric's Reactor Division in San Jose, California, whose departures from the industry were also impelled by safety concerns. That had proved quite a success: it became the number one national story in the Sunday *New York Times* Week in Review and was picked up by the press across the country. The country had been alerted to safety problems in the industry.

Giving *60 Minutes* an exclusive story helped the resignation gain even greater national coverage—"Government not doing its job"—with resulting major press attention. It was a front-page story in the *New York Times* for three days in a row. Three years later, at the time of the Three Mile Island nuclear plant accident, *60 Minutes* rebroadcast the segment.

Pollard did come to work for UCS and retired in 1996 after 20 years with the organization. He became a uniquely powerful nuclear critic owing to his patience with the press, the fact that he always had documents available to substantiate his conclusions, his links to numerous other whistle blowers within the NRC and the industry, and his uncanny ability to ferret out important safety problems. His major contributions over those years were made possible by UCS's capacity to offer him a job and to protect him from the retribution that is otherwise visited on whistle blowers. Shortly after his retirement, UCS was able to fill the vacancy with a capable replacement so that our nuclear watchdog role would not end.

In the years since the ECCS issue was center stage, other troubles with nuclear power have arisen. Indeed, the industry never seems able to shake off its difficulties.[4] In 1975 there was the near-miss fire at Brown's Ferry nuclear plant, in which all of the plant's emergency systems were disabled. Four years later, there was the near-catastrophic accident at Three Mile Island. That crisis gripped the country and UCS membership soared. Dan Ford's experience in unearthing hidden nuclear secrets was exploited again as he delved into the background and development of what remains the country's worse nuclear accident.[5]

In 1975, we prepared a declaration on nuclear power, which was released on the thirtieth anniversary of the atomic attack on Hiroshima. It was signed by more than 2000 scientists, including some of the most senior in the country. Among the signers was James Bryant Conant, who had been chair of the National Defense Research Committee that guided the United States during World War II, a member of the Manhattan Project Steering Committee, a member of the General Advisory Committee of the AEC, and had received the Atomic Pioneer's Award from President Nixon. The declaration and a portion of the accompanying news release are reprinted in this part.

Since the Three Mile Island accident, the economics of nuclear power have continued to worsen, leading to the closure of many plants. Wall Street lost confidence and faith in the technology, and deregulation of electric generating facilities has made competition so intense that now nuclear power's survival is in question.[6] Freeman Dyson[7] has pointed out that nuclear power, as an ideologically driven technology, was not allowed to fail. When it commenced to fail, rules were rewritten, costs were omitted from the accounting, and the incipient failure was denied by the promoters, who thereby deceived themselves as well as the public. After the failure, the future of the technology was clouded by a "deep and justified public distrust."

The materials that follows sets out both the safety issues that had arisen to plague nuclear power and the political and administrative failures that underlay their development and persistence, all of which contributed heavily to the decline. First is the news release that opened the nuclear power controversy. It is followed by the document that accompanied the release at the press conference that presented the consequences of a reactor accident and our early information on ECCS weaknesses. The third document, "Cooling Water," is a full discussion of the ECCS test failures and how the AEC got into their troubles. The next, "Nuclear Safety," sets out the course of the ECCS hearings themselves and the consequences of the whistle blowers' revelations. The last is a few excerpts from a long article I wrote for Martin Shubik, of Yale's School of Organization and Management, following my participation in a seminar on Risk and Organization that he had assembled in 1990.

References and Notes

1. Brian Balogh, *Chain Reaction: Expert Debate and Public Participation in American Commercial Nuclear Power, 1945–1975* (Cambridge University Press, Cambridge and New York, 1991). Contains an account of the UCS role in the nuclear safety debate.
2. *Reactor Safety Study: An Assessment of Accident Risks in U.S. Commercial Power Plants* (U.S. Nuclear Regulatory Commission, Washington, DC, October 1975). Document designation WASH-1400 (NUREG 75/014).
3. *The Risks of Nuclear Power Reactors: A Review of the NRC Reactor Safety Study WASH-1400*. H. W. Kendall, Study Director; Richard B. Hubbard and Gregory C. Minor, Editors; W. Bryan, D. Bridenbaugh, D. Ford, R. Finston, J. Holdren, T. Hollocher, R. Hubbard, D. Inglis, M. Kamins, H. Kendall, A. Lakner, G. Minor, J. Primack, contributors (Union of Concerned Scientists, Cambridge, MA, August 1977).
4. *Union of Concerned Scientists, Safety Second: The NRC and America's Nuclear Power Plants*. Contributors: Michelle Adato, principal author, James Mackenzie, Robert Pollard, Ellyn Weiss (Indiana University Press, Bloomington and Indianapolis, 1987).
5. Daniel F. Ford, *Three Mile Island: Thirty Minutes to Meltdown* (Penguin Books, New York, 1982).
6. See Joseph P. Tomain, *Nuclear Power Transformation* (Indiana University Press, Bloomington and Indianapolis, 1987).
7. Freeman J. Dyson, *Imagined Worlds* (Harvard University Press, Cambridge, MA, and London, England, 1997).

PRESS RELEASE

Union of Concerned Scientists, July 26, 1971

Nuclear Power

Two nuclear physicists, a nuclear engineer, and an economist—members of the Union of Concerned Scientists—released a report today that examined a new problem with nuclear reactor safety. According to the report, recent tests carried out under Atomic Energy Commission sponsorship have indicated that presently designed reactor safety systems are likely to fail if the systems supplying cooling water to a reactor should rupture or break. The reactor core is then expected to melt down and ultimately lose appreciable quantities of radioactive material into the environment. In the words of the report, "The resulting catastrophe and loss of life might well exceed anything this nation has seen in time of peace."

The authors analyzed the test failures and their implications in a detailed technical assessment. They studied the course of a meltdown accident in a modern reactor and estimated that released radioactivity, transported and dispersed by the wind, might still be lethal to humans as far as 75 miles from the accident, with the possibility of damage or injury to several hundred miles. They concluded that "presently installed emergency core-cooling systems would very likely fail to prevent such a major accident."

They call on the AEC to enforce "a total halt to the issuance of operating licenses for nuclear power reactors under construction, until safeguards of assured performance can be provided." They recommend a prompt technical and engineering review of presently operating power reactors to determine if they offer unacceptable risks to the public.

The report is sharply critical of the AEC for delaying adequate engineering tests of reactor safety features until after 21 power reactors have come into operation.

The report's assessment indicated that "the gaps in basic knowledge concerning the effectiveness of the safety features of large power reactors, the surprising scarcity of adequate tests—amounting nearly to a total lack—has astonished our group, especially in view of the large number of reactors of apparently hazardous design that are already operating."

The group believes that a major program will be required to develop information adequate to design safety features that will ensure protection against an accident. The country must not be allowed to "accumulate large numbers of aging reactors with flaws as critical as those we now see. It is past time that these flaws be identified and corrected so that the nation will not be exposed to hazards from the peaceful atom."

This report is the first technical review of emergency core-cooling systems and their reliability carried out independent of AEC sponsorship by scientists and engineers. The report's conclusions and recommendations are more far-reaching than anything that the AEC has released on the basis of its own reviews.

Nuclear Reactor Safety: An Evaluation of New Evidence

> ... Our safety lies, and lies exclusively, in making public decisions subject to the test of public debate. What cannot survive public debate—we must not do.
> —John Kenneth Galbraith

Introduction

The nation faces a growing electric power shortage whose roots lie in an ever-increasing per capita demand for electricity to support our rising standard of living and in a burgeoning population whose ize is expected to grow by 50% in the next 30 years. Our electric power problems are aggravated by an increasing reluctance on the part of the public to permit the construction of environmentally damaging fossil-fuel plants. To help alleviate this power shortage, the nation is turning to nuclear power: the controlled release of the enormous energies available from the atomic nucleus. Widely heralded as "clean power," because its production generates no sulfur oxides, smoke, or other visible pollutants so much in evidence from coal or oil generating plants, it has come, so its proponents say, just in time to augment and ultimately dominate as the national source of electricity.

Recent tests carried out under the sponsorship of the United States Atomic Energy Commission, however, have indicated that emergency systems incorporated in presently operating and planned reactors might not function adequately in the event the reactor should lose its cooling water (loss-of-coolant accident). In such a circumstance the reactor core would be expected to melt down and breach all the containment structures, very likely releasing some appreciable fraction of its fission

---

Ian A. Forbes, Daniel F. Ford, Henry W. Kendall, James J. MacKenzie, Union of Concerned Scientists, 1971.

product inventory. The resulting catastrophe and loss of life might well exceed anything this nation has seen in time of peace. The nation cannot move to dependence on nuclear power until the possibility of such an accident is totally negligible.

Our deep concern over this situation has led us to study the technical nature of a loss-of-coolant accident and the emergency core-cooling systems designed to prevent meltdown. We have evaluated the possible consequences of a major fission product release and have assessed in some detail what the lethality and range of the radioactive cloud might be from an accident involving a large modern reactor.

In the United States, the reactors in almost all nuclear power plants are cooled by the circulation of ordinary water. These reactors are called "light-water reactors," and it is this type that we shall be concerned with in this document. There are presently 21 operating light-water power reactors in the United States, 56 more under construction, and 44 others are in planning and design stages. By the year 1980 there will be more than 100 reactors operating singly and in clusters across the country, with a total electrical generating capacity of 150,000 megawatts of electricity.

Nuclear reactors have some unique problems associated with their operation that are unlike those accompanying any other source of energy. These problems are generally related to containing the potentially lethal nuclear radiation that accompanies the power developed by the fissioning (splitting) of uranium atoms. There are two major sources of this radiation. The first accompanies the release of energy by the atomic nuclei as they are split or fissioned. This energy is the source for generating the steam to drive the turbine. If the reactor were to be deliberately shut down or if its cooling water were lost through a pipe rupture or other accident, the levels of this radiation, and the principal energy release, would be quickly reduced.

The second source of radiation is the spent fragments of the fissioned atoms. These fission products are the nuclear ashes that result from the "burning" of uranium. They are intensely radioactive and accumulate in great abundance as a reactor continues to run. This source of radiation persists at very high levels even when the reactor has been shut down. Under normal circumstances these fission products remain within the fuel elements where they are formed.

Fuel elements, composed of small ceramic-like cylinders of uranium dioxide enclosed in long thin tubes of an alloy of zirconium (zircaloy), must be periodically replaced. The highly radioactive waste products that have accumulated in them must then be removed by chemical processing and stored safely away for thousands of years until the radioactivity has died away.

A nuclear reactor employs heavy shielding of steel, water, and concrete to absorb the radiation from the fissioning nuclei and from the fission products. Very substantial precautions are taken to prevent the escape of more than the most minute quantities of radiation or radioactive materials. Multiple sensor systems detect improper operation and initiate emergency procedures designed to prevent an unwanted release. Should one level of reactor containment be breached, others, still effective, are there to prevent major releases.

Nevertheless, as we shall see, there is an enormous potential for damage from a large release of fission products. Accordingly, there is a great burden on the designers of reactors to understand fully all aspects of potential accidents and have known and proven facilities to prevent or largely mitigate the consequences of a major accident. We have grave concerns that this is not presently the case with regard to the consequences of a loss-of-coolant accident.

In the following sections we describe first, the consequences of a loss-of-coolant accident that releases a substantial amount of radioactive material into the environment; second, how such an accident could occur through the failure of inadequate safeguards; and finally, our urgent recommendations. The aim of these recommendations is to lead to nuclear power generation that is as secure from a hazardous accident as human ingenuity and care can provide.

The Consequences of a Major Reactor Accident

If the emergency cooling system did not function at all, the core would melt and the molten mass of Zircaloy and UO_2 would collapse and probably melt through the pressure vessel in 30 minutes to 1 hour.
—C.G. Lawson, *Emergency Core-Cooling Systems for Light-Water-Cooled Power Reactors*, Oak Ridge National Laboratories, 1968 (ORNL-NSIC-24), p. 52

It seems certain that melt-through would be a catastrophic event in that large quantities of molten material will be discharged suddenly.
—*Emergency Core Cooling*, report of the Task Force Established by the U.S. AEC to Study Fuel Cooling Systems of Nuclear Power Plants, 1967, p. 145

The emergency core-cooling system is the reactor safety feature whose reliability is our pressing concern. Although the nature and sequence of events accompanying reactor accidents is subject to much speculation and conjecture, owing both to the fortunate lack of experience with major reactor accidents and the unfortunate lack of extensive experimental data on reactor safety systems, it is clear that the emergency core-cooling system is a vital safety system in currently operating reactors. As the statements from the two AEC reviews of emergency core cooling quoted at the beginning of this section affirm, emergency core-cooling performance determines whether a loss-of-coolant accident can be terminated without a major catastrophe. To facilitate an understanding of the seriousness of the recent indications of emergency core-cooling system unreliability, we describe in this section the consequences of emergency core-cooling system inadequacy in a loss-of-coolant accident.

If, through equipment malfunction or failure, human error, or an externally initiated event such as sabotage or severe earthquake, one of the major cooling lines to a large reactor core were ruptured, the water circulating through the primary cooling system would be discharged from the system through the rupture and the reactor core would be without coolant. As stated earlier, the absence of water (which normally serves as a "neutron moderator" as well as coolant), plus emergency insertion of the control rods, would prevent the continuation of uranium fission. That is, the reactor would become, and would remain, subcritical and the primary source of reactor energy would be removed.

There is, however, the other source of heat, which could not be turned off by reactor shutdown—the heat generated by the intense radioactivity of the fission products in the fuel rods. In a 2000 MW (thermal) reactor (allowing generation of about 650 MW of electricity), which is typical of many now operating, the heating provided by this source 3 seconds after control rod insertion amounts to approximately 200 MW, after 1 hour 30 MW, after a day 12 MW, and it would still be appreciable for some months.

Under normal reactor operating conditions, the external surfaces of the fuel cladding is at a temperature of about 660°F, while the interiors of the fuel pellets are very much hotter, typically 4000°F, near the melting point of the material. After coolant loss, the pin surfaces begin to heat rapidly both from the higher temperatures inside and from the continued heating by the fission products. In 10 to 15 seconds the fuel cladding begins to fail and within 1 minute the cladding has melted and the fuel pins themselves begin to melt. If emergency core cooling is not effective within this first minute, the entire reactor core, fuel and supporting structure, begins to melt down and slump to the bottom of the innermost container. Emergency cooling water injected at this stage may well amplify the disaster as the now molten metals can react violently with water, generating large quantities of heat, releasing steam and hydrogen in amounts and at pressures that can themselves burst the containers (see Battelle Memorial Institute Report BMI-1825). Approximately 20% of the fission products are gaseous and the meltdown has released them entirely from the now fluid core. If the containment vessels do not burst, the molten mass of fuel and entrained supporting structure continues to melt downward, fed by the heat generated by fission-product radioactivity. At this point in the accident there is no technology adequate to halt the meltdown—it is out of control (see *Emergency Core Cooling*, Report of the Task Force Established by the U.S. Atomic Energy Commission to Study Fuel Cooling Systems of Nuclear Power Plants, 1967). This meltdown situation has been referred to as a "China Accident." How far down the core would sink into the earth and in what manner the material would ultimately dissipate itself are not entirely understood, but what is close to certain is that virtually all the gaseous fission products and some fraction of the volatile and nonvolatile products would be released into the atmosphere.

The released radioactive material would be transported and dispersed in the atmosphere. Unlike the fission products of a nuclear weapon explosion, which are distributed in an immensely hot bubble of gas that rises rapidly, the gases here are relatively cool. Accordingly, they rise little at best and may easily be trapped under temperature inversions common at night. The cloud can be lethal at dozens and, in some circumstances, at close to one hundred miles.

In 1957 the Atomic Energy Commission published a detailed and thorough report entitled *Theoretical Possibilities and Consequences of Major Accidents in Large Nuclear Power Plants*, frequently referred to by its document code designation, WASH-740. It presented the results of a detailed analysis of the dispersion of radioactivity, amounting in one of their examples to 50% of the fission products contained in a reactor, and of the human exposure to radiation that would result if these fission products were released from a reactor in forms easily dispersed by the winds. We have studied the report in detail and it forms the basis for our evaluation of the accident hazards.

The implications and conclusions of WASH-740 are generally discounted by the Atomic Energy Commission and the nuclear power industry on the grounds that (1) no mechanisms exist that would disperse 50% of the reactor fission product inventory through a breach in all the containment vessels, as was assumed in the report, and (2) it is essentially impossible to breach all the containment vessels anyway, in view of the elaborate safety equipment incorporated in modern reactors. We discuss these two criticisms next.

The fission product inventory of a modern 2000-MW(t) reactor after a long period of operation can be as much as 15 times greater than that assumed in WASH-740. Accordingly, the consequences of the worst accident discussed in WASH-740 could be duplicated by the release of barely more than 3% of the radioactive material in a modern reactor. This much material represents less than 1/7 of a reactor's volatile or gaseous fission products, all of which (and more) could well be expected to be released after a complete core meltdown.

With reference to the second criticism of WASH-740, we shall see in the next section in more detail that it is not at all impossible to breach the containment vessels.

There are other assumptions in WASH-740. They concern the mechanisms of radiation exposure from the cloud of radioactivity released by the reactor; the deposition of radioactive materials on the ground (fallout); and the mechanisms of transport and dispersion by the atmosphere for a variety of commonly occurring meteorological conditions. We have examined these assumptions and the calculations based on them and have concluded that they represent an adequate basis for estimating the consequences of a major accident.

We have considered in this example an accident just prior to refueling in a reactor generating 2000 MW of thermal power, typical of many now operating. We have assumed that one-third of the fuel is removed for reprocessing each year, as is customary, and replaced with fresh fuel. The quantity of fission products in the reactor core reaches a maximum just before refueling. We consider an accidental release of a portion of these products to the atmosphere. The manner and speed with which the material spreads will depend on a number of parameters. These include the particle size (for the non-gaseous fission products), the temperature of the emitted materials, and the then-existing weather conditions: wind speed and direction, rain, temperature stratification, and so forth.

A release that involves primarily highly volatile or gaseous materials is more likely than one that involves dispersal of substantial solid material as a smoke or dust, although the latter may certainly occur. The volatile and gaseous radioactive materials are the halogens and noble gases and include the xenons, kryptons, bromines, iodines and strontiums and comprise some 20% of the fission product inventory. These and the semivolatiles—ruthenium, cesium and tellurium—are released in an uncontrolled core meltdown. The iodines and strontiums are biologically the most hazardous.

As in WASH-740, we have considered hot clouds (3000°F), cold clouds, fumes, dust and gaseous emissions, and a range of commonly occurring meteorological conditions including rain, no-rain, daytime and night-time lapse rates (that is, the presence or absence of temperature inversions), and various cloud rise distances from zero to several hundred meters. The larger cloud rises depend on initial temperatures that appear improbably high, and the most probable situation appears to be the emission of a ground-level cloud. This latter case unfortunately results in the greatest damage.

In the case of emissions from the reactor of purely gaseous radioactive products, the biological effects are dominated by radiation exposures from cloud passage. If, in addition, there is radioactive particulate matter, as dust or smoke, then one expects deposition of radioactive materials quite similar to nuclear weapons fallout. Unlike weapons fallout, which decays rapidly from the first, reactor "fallout" decays rather slowly for some time. Some of the radioactive materials have half-lives in the years or tens of years, and evacuation, land denial, or land restrictions may thus persist for appreciable periods of time. Substantial uptake and ingestion of radioactive materials would occur in humans exposed to clouds containing radioactive dusts or smokes.

We shall summarize a number of cases from our study in terms of maximum distances at which various consequences of cloud passage and radioactive deposition might occur. The effects refer to a person exposed to cloud passage with no shielding or protection. The cloud would be increasingly difficult to see after it had moved away from the accident site and would be invisible long before it had lost its lethality. Persons inside structures at the time of cloud passage would receive substantially less initial exposure from cloud passage but would receive continuing exposure from material finding its way into the structures and only slowly dispersing.

Under daytime conditions, with no temperature inversion, a cold, ground-level cloud, and an 11.5 mph wind, lethal injuries might be expected to a distance of 2.5 miles, injuries likely to 5 miles, and possible but unlikely effects no further than 20 miles, assuming 20% of the fission product inventory is released. If 5% is released, these distances become 1 mile, 2.5 miles, and 12 miles, respectively. The region affected is a strip, extending downwind and as wide as 3/4 mile across.

Injury here means radiation sickness and other prompt effects in addition to increased susceptibility to many other diseases throughout the lifetime of the exposed persons. In particular, enhanced incidence of leukemia and other cancers would be expected in addition to genetic damage.

If, however, the radioactive materials are released under a temperature inversion, by no means an uncommon nocturnal condition, with a 6.5 mph wind, the meteorological trapping greatly extends the range of possible death or damage. Assuming a 20% release, lethal effects can extend 75 miles downwind in a strip of maximum width up to 2 miles. Injuries would be likely at up to one to two hundred miles, the presence of moderate rain yielding the lower figure. For release of 5% of the fission products, these numbers are reduced to 40 miles and 80 to 100 miles, respectively, in a strip up to 1 mile wide.

Non-gaseous particulate fission products might be generated in chemical explosions arising should molten reactor core material contact water in appreciable quantity or if the core meltdown generated high temperatures. These products will, in part, be deposited on the ground as fallout. If they are initially borne aloft in a hot cloud, the point of maximum deposition may be some distance down wind, but if emitted in a cloud at essentially ambient temperatures, deposition would be expected to be a maximum at the reactor site and decrease downwind. Rain, depending on its rate, can substantially increase the rate of deposition, and by depleting the cloud, much reduce the range of fallout, at the cost of increased local concentrations. Non-volatile particulate material would include entrained amounts of cesium-137, with a half-life of 30 years, and of the partially volatile strontium-90, with a half-life of 28 years. The latter material is an especially noxious contaminant for, in addition to its long half-life, it is chemically similar to calcium and, when ingested, lodges in the bones. Low concentrations of strontium-90 on arable land require agricultural restrictions, for it can move through various food chains into humans.

The estimates of the distances at which evacuation (urgent or merely necessary), severe land restrictions, or agricultural use restrictions would occur are particularly sensitive to the particle size, emission mechanism from the core, and meteorological conditions during transport. It is presently not possible to refine these estimates and select among them owing to a lack of knowledge of the nature of possible accidents. If we assume 4% of the non-volatile core products are released, then urgent evacuation might be required out to 15 to 20 miles, within 12 hours, or might not be required anywhere. Necessary evacuation might be required on a less urgent basis out to 80 to 100 miles, or possibly not at all. Land use restrictions would be inevitable, however, for this postulated release, and would extend a minimum of fifteen miles and could reach to distances of 500 miles or more. Denial or land use restriction might persist for some years.

Some idea of the hazardous nature of the materials involved comes from recognizing that the strontium-90 inventory in the reactor we are considering is sufficient to contaminate more than 1000 *cubic miles* of water in excess of permitted AEC tolerance levels.

Many of the power reactors now operating are sited near metropolitan areas, and a very large number are planned to be similarly sited. One reactor under construction near Boston is within 50 miles of close to 3,000,000 persons, and is 6 miles from a town of 15,000 persons.

We should emphasize that in making these estimates, we have considered only commonly occurring meteorological conditions and the release of no more than one-fifth of the hazardous materials from the core. Uncommon, but by no means rare, weather conditions might appreciably extend the range of lethal or damaging effects.

G. W. Keilholtz, C. E. Guthrie, and G. C. Battle, Jr., in *Air Cleaning as an Engineered Safety Feature in Light-Water-Cooled Power Reactors* (ORNL-NSIC-25, September 1968, page 4) state that:

> Several power reactors with outputs of over 1000 MW(e) each are scheduled for startup in 1970–71, and reactors of about 1500 MW(e) are being designed. The increase in numbers is increasing the probability that a major accident might occur, the increase in size is increasing the fission-product inventory in the average power reactor, and increases in both number and size are increasing the total amount of fission products that must be contained, within reason, to prevent excessive contamination of the earth's atmosphere.

Reactors of 1000 MW and 1500 MW electrical power are respectively 1 1/2 times and 2 1/4 times larger than the reactor we considered in our calculations on the consequences of reactor accidents. For these larger reactors the ranges of accident effects would be greater than our estimates.

It is certainly conceivable that more than one-fifth of the fission products might be released in a wholly uncontrolled core meltdown. The range estimates for lethal or damaging effects we have made should be regarded as conservative and by no means considered extreme distances that could not be exceeded in particularly unfortunate circumstances.

Quite aside from the direct and indirect damage and loss of life that might ensue from a drifting, lethal radioactive cloud, there is the fear and anxiety that its existence would entail. It is probably not possible to evacuate a major city near the accident site exposed to the threat of cloud passage and radioactive fallout, and little could be done aside from warning the majority of the inhabitants to seek shelter from the invisible materials moving toward them. It is hard to doubt that the warning alone would initiate a great urban dislocation and panic even if the cloud itself were diverted by the wind.

It is abundantly clear from our study that a major nuclear reactor accident has the potential to generate a catastrophe of very great proportions. The full scale and consequences of a catastrophe cannot fully be reckoned, yet it is against such an ill-understood but awesome event that the scale of and confidence in the reactor safeguards must be weighed.

Emergency Core Cooling and a Loss-of-Coolant Accident

We have already pointed out that in the unlikely event that the primary coolant flow is interrupted, as through a break in one of the recirculating lines, causing ejection of the primary coolant from the core, a simple shutdown of the reactor (that is inserting the control rods and halting the chain reaction) may be insufficient to prevent a release of core radioactive materials into the environment. Even though fissioning ceases when the reactor "scrams" or shuts down, the heat generated by the decay of fission products produced during normal operation is sufficient, in a large reactor, to cause meltdown of the core within a minute or so.

The function of the emergency core-cooling system is to resupply the core with coolant water in the event of the loss of primary coolant in a loss-of-coolant accident. If the emergency cooling system did not function at all, the core would melt and the molten mass of fuel cladding and uranium dioxide fuel, weighing a few hundred tons for a large reactor, would collapse to the bottom of the inner containment vessel and would be expected to melt through the vessel in 30 to 60 minutes. Other containment might still prevent loss of radioactive material (if none had not already escaped through the rupture) but, driven by fission product heating, the mass would continue to melt downward, through all man-made structures. The possibility of avoiding the release of radioactive materials once this happens appears negligible and the stage is then set for the kind of catastrophe discussed earlier.

In the course of reviewing this paper, Professor James A. Fay, of MIT's Mechanical Engineering Department, made some preliminary estimates of the rate of descent of the molten mass of fuel, fission products, and supporting structure following core meltdown. From this we have made some crude estimates of the nature of last stages of meltdown accident. Although melting of the fuel elements themselves requires very high temperatures (of order 5000°F), it would appear likely that the core and supporting material after meltdown and after melt-through of the steel containment structures would be appreciably cooler. If the mass were at a temperature of 3000°F it would sink through rock at nearly 200 feet per day. As it continued to cool from melting and entrainment of rock, from loss of its heat source through radioactive decay, and by loss to the environment, its rate of descent would slow. At 2100°F it would be melting downward at 50 feet per day and when at 1300°F descending no more than 1 foot per day. Radioactive decay alone diminishes the heat source by about 90% by the end of a week, compared to its strength at the time of meltdown, so the sinking would be nearly arrested after a few weeks or a month, with the hot mass expected to have descended no more than a fraction of a mile. Ejection of substantial amounts of radioactive materials would have ceased much before this time, unless the mass encountered groundwater on its trip down. Chemical reactions of explosive violence could, in this instance, cause more hazardous material to be spewed forth. Ultimately the mass would cool off slowly, entombed well underground. Extensive deposition of fallout at and near the accident site and possible contamination of groundwater would be a reminder for many tens of years of the core meltdown accident.

It is clear why a reliable emergency core-cooling system is, therefore, such an important component of nuclear reactor safety equipment.

A series of tests of a simulated emergency core-cooling (ECC) system were conducted during the period of November 1970 through March 1971 at the National Reactor Testing Facility in Idaho. In these tests, the simulated emergency core-cooling system failed in a particularly disturbing manner, and, accordingly, we have carried out a detailed examination of the test results and a systematic evaluation of the issues thus posed. We have examined the technical documents pertaining to the Idaho experiments, talked with some of the investigators who performed the tests and to other reactor experts, and examined data on reactor safety systems contained in internal AEC reviews of the issues, in an effort to form a sound judgment as to the Idaho tests' implications for the safety of nuclear reactors.

The tests of the ECC system, part of the Loss-of-Fluid Test (LOFT) program, were performed on a small semi-scale reactor mockup. The ECC system on the mockup was of the type common to pressurized-water reactors, i.e., it was designed to flood the core from the bottom.

The mockup consisted of a small, cylindrical core (about 9 inches high and 9 or 10 inches in diameter), with upper and lower plenum, inside a pressure vessel (see figure) and a single coolant loop.[1] The core contained zircaloy or, in other tests, stainless-steel rods, which were electrically heated.

Operating power was about 1.1 MW, yielding a typical commercial reactor power density. Core pressure and temperature were likewise typical of a large pressurized-water reactor, about 2180 psig and 575°F, respectively. All but the first two tests used orifices on the core inlets to simulate the pressure drop across a large core.

The tests were intended to study the effects of ECC after a rupture of the primary coolant loop piping. The system break was generally made in the cold (inlet) leg at a point higher than the core top. The break area-to-system volume ratio was varied between 0.007 ft^{-1} and 0.0007 ft^{-1}. In one test, a break was made in the hot (outlet) leg. The emergency coolant was introduced into the inlet plenum. Between 3 and 10 seconds (depending on break size) were required to depressurize the system to the 660 psig level required for ECC flow to commence.[2]

Only a negligible fraction of the emergency core-cooling water reached the core in any of these tests, even in those tests where a quick-closing valve on the ECC line or other fast-acting bypass systems were added. In fact, no appreciable difference in post-break core temperatures was observed when the tests were conducted *without* the emergency core cooling utilized.

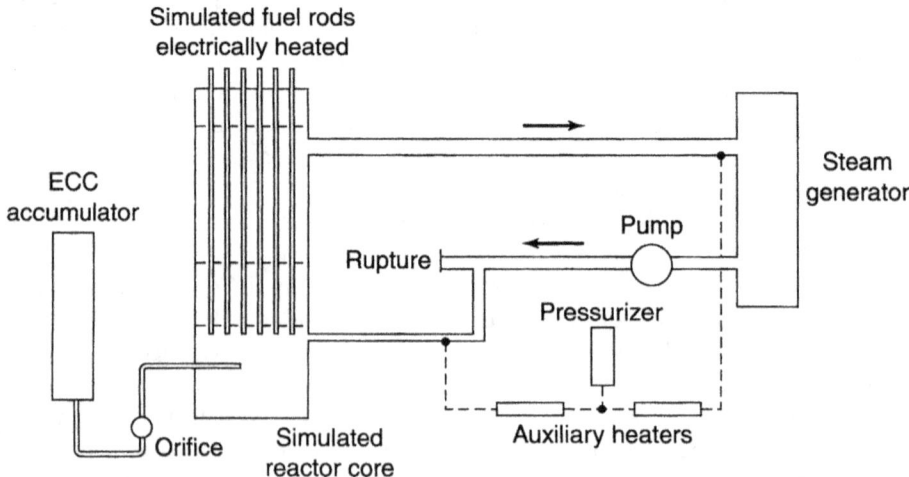

Figure 1. Schematic drawing of the semiscale reactor mockup used in the emergency core-cooling tests of the National Reactor Testing Facility in Idaho, November 1970–March 1971.

The official reports of the Idaho Nuclear Corporation, which performed the experiments for the Atomic Energy Commission, express the test results: "The first semiscale test involving emergency core cooling (ECC), test 845, was performed in which ECC was injected into the annulus between the flow skirt and vessel wall. Early analysis of test data indicates that essentially no emergency core coolant reached the core." Subsequent tests were performed with ECC fluid injected directly into the inlet plenum to reduce the "potential for entrainment of ECC coolant in the escaping primary fluid." Similar behavior of ECC fluid was observed in these subsequent tests: "no significant amount of ECC fluid reached the core." A further test was conducted "with a high-inlet break configuration with a nozzle in place of an orifice in the blowdown line. A quick-opening valve was installed in a bypass line around the system low point to determine whether pressure buildup due to a water seal was responsible in previous tests for emergency coolant expulsion through the break. ECC liquid was ejected from the system in test 849 as in previous tests and at no time did ECC liquid reach the core."

The ECC fluid was observed in the tests to escape through the same break in the primary coolant loop as the primary coolant. As noted in the official test reports, "On the basis of these data, the conclusion was reached that the ECC fluid entered the inlet annulus and was swept out of the vessel to the inlet nozzle and external downcomer, and out the break." Thus the ECC system would fail at the same time as the primary cooling system and would be, therefore, of no assistance in a loss-of-coolant accident.

The Idaho test results raise a major prima facie uncertainty about the reliability of the emergency core-cooling system. To develop a full interpretation of the experimental data, we have carefully considered numerous technical issues related to the use of the semiscale test data in the analysis of the emergency core-cooling system of commercial-size power reactors. On the basis of our analysis we conclude that there are no convincing grounds upon which the prima facie problem with the ECC system indicated by the Idaho data can be dismissed. In our judgment the results of the Idaho tests clearly demonstrate the basic inability of current pressurized-water reactor emergency core-cooling systems to perform the function for which they were designed.

The fact that the mockup used in the tests was very small detracts little from the apparent unreliability of the emergency core-cooling system for pressurized-water reactors. It is said that similar tests will be run with a larger core: the results will shed light on this question.

The situation for boiling-water reactors is somewhat different, inasmuch as the ECC system for these reactors consists of a spray arrangement located at the top of the core (sometimes in conjunction with a lower plenum flooding arrangement—which might now be considered useless), rather than of the bottom-flooding system used in the semiscale test system. The spray arrangement *ought* to provide at least *some* core cooling in the event of a break. It is not apparent, though, whether it could be expected to provide *sufficient* cooling in view of the facts:

a. The Idaho tests, the only ECC tests under simulated accident conditions conducted to date, failed completely.

b. The Idaho tests demonstrated a severe lack of knowledge of the mechanics of emergency core cooling.

c. The Idaho reports (February 1971) include the statement that "parameters used in the G.E. (General Electric) model for top spray cooling used for accident evaluation require modification."

We have concluded, from an extensive study of data available on various aspects of the problems of emergency core cooling, that several mechanisms exist in a loss-of-coolant accident situation that would very likely render the emergency core-cooling system ineffective. These mechanisms tend to force emergency core coolant away from the core hot spot (at the center) and propel coolant around, rather than through, the center of the core to the cold leg break so that there is a loss of emergency coolant through the same break as the primary coolant.

Any tendency for coolant to be forced away from the center of a reactor core would greatly diminish the efficiency of the emergency core coolant since it is the center of the core which is hottest and most requires emergency coolant in an accident situation.

The following mechanisms could produce this kind of effect:

a. **Flow blockage:** With a loss of primary coolant as the result of a break, the temperature of the fuel rods will rise. For temperatures above about 1400°F, the fuel rods will swell and buckle, restricting or preventing normal flow. The greatest swelling and blockage would occur in the hottest part of the core and would force diversion of ECC around the hot spot through the cooler core region.

b. **Steam expansion:** With loss of pressure as the result of a break, much of the primary coolant in the reactor will flash to steam. Expansion of the steam will tend to prevent emergency core coolant from entering the core. Also the steam expansion will be greatest in the hottest core region, again preventing the emergency core coolant from reaching the point where it is most needed.

c. **Leidenfrost migration:** During emergency core-coolant injection, steam-entrained water droplets, behaving like drops of water in a hot frying pan, can cool the fuel rods many feet above the water surface. These droplets are forcibly repelled from host surfaces by a rapid buildup of steam between the droplet and the surface. Since the forces exerted on the droplets by cool surfaces are less than the forces exerted by hot surfaces, there is a tendency for the droplets to migrate from the hot central core regions to the cooler periphery (Liedenfrost migration). This produces hot-spot flow starvation.

The mechanisms described above may diminish drastically the amount of emergency coolant reaching the reactor core's center. Far more serious than this, however, is the possibility that only part or none of the emergency core coolant reaches any portion of the core. Accumulation of steam in the core or steam generators after a break may form a pressure head which the emergency core-coolant injectors cannot overcome (i.e., a steam "bubble," which prevents the emergency core coolant from entering the core). In this event, part or perhaps all of the emergency core coolant will bypass the core and merely flow out the ruptured pipe.

A good *experimental* knowledge of both emergency core coolant flow starvation and bypass effects is required if the ability of emergency core cooling to prevent core meltdown is to be relied upon. Our review of AEC and industry safety research, however, indicates that almost no attempt has been made, up until the last few months, to obtain this information, even though, of course, it relates in a vital way to the safety of currently operating and planned reactors.

This palpable lack of engineering knowledge obtained through experiments under realistic accident simulations was emphasized in the Oak Ridge National Laboratory report, *Emergency Core-Cooling Systems for Light-Water Cooled Power Reactors*, by C. G. Lawson (ORNL-NSIC-24, 1968). The purpose of the report was "to identify inadequacies in assumptions, available data, or general basic knowledge" with regard to emergency core cooling so that "areas of meaningful research and development" can be determined.

The Lawson Report's major findings were:

- "Tests of emergency cooling equipment have never been performed in the environmental conditions of the dynamic pressure, temperature, and humidity that might prevail in a loss-of-coolant accident." (p. 6)

- Parameters used to calculate the design-basis criteria for ECCS safety margins were based on "insufficient data" and were to be judged "tentative and arbitrary." (pp. 7–8)

- So little engineering information existed on the behavior of the zircaloy-clad fuel rods during a loss-of-coolant accident "that a positive conclusion of the adequacy of these emergency systems would be speculative." (p. 9)

- The swelling of zircaloy cladding during blowdown was indicated by existing data, but "additional information is required to determine how much and whether cooling would be prevented." (p. 59)

- Emergency power for the ECCS was found unreliable ("The emergency diesels failed to operate 1% of the time they were tested"), from which it is concluded, "The relatively poor showing of the emergency power supply makes the effectiveness emergency cooling systems questionable." (p. 63)

- "The emergency core-cooling systems of several boiling- and pressurized-water reactors were reviewed, the design basis and backup data were examined, and the need for certain additional data was established. Generally, the design approach used by the manufacturers is conservative when evaluating the energy released or the cladding temperature. Occasionally there is an absence of experimental data that is inconsistent with the apparent sophistication of the calculation procedures." (p. 88)

The basic criticisms in Lawson's report are not out of date, as the remarks of George M. Kavanagh, AEC Assistant General Manager for Reactors on May 13, 1971 before the Joint Committee on Atomic Energy, demonstrate:

Heavy reliance has been placed on engineering safety features such as the ECCS, where technology is complex. . . . Some of the information needed to confirm convincingly the adequacy of such systems, which are intended to arrest the course of hypothetical large primary system failures, is not yet available. (Quoted in *Science*, May 28, 1971, p. 191)

Further comment on the basic lack of engineering data pertaining to ECCS reliability was made in February 1970 by the AEC Division of Reactor Development and Technology:

> If the primary coolant system fails at elevated temperature and pressure, there will inevitably be a loss of some fraction of the primary coolant by the blowdown, or depressurized process. Forces resulting from the depressurization may cause mechanical core damage, and the loss of primary coolant may lead to a loss, or at least a significant lowering, of cooling capacity. The safety features interposed at this point include adequate design of the core structure to resist such blowdown forces and the use of emergency core cooling systems such as sprays or core-flooding systems to ensure that the cooling capacity remains great enough that fuel cladding will remain intact. *Since little experience with such systems is available*, the principal problems are those related to the performance of emergency core cooling systems, i.e., *whether* they will operate as planned under the postulated accident situation, with respect to the time of initiation and the rate of flow. (p. I-10; emphasis added)

The Division of Reactor Development and Technology's comment was made as it presented its *Water-Reactor Safety Program Plan* February, 1970 (AEC Document WASH-1146) whose general objective was to present a "comprehensive plan for the timely and effective solution of water-reactor safety problems" (p. I-11) The report identified "major problem areas" in nuclear power reactor safety and emphasized that determination of "all the factors affecting the performance and reliability of ECCS" as "the most urgent problem area in the safety program today." The *Safety Program Plan* gave priority ratings to dozens of items on its safety research agenda. What was surprising was the length of the list of "very urgent, key problem areas, the solution of which would clearly have great impact, either directly or indirectly, on a major critical aspect of reactor safety."

These AEC documents and others we have examined indicate the palpable lack of experimental proof of ECCS reliability. The *Safety Program Plan* indicates that basic safety research has yet to be completed and is, instead, scheduled over the next decade. These reports bring to light the AEC's manifest failure to adhere to the vital and important procedure of establishing the safety of nuclear power plants *before* initiating its full-scale program for nuclear power plant construction, a program that today has produced 21 operating plants and 53 under construction.

Conclusions and Recommendations

The grave weaknesses apparent in engineering knowledge of emergency core-cooling systems and the strong implications that these systems would fail to terminate safely a loss-of-coolant accident makes it clear that in the event of a major reactor accident the United States might easily suffer a peacetime catastrophe whose scale, as we have seen, might well exceed anything the nation has every known.

The gaps in basic knowledge concerning the effectiveness of the safety features of large power reactors, the surprising scarcity of adequate tests—amounting nearly to a total lack—has astonished our group, especially in view of the large number of apparently hazardous designs that are already operating. Not until 14 years after the publication of WASH-740 do we see experimental tests of an emergency core-cooling system, tests carried out on nothing larger than a 9-inch model, described by the AEC as not meant to simulate a reactor fully. It is now more than 11 years since the first reactor for the commercial production of power was brought into operation.

The hazards inherent in the present situation have not gone entirely unnoticed by the AEC; the Commission was evidently disturbed by the Idaho test results and appointed a Task Force to assess them. The Task Force report has not yet been released, but "interim" criteria for determining the adequacy of emergency core-cooling systems were published in the *Federal Register*. In an unusual move, the AEC waived the normal 60-day waiting period, noting, "In view of the public health and safety considerations . . . the Commission has found that the interim acceptance criteria contained herein should be promulgated without delay, that notice of proposal issuance and public procedure thereon are impracticable, and that good cause exists for making the statement policy effective upon publication in the *Federal Register*."

In the delicately chosen words of AEC Chairman Glenn T. Seaborg, "The use of recently developed techniques for calculating fuel cladding temperatures following postulated loss-of-coolant accidents, and the results of recent preliminary safety research experiments, have indicated that the predicted margins of ECCS (Emergency Core-Cooling System) performance may not be as large as those predicted previously."

The Chairman has failed to indicate why obviously critical preliminary safety tests have not, until this year, been carried out in view of the potential hazards associated with the 21 power reactors authorized by the AEC and now operating.

We have concluded that there are major and critical gaps in present knowledge of safety systems designed to prevent or ameliorate major reactor accidents. We have further concluded that the scanty information available indicates that presently installed emergency core-cooling systems would very likely fail to prevent such a major accident. The scale of the consequent catastrophe which might occur is such that we cannot support the licensing and operation of any additional power reactors in the United States, irrespective of the benefits they would provide to our power-shy nation.

We do not believe it is possible to assign a reliable numerical probability to the very small likelihood of a loss-of-coolant accident in a power reactor. There are too many sources of uncertainty whose importance cannot be adequately assessed. The acquisition of this information by trial and error, in the absence of safeguards that would mitigate or prevent core meltdown, could be extremely costly to the nation.

While it appears that the probabilities are not very large, we do not believe that a major reactor accident can be totally or indefinitely avoided. The consequences of such an accident to public health are too grave to assume anything more than a very conservative position. Accordingly we have concluded that power reactors *must have assured safeguards* against a core meltdown following a serious reactor accident.

We have grave concern that reactors now operating may at present offer unacceptable risks and believe these risks must be promptly and thoroughly assessed.

Accordingly, we recommend:

1. A total halt to the issuance of operating licenses for nuclear power reactors presently under construction until safeguards of assured performance can be provided.
2. A thorough technical and engineering review, by a qualified, independent group, of the expected performance of emergency core-cooling systems installed in operating power reactors to determine whether these reactors now constitute an unacceptable hazard to the population.

It is apparent that a major program will be required to develop, through both theoretical studies and experimental measurement, information adequate to design reactor safety features that will ensure protection against core meltdown following a loss-of-coolant accident.

We believe that a complete and adequate understanding of loss-of-coolant accidents can be gained. Moreover, there appear to be no technical difficulties so acute that adequate protection from the consequences of a major accident cannot be assured. The United States will become increasingly dependent on nuclear power. Nuclear power *can* be both clean and safe, but it will not be, in years to come, if the country is allowed to accumulate large numbers of aging reactors with flaws as critical and important as those we now see. It is past time that these flaws be identified and corrected so the nation will not be exposed to hazards from the peaceful atom.

Notes

1. Three coolant loops are generally employed in power reactors.
2. This is probably fast enough, but the topic warrants further examination.

Cooling Water

Unique and very substantial hazards are associated with nuclear power reactors. They contain enormous quantities of radioactive materials, the "ashes" from the fission (splitting) of uranium, whose accidental release into the environment would be a catastrophe. Great reliance is placed on engineered safety systems to prevent or mitigate the consequences of such accidents. Foremost among safety systems are the emergency fuel-core cooling systems which, should normal cooling systems accidentally fail, are designed to prevent an overheating and melting of the reactor fuel and subsequent release of lethal radioactivity into the environment.

Recently developed evidence from various experiments conducted under the auspices of the United States Atomic Energy Commission (AEC) suggests that emergency cooling systems may well be unable to perform the functions for which they are designed and that the margin of safety previously throught to exist in these systems' operations during an accident is very much smaller than has been expected and may, in fact, be nonexistent. A review of the technical literature indicates that there may be still further problems with emergency fuel-core cooling systems (also referred to as emergency core-cooling systems—ECCS) than those already uncovered by recent experiments.

The principal difficulty continues to be lack of adequate information, as indicated by George Kavanagh, assistant general manager for reactors, U.S. Atomic Energy Commission:

> Heavy reliance has been placed on engineering safety features such as the ECCS, where technology is complex. . . . Some of the information needed to confirm convincingly the adequacy of such systems, which are intended to arrest the course of hypothetical large primary system failures, is not yet available.[1]

These problems are a matter of serious concern, for nuclear power reactors are expected to play an increasingly important role in supplying electric power to the nation.

By Ian A. Forbes, Daniel F. Ford, Henry W. Kendall, and James J. MacKenzie, Union of Concerned Scientists, 1972. From *Environment* 14(1) 40–47 (January/February 1972). Reprinted with permission of the Helen Dwight Reid Educational Foundation. Published by Heldref Publications, 1319 18th St. N.W., Washington, DC 20036-1802. Copyright © 1972.

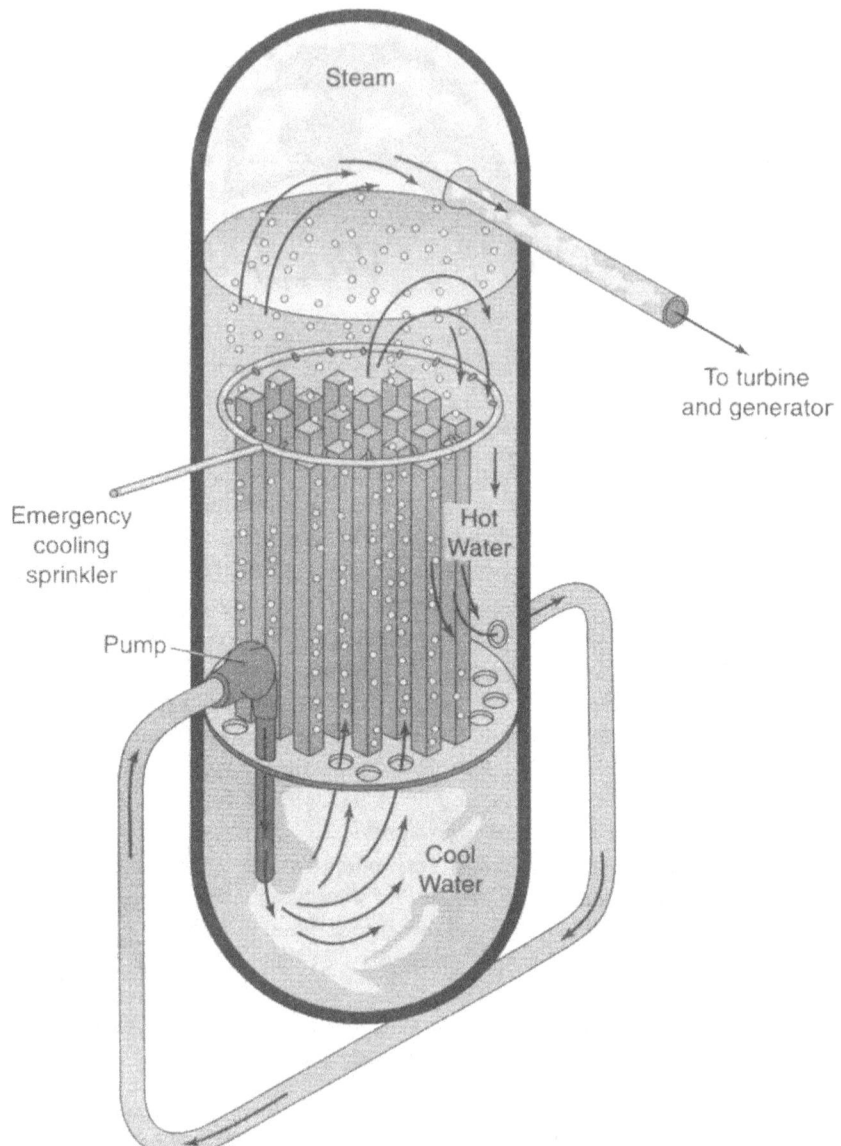

Figure 1. A nuclear reactor, the heart of a nuclear electric power plant, is shown in simplified form. The reactor's fuel, or core, is represented by the long oblongs in the center of the drawing. The core is about twelve feet high and fifteen feet across. Each oblong fuel assembly, roughly 6 inches across, is actually a bundle of 49 thin rods filled with uranium oxide pellets. In a large reactor there may be more than 700 such assemblies, containing 150 tons or more of uranium fuel. In the simplest design, the "boiling-water" reactor shown here, cool water is pumped upward among the fuel assemblies, which are heated by nuclear reactions. The water boils and steam is drawn off to generate electricity, while the remaining water is recirculated. In actual reactors of the type shown here, there are actually two recirculation loops and, of course, systems for replenishing the cooling water, which is often referred to as the "pri-

Electric power consumption is increasing at a rate many times that of the U.S. population, and, with increasing difficulties in obtaining clean fossil fuels, such as natural gas, the pressures to tap the energies locked in the atomic nucleus will soon become very great. Twenty-two nuclear power stations are already operating, 55 are under construction, and 44 more reactors have been ordered.

Possible Accidents

Although the nature and sequence of events accompanying major reactor accidents are subject to much speculation and conjecture, owing to both the fortunate lack of experience with major accidents and the unfortunate lack of extensive experimental data on reactor safety systems, it is clear that the emergency core-cooling system is a vital safety system in currently operating reactors. As the following statements from two AEC reviews of emergency core-cooling affirm, emergency core-cooling performance determines whether a loss-of-coolant accident can be terminated without a major catastrophe.

> If the emergency cooling system did not function at all, the [fuel] core would melt and the molten mass of zircaloy and UO_2 [uranium oxide] would collapse and probably melt through the pressure vessel in 30 minutes to one hour.[2]
>
> It seems certain that melt-through will be a catastrophic event in that large quantities of molten material will be discharged suddenly.[3]

If through equipment malfunction or failure, human error, or an external event such as sabotage or severe earthquake, one of the major cooling lines to a large reactor core were ruptured, the water circulating through the primary cooling system would be discharged from the system through the rupture and the reactor core would be without coolant. In the absence of water, and with operation of emergency systems, the power generation by fission would terminate.

There is, however, a source of heat which could not be turned off by reactor shutdown—the heat generated by the intense radioactivity of the "ashes" of nuclear fis-

---

mary coolant." Because the whole system is under pressure, a break in the cooling-water pipes would lead to rapid loss of the coolant. This would shut down the reactor, but radioactive wastes accumulated in the fuel would continue to generate heat. Unless an alternative cooling system were available, this heat would rapidly lead to melting of the fuel and supporting structures and ultimately to release of radiation to the outside air. An emergency cooling spray is therefore provided in case of emergency. As a further safeguard the entire reactor is enclosed in a massive concrete and steel "containment" structure. . . . If the primary cooling water were lost, overheating of the reactor fuel would cause blockages of the channels among and within the fuel assemblies, keeping out the emergency spray, in which case the fuel might continue to heat and to melt. In such an accident there is considerable doubt as to whether the containment structures would prevent release of radiation to the outside air.

—Sheldon Novide
Editor, *Environment*

sion, the fission products in the fuel. In a reactor generating 650,000 kilowatts of electricity (a small plant by today's standards), the heating provided by fission products seconds after shutdown amounts to approximately 200,000 kilowatts; after one hour, 30,000 kilowatts; after a day 12,000 kilowatts; and would still be appreciable for some months.

Under normal reactor operating conditions, the sheathing of the fuel is at a temperature of about 660 degrees F., while the interiors of the fuel pellets are typically at 4,000 degrees F. near the melting point of the material. After loss of cooling water (primary coolant) in an accident, the fuel surfaces would begin to heat rapidly, both from the higher temperatures inside and from the continued heating by the fission products. In ten to fifteen seconds the fuel sheathing would begin to fail and within one minute would melt, and the fuel itself would begin to melt. If emergency cooling were not effective within this first minute, the entire reactor core, fuel, and supporting structure would then begin to melt down and slump to the bottom of the reactor vessel. Emergency cooling water injected at this stage might well amplify the disaster because the now-molten metals could react violently with water, generating large quantities of heat, releasing steam and hydrogen in amounts and at pressures that could themselves burst the containment vessels.[4] Approximately 20 percent of the fission products would be gaseous, and the meltdown would release them entirely from the now fluid core. If the containment vessels did not burst, the molten mass of fuel and entrained supporting structure would continue to melt downward, fed by the heat generated by fission product radioactivity. At this point in the accident there would be no technology adequate to halt the meltdown—it would be out of control.[3] How far down the core would sink into the earth and in what manner the material would ultimately dissipate itself are not entirely understood, but it is probable that virtually all the gaseous fission products and some fraction of the volatile and nonvolatile products would be released to the atmosphere.

The released radioactive material would be transported and dispersed into the atmosphere. Unlike the radioactive materials produced in a nuclear weapon explosion, which are carried in an immensely hot bubble of gas that rises rapidly, the gases here would be relatively cool. Accordingly, they would rise little at best and might easily be trapped under temperature inversions common at night. The cloud could be lethal at dozens and, in some circumstances, at close to 100 miles, and grave consequences might occur at many hundreds of miles.

Safety Systems

A completely reliable emergency core-cooling system is, then, an indispensable component of nuclear reactor safety equipment. Recent tests conducted for the AEC, however, have called into question the design of emergency cooling systems being installed in large nuclear power plants. A series of tests of a simulated emergency core-cooling-system was conducted during the period November 1970 through March 1971 at the National Reactor Test Station in Idaho. Preliminary reports indicated that the test system had failed to operate as expected. We have carried out a detailed examination of the test results and a systematic evaluation of the issues thus posed. In an effort to form a judgment as to the Idaho tests' implications for the safety of nuclear reactors, we have examined the technical documents pertaining to the Idaho experiments, have talked with some of the investigators who, with other reactor experts, performed the tests, have examined data on reactor safety systems contained in internal AEC reviews of the issues, and have reviewed the experiments in the safety research programs.

The tests, part of the Loss-of-Fluid Test (LOFT) program, were performed on a small semi-scale reactor mockup. The ECC system on the mockup was of the type common to pressurized-water reactors—i.e., it was designed to flood the fuel core from the bottom with water.

The mockup consisted of a small cylindrical core (about nine inches high and nine or ten inches in diameter) and a single cooling-water inlet and outlet. The core contained zircaloy or, in other tests, stainless-steel-clad simulated fuel rods that were electrically heated.

In all of these tests, only a negligible fraction of the emergency core-cooling water reached the core, even in those tests where a quick-closing valve on the ECC line, or another fast-acting bypass system, was added. In fact, no appreciable difference in post-break core temperatures was observed when the tests were conducted *without* use of the emergency core cooling.

The official reports of the Idaho Nuclear Corporation, which performed the experiments for the AEC, express the test results:

> The first semi-scale test involving emergency core cooling (ECC), test 845, was performed in which ECC was injected into the annulus between the flow skirt and vessel wall. Early analysis of test data indicates that essentially no emergency core coolant reached the core. Similar behavior of ECC fluid was observed in subsequent tests: no significant amount of ECC fluid reached the core.

The ECC water was observed in the tests to escape through the same break in the primary coolant inlet and outlet as did the primary coolant. As noted in the official test reports, "On the basis of these data, the conclusion was reached that the ECC fluid entered the inlet annulus and was swept out of the vessel to the inlet nozzle and external downcomer, and out the break." Thus the ECC system would fail at the same time as the primary cooling system and would be, therefore, of no assistance in a loss-of-coolant accident.

The Idaho test results raise a major *prima facie* uncertainty about the reliability of the emergency core-cooling system. To develop a full interpretation of the experimental data, we have carefully considered numerous technical issues related to the use of the semi-scale test data in the analysis of the emergency core-cooling system of commercial-size power reactors. On the basis of our analysis we conclude that there are no convincing grounds upon which the *prima facie* problem with the ECC system indicated by the Idaho data can be dismissed. In our judgment the results of the Idaho tests clearly indicate the basic inability of current pressurized-water reactor emergency core-cooling systems to perform the function for which they were designed.

The fact that the mockup used in the tests was very small detracts little from the view that the emergency core-cooling system for pressurized-water reactors is apparently unreliable. It is said that similar tests will be run with a larger core; the results will shed light on this question.

There are differences between the emergency coolant systems of pressurized-water and boiling-water reactors. The LOFT program results are directly applicable only to pressurized-water reactors. Equivalent tests applicable to boiling-water reactors have not yet been performed. We have, however, found that similar important uncertainties are associated with the emergency systems of both types.

The failure of the emergency cooling systems in these preliminary tests indicates the palpable lack of experimental proof of ECC's reliability. The AEC's Safety Program Plan indicates that basic safety research has yet to be completed and is, instead, scheduled over the next decade. These reports bring to light the AEC's manifest failure to adhere to the vital and important procedure of establishing the safety of nuclear power plants *before* initiating the agency's full-scale program for nuclear power plant construction.

Further testing of emergency cooling systems is important for other reasons. Even if these systems function as designed—and the tests just described indicate they may not—there are reasons to believe a major release of radiation would still result. The natural course of a loss-of-coolant accident can be halted if adequate emergency cooling of a core can be initiated within the time (a fraction of a minute) before the irreversible event of core meltdown has begun. Such irreversible events are started when alterations in core geometry prevent or substantially constrict coolant flow through a major portion of the core, especially around a core hot-spot, or when temperature increases have initiated abundant chemical reactions between metal and water or between metal and steam. Severe coolant flow reduction, from any cause, may induce such occurrences. It is known that appreciable metal-steam or metal-water reactions, once started, can generate large quantities of heat and sufficient pressure to rupture or burst reactor containment vessels. Metal-water reactions may very likely follow on the heels of major reduction of coolant flow and can be considered as successive aspects of the development of a single accident, one causing the other.

Standards Uncertain

Emergency core-cooling system performance in a loss-of-coolant accident would be considered successful if the value of the relevant variables were kept below the thresholds defining the onset of the irreversible event of core meltdown. The Atomic Energy Commission has recently established regulations specifying the minimum design standards for emergency core-cooling systems. In the absence of full-scale testing or experience with accidents, such standards are difficult to establish.

Basic research on the susceptibility of altered core geometries to cooling, and on what core geometry changes are induced during a loss-of-coolant accident, irrespective of their effects on cooling, has only recently started. No adequate fraction of the program has been completed. Although some preliminary theoretical calculations indicate that coolant flow reductions of up to 90 percent (a possible result of swelling of fuel rods, which has been known to occur) may have no deleterious effect on emergency cooling adequacy, there are other engineering reports that appear to contradict this. Moreover, the predictions themselves are of uncertain validity. Tests of adequate scope (involving flow resistance in distorted core geometries) to settle this important question have not been carried out, but like many other tests, are planned. It is absolutely clear that complete blockage of cooling-water flow cannot be tolerated, but there is insufficient assurance, again resulting from lack of tests of adequate scope, that core damage in an accident will not lead to nearly total flow stoppages in a large enough portion of a reactor core so that a major release of radioactivity cannot be prevented.

One of the Oak Ridge National Laboratory's principal investigators of changing core geometry during a loss-of-coolant accident, R. D. Waddell, Jr., summarized current knowledge concerning the possibility of cooling altered core geometries when he flatly stated: "It would be presumptuous at this time to predict what level of coolant area reduction could be tolerated in a LOCA (loss-of-coolant accident)."[5]

The AEC's recent design criteria would allow fuel surface temperature to reach 2,300 degrees F. Two recent experiments indicate that gross swelling and even rupture of fuel tubes can take place at temperatures very much less than 2,300 degrees F. Both swelling and rupture represent irreversible changes in core geometry that could lead to complete core meltdown.

The first of the two experiments carried out at the Oak Ridge National Laboratory, Oak Ridge, Tennessee, was a test of fuel-rod failures in a simulated loss-of-coolant accident.[6] The experiment employed a special test reactor (called TREAT) using actual fuel rods exposed to steam flow, simulating conditions after accidental loss of cooling water. The test reactor was operated at steady power for 28 seconds so as to bring the maximum measured fuel cladding (fuel sheathing) temperatures to 1,770 degrees F. Heating of the fuel rods was halted two seconds later. (As we have said, in a real accident, fission product heating would continue, not subject to control.) Fuel-rod ruptures had by then become extensive. *All* fuel elements were swollen and bowed. One of the elements, previously irradiated in another reactor in an amount equivalent to a few percent of its normal fuel burnup, ruptured, releasing parts of its uranium dioxide fuel and fission products into the core and steam. Its failure initiated what appears to have been the start of an unexpected propagating process of fuel damage. Hydrogen generated by zirconium-steam reactions was identified. In the words of the report,

> The zircaloy cladding swelled and ruptured resulting in a 48 percent blockage of the bundle coolant channel area at the location of maximum swelling . . . examination revealed ductile ruptures and significant oxygen pickup.

The relevance of these results derives from the fact that the test "was conducted under the most realistic loss-of-coolant accident conditions of any experiment to date." The investigators found that fuel-rod ruptures were close together and concluded,

"This indicates a high sensitivity to temperature. . . ." This sensitivity is verified in other tests carried out by R. D. Waddell, Jr. at Oak Ridge.

On wonders what would occur in an accident that developed a cladding temperature of 2,300 degrees F. (500 degrees F. greater than in this test), the temperature generally expected in a major accident and acceptable under present design criteria.

Another test of the TREAT series has been completed, although only a preliminary and not a final report is so far available. In this test, run at somewhat higher temperatures, portions of the fuel-rod assembly were 100 percent blocked. The report states that

> the fuel rod swelling obtained was sufficient to block more than 85 percent of the coolant channel area across the bundle in the second experiment. Although not yet demonstrated, it is thought that obstruction of the coolant channels to this extent could significantly deter effective application of emergency coolant to the region affected by the swollen cladding.[7]

A planned third test was canceled after funds ran out.

In these two unique in-reactor simulations of the important portion of a loss-of-coolant accident, a total of fourteen fuel rods was tested. Only two of the fourteen had been previously irradiated, the remainder being fresh rods. There are 30,000 or more fuel rods in a modern power reactor. A comparison of these numbers gives some measure of the thinness of the experimental base of information.

One consequence of this temperature during an accident, independent of the effects we have just discussed, is the likelihood of metallurgical reactions that can cause damage to a core and to its containment vessels so severe as to itself constitute the beginnings of uncontrolled meltdown. The second of the recent experiments on reactor accidents sheds light on this issue. The tests Zr-3, Zr-4, and Zr-5 were devised and run by the Idaho Nuclear Corporation (INC) under contract with the Atomic Energy Commission to determine the response of zircaloy-2 fuel cladding to emergency core-cooling conditions. The fuel cladding was subjected to a temperature cycle in the presence of steam.[8] These tests showed that unexpected chemical reactions among the materials of the fuel, supporting structures, and water and steam, might occur at temperatures below those expected in an accident.

The tests Zr-3, 4, and 5 were three of over one hundred tests run under the boiling-water reactor portion of the Full Length Emergency Cooling Heat Transfer (FLECHT) program.[9] This portion forms the principal body of experimental information on which boiling-water reactors' assurance of safety is based. Only five of the tests employed zircaloy-clad rods and were intended to simulate accident conditions in a reactor. Only one of these, Zr-2, employed pressurized rods to simulate fission gas pressure, and so alone was able to reproduce the fuel-rod swelling characteristic of a real accident. This critical test was deeply flawed and little could be usefully concluded. The remainder of the series of tests, especially the stainless steel series, was marred by poor instrumentation, occurrence of unexplained phenomena, and was largely inapplicable to the expected conditions of a reactor accident.

The implications of tests Zr-2, 3, and 4 and the TREAT measurements are that, as a core heats to above 1,600 degrees F. during the depressurization (or blowdown) phase of a loss-of-coolant accident, extensive fuel-rod swelling starts, with consequent constriction of the coolant flow channels. At 1,800 degrees F. one expects coolant channels to be between 50 percent and 100 percent blocked (depending on the internal pressures of the individual rods) and extensive fuel-element rupture to have occurred. As the temperature continues to increase toward 2300 degrees F., with impeded coolant flow (and the emergency core-cooling results suggest that the temperature *will* rise close to this value, even with *unimpeded* coolant flow) it is a reasonable and indeed likely conclusion that unexpected cladding and fuel reactions with steam will develop unusually high and sharply defined maximums of temperature, increased melting of cladding, further fuel-rod perforation, and extensive additional damage to the core. Most of these phenomena occur through mechanisms not considered in present safety standards. Possibly, the recently discovered propagating fuel element failures would cause even more rapid and widespread core destruction than we can now foresee. If such destruction occurred, substantial amounts of fission products would be released into the containment. It should be recognized that these are the last developments that precede uncontrolled core meltdown. It appears possible, although not certain, that the development of an accident could not be arrested in a large, badly damaged reactor in which some fraction of the core was at 2,300 degrees F., and that, at this temperature, emergency core-cooling water would then serve only to aggravate the accident.

In addition to coolant flow interference from mechanisms related to temperature changes and metallurgical phenomena, there are other mechanisms that might alter core geometry during a loss-of-coolant accident. The loss of primary coolant in a pressurized water reactor, with attendant shock waves and waterhammer effects (sharp but quickly fading impacts), could well be a brusque and destructive event, with violent coolant flow conditions that a reactor core might not be able to withstand. In addition to shock-wave and waterhammer damage to the core in an accident, there is also a presumption that thermal shock to the fuel rods from contact with emergency core coolant may also prove highly damaging. The AEC's present design criteria for emergency cooling fail to consider all such mechanisms that might render the coolant flow ineffective. Indeed, owing to the serious shortcomings of the reactor safety research program, little is known about the magnitude and consequences of destructive forces and abrupt cooling on the core.

The areas where information is lacking can be most easily identified by a study of the proposed safety engineering research program at the National Reactor Test Station.[10] In requesting supplementary funding for fiscal year 1972 for light-water reactor safety studies, a spokesman said, "And this is largely with recognition that over the last few years we may have undershot somewhat over all." The proposed program includes studies of core power decline (following a loss-of-coolant accident), mechanical responses of the core, emergency coolant initiation and delivery, melting of the fuel cladding and of the fuel itself, and cladding temperature rise reversal, to name a few.

Confidence in the successful operation of reactor safety systems can come only after the completion of a major program of engineering research and study, including closely controlled, large-scale, "near-real" accident simulations. Such a program has not been carried out. There is presently available only an exceptionally thin base of engineering experience with emergency core-cooling system behavior. None of the engineering studies have been conducted to date under near real conditions.

Computer Predictions

For situations in which direct experience is difficult to obtain, computers can sometimes be used to predict the behavior of systems, but this requires that the underlying behavior of the systems be well understood. This technique is often referred to as "mathematical modeling," and is extensively used in nuclear power plant design. Yet mathematical models cannot be used reliably to span large gaps in engineering knowledge, owing to the very great uncertainties that accumulate in long and unverified chains of inference. The quality of the mathematical simulation is influenced in a vital way by the extent to which the modeling procedure has been based upon experimentally warranted assumptions and parameters and confirmed by suitable tests. Without presently lacking engineering experience it is difficult, if not impossible, to determine whether these models are truly accurate and useful. The lack of critical data is likely to lead to elegant but empty theorizing. It is possible, however, to determine whether there exist in a computer program obvious weaknesses that reduce or eliminate confidence in its application.

Accordingly, experimental confirmation of the accuracy of the computer techniques presently used to predict the performance of reactor safety systems is a compelling prerequisite for their use in determining the suitability of nuclear power plants for safe operation.

The Atomic Energy Commission, nevertheless, has taken the position that certain mathematical models may be used to span the very substantial gap between the meager conclusions one can draw from available experimental information and credible assurance of the satisfactory performance of untested emergency cooling systems. The AEC places its reliance on the predictive capabilities of mathematical models embodied in computer codes being developed by its own researchers and by reactor manufacturers.

A reliable estimation of the expected performance of the emergency core-cooling system during a loss-of-coolant accident in a large reactor would require a most sophisticated mathematical model together with the resources of a very large computer. It is likely that no modern computer is of adequate size. This mathematical model, embodied in a computer program, or code, would have to simulate accurately the complex phenomena expected to occur in a loss-of-coolant accident. To construct a code that would do all this reliably is beyond the present capabilities of engineering science. Feasible computer programs that attempt to represent these events will be forced to make very substantial compromises in their description of events because of the complexity of the mathematics and uncertainty in, or total lack of, important engineering data. Many of the phenomena must be omitted or only approximately described. To limit the scale and complexity to manageable proportions, approximations will have to be made. The validity of the models—their ability to make reliable predictions about the consequences of a loss-of-coolant accident—will be greatly influenced by these mathematical simplifications, computational approximations, and by neglect or oversimplified treatment of many processes.

There is abundant evidence that the computer codes presently in use are insufficiently refined from a mathematical point of view, inadequately tested in a suitable experimental program, and embody unsatisfactory simplifications of the real situations. They cannot provide a secure base for predicting the performance of presently designed emergency core-cooling systems during loss-of-coolant accidents. Despite all these difficulties, computer predictions rather than actual experiments form the basis of design of emergency cooling systems.[11]

The failure of recent semi-scale tests described above would seem to emphasize difficulties with present computer codes and predictions. The AEC, however, takes the following position:

> The recent small mockup tests at NRTS [National Reactor Test Station] were not designed to represent the response of an actual operating nuclear power plant to a loss of coolant accident. There were significant differences in the experimental system which was tested as compared to an operating reactor.[12]

Science reported that AEC Director of Reactor Development and Technology, Milton Shaw, "insists these findings [Idaho Semiscale Test results] have little direct bearing on the safety of nuclear reactors."[13] The implication is that computer predictions of full-scale accident conditions are not put to question by less realistic small-scale tests.

One disturbing feature of any attempt to study the quality (or lack of it) in safety system performance is that a number of computer codes, developed by reactor manufacturers to analyze the performance of their systems during accident conditions, are not available for public examination. The vendors consider these codes to be proprietary information. However, because sufficient public data have been provided on the assumptions employed in the codes and on the basic equations they incorporate, we have been able to conduct a partial review of the codes' structures and adequacies. We have concluded that the proprietary codes are not more acceptable than those in the public domain. We believe it is very wrong to conceal, even partially, from public view material which so evidently affects the public health and safety, especially since there is a deep suspicion of the adequacy of the material.

In summary, the gaps in engineering knowledge of safety system performance are simply too great to be bridged adequately by present mathematical models.

Conclusions

The grave weaknesses apparent in engineering knowledge of emergency core-cooling systems and the strong implications that these systems would fail to terminate safely a loss-of-coolant accident make it clear that in the event of a major reactor accident the United States might easily suffer a peacetime catastrophe, the scale of which might well exceed anything the nation has ever known.

The gaps in basic knowledge concerning the effectiveness of the safety features of large power reactors and the surprising scarcity—nearly amounting to a total lack—of adequate tests have astonished our group, especially in view of the large number of reactors of possibly hazardous design that are already operating.

We have concluded that there are major and critical gaps in present knowledge of safety sytems designed to prevent or ameliorate major reactor accidents. We have further concluded that the scanty information available indicates that presently installed emergency core-cooling systems could well fail to prevent such a major accident. The scale of the possible consequent catastrophe is such that we cannot support the licensing and operation of any additional power reactors in the United States, irrespective of the benefits they would provide.

References and Notes

1. *Science*, May 28, 1971, p. 191.
2. Lawson, C. G. *Emergency Core-Cooling Systems for Light-Water-Coled Power Reactors*, Oak Ridge National Laboratories, ORNL-NSIC-24, Oak Ridge, Tenn., 1968, p. 52.
3. Report of the Task Force Established by the USAEC to Study Fuel Cooling Systems of Nuclear Power Plants, *Emergency Core Cooling*, 1967, p. 145.
4. Battelle Memorial Institute Reports, BMI-1825 and BMI-1910.
5. Waddell, R. D. "Measurement of Light Water Reactor Coolant Channel Reduction Arising from Cladding Defermation During a Loss-of-Coolant Accident," *Nuclear Technology* 11(4): 491, (August 1971).
6. The report on these tests: Lorenz, R., D. Hobson, and G. Parker, *Final Report on the First Fuel Rod Failure Transient Test of A Zircaloy-Clad Fuel Rod Cluster in TREAT*, Oak Ridge National Laboratory, (ORNL-4635). Oak Ridge, Tenn., Mar. 1971.
7. ORNL TM 3263.
8. The results are recent and are reported in an Idaho Nuclear Corporation document: M. J. Graber, W. E. Zelensky, and R. E. Schmunk, *A Metallurgical Evaluation of Simulated BWR Emergency Core-Cooling Tests* (IN-1453), Feb. 1971.
9. GEAP-13197.
10. Presentation to AEC Atomic Safety and Licensing Board, Idaho Falls, Idaho, July 1, 1971.
11. For detailed documentation of the extensive gaps in our experimentally derived understanding of reactor accident phenomena and of the performance capabilities of emergency systems, refer to: Lawson, C. G. *Emergency Core-Cooling Systems for Light-Water Cooled Reactors*, ORNL-NSIC-24 Oak Ridge National Laboratory, Oak Ridge, Tennessee, 1968. USAEC, *Water-Reactor Safety Program Plan*, WASH-1146, Feb. 1970. Committee on Reactor Safety Technology (CREST), European Nuclear Energy Agency, *Water-Cooled Reactor Safety*, OECD, Paris, May 1970, Forbes, Ian A., Daniel F. Ford, Henry W. Kendall, and James J. MacKenzie, *Nuclear Reactor Safety: An Evaluation of New Evidence*, The Union of Concerned Scientists, Cambridge, Mass., July 1971, reprinted in *Nuclear News*, Sept. 1971.
12. AEC News Release, May 27, 1971.
13. *Science*, May 28, 1971, p. 919.

Nuclear Safety

In January of this year a public hearing[1] began on the adequacy of the Atomic Energy Commission's design standards for a key safety system in nuclear electric power plants. The hearing was the result of a public controversy that Ian Forbes, James MacKenzie, and the present authors, of the Union of Concerned Scientists (UCS), opened in the summer of 1971 with a critique of presently designed emergency core cooling systems for commercial nuclear power plants. (See "Cooling Water," *Environment*, January/February 1972.) Emergency cooling systems are needed to prevent overheating, melting, and release of radiation from the massive uranium fuel "core" of such plants. The hearing has produced an illuminating record: The community of reactor safety experts has been shown to be widely divided on this vital issue. Many of the AEC's own experts—in the National Laboratories, the AEC contractor laboratories, and within the AEC itself—consider highly inadequate the AEC criteria which now define an acceptable emergency cooling system; experts who work for nuclear power plant (reactor) manufacturers, on the other hand, claim the criteria are too stringent. The AEC licensing division sided with the reactor manufacturers against the AEC's own experts when it promulgated the present safety criteria in June 1971. Disturbing evidence has accumulated to reveal that the AEC has moved systematically to suppress internal dissent and to disregard the mass of information adverse to the industry position. Important safety research programs were cut back or terminated abruptly, with the consequence of diminishing the flow of additional information.

Possible troubles with the emergency systems and the consequences of an accident have been discussed in articles by Sheldon Novick[2] and by Ralph Lapp.[3] The UCS issued two reports in 1971[4] following the unexpected failure of some reactor emergency system tests and the AEC's hasty promulgation of Interim Criteria for reactor safety; UCS raised the safety question in legal proceedings to prevent the licensing of several nuclear plants. By December 1971 one of the AEC's reactor licensing boards, hearing the Indian Point 2 Reactor controversy, informed the AEC

By Daniel F. Ford and Henry W. Kendall, Union of Concerned Scientists. From *Environment* **14**(7), 2–9, 48 (September 1972). Reprinted with permission of the Helen Dwight Reid Educational Foundation. Published by Heldref Publications, 1319 18th St. N.W., Washington, DC 20036-1802. Copyright © 1972.

that the board had serious questions about the technical and legal validity of the AEC safety system design standards. The issue was meanwhile being raised in other licensing proceedings across the country.

The AEC agreed at the urging of the UCS and others to hold national hearings on emergency core cooling in an attempt to extricate itself from cascading difficulties in local hearings connected with the licensing of individual nuclear power plants. In this single national hearing the AEC said that it would make available for questioning, for the first time, its own reactor safety experts from within the AEC Regulatory Staff, the Oak Ridge National Laboratory, and the National Reactor Testing Station in Idaho, operated for the AEC by the Aerojet Nuclear Company and the AEC's prime reactor safety research center.

The AEC's opponents in local licensing hearings joined with the UCS to form an intervenor's group to participate in the hearings as the Consolidated National Intervenors. The hearings began in January 1972 with the Intervenors, the U.S. reactor manufacturers (Westinghouse, General Electric, Combustion Engineering, and Babcock and Wilcox), several electric utilities, and the AEC Regulatory Staff as participants. Principal counsel for the Intervenors were Myron M. Cherry, well-known for his role in shaping AEC policy and opening to public scrutiny information secreted by the AEC, and Thomas B. Arnold, a Boston environmental lawyer.

Over strenuous objections the Intervenors and all other participants were denied the power of subpoena normally provided to intervenor groups in reaction interventions. Potentially important testimony by a number of reactor safety personnel therefore became unavailable, for many experts were reluctant to come forward voluntarily. In addition, "discovery" was forbidden, that is, the right to examine underlying information in the possession of the reactor manufacturers or the AEC (information which each relies upon).

The AEC was very reluctant to make public all of its internal documents concerning the adequacy of the safety systems, but, under threat of lawsuit under the Freedom of Information Act, made a token release of 60 documents in February 1972. These have been supplemented as new memoranda, reports, and correspondence were identified and requested.

Despite all of the obstacles to a full and complete record, the hearing has presented an unparalleled opportunity to see how this large agency, the AEC, has dealt with a most critical public safety issue.

The Possibility of Catastrophe

The basic issue revolves around the possibility of a catastrophic reactor accident with the consequent release of large amounts of radioactivity to the environment. Reactors accumulate prodigious quantities of radioactive materials as they continue in operation: A large power reactor can contain fission products equivalent to the fall-out from many dozens of nuclear weapons in the megaton range. The uncontrolled release of even 5 or 10 percent of this inventory could bring death to persons from 60 up to 100 miles from the reactor. Persons hundreds of miles distant

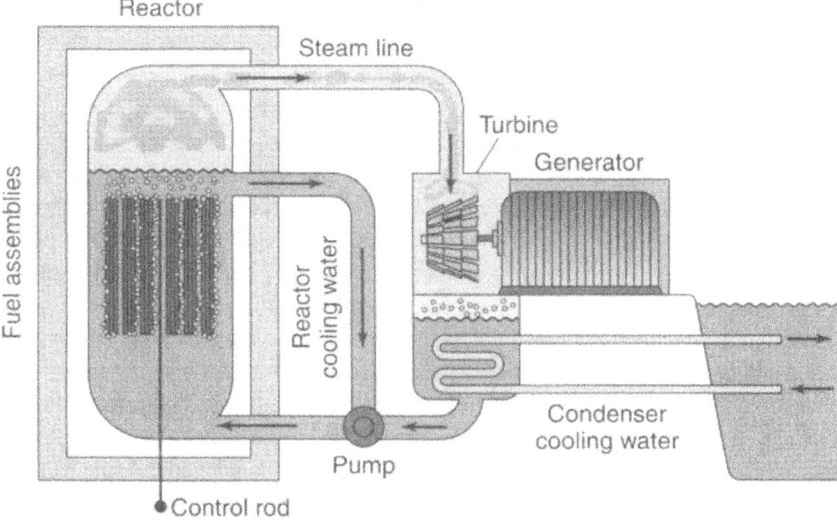

Figure 1. A highly simplified diagram showing the essential parts of a nuclear power plant (of the type called a "boiling-water reactor"). The heart of the plant, or its "core," is the array of long, thin rods filled with pellets of uranium fuel. These become hot during reactor operation, and water is pumped through the fuel and back again in an enclosed circuit. Steam is drawn off from this circuit and carried to a turbine, which spine to produce electricity. The Atomic Energy Commission believes the worst possible accident which could occur in such a plant is the breaking of the pipe which carries water to the fuel. With cooling-water lost, the fuel would quickly overheat and melt, releasing radioactivity. To forestall this, various spray or flooding systems are installed to douse the fuel in the event of a pipe breaking. It is the performance of these emergency cooling systems which is questioned in the accompanying article.

could suffer radiation sickness, genetic damage, and increased incidence of many diseases including cancer. Large-scale and long-lasting contamination of the world's oceans could result from the accidental release of fission products from proposed off-shore nuclear power plants. Several reactors now operating or under construction are sited close enough to large urban areas so that each could put more than ten million persons within range of a lethal plume of radioactivity. One reactor, proposed for a site near Philadelphia, will, if built, be within *one mile* of a planned community of 25,000 persons. The stage is set for an accident of devastating proportions.

Such an accident is a possibility because of the accumulation of radioactive wastes in the fuel of a nuclear power plant during operation. These wastes generate enough heat to melt the fuel and its supporting structures unless steadily cooled. During normal operation, large volumes of water are pumped through the fuel core to cool it. The hot steam which results is used to generate electricity, but when the plant is not in operation the flow of water must still be maintained to prevent overheating. Emergency cooling systems generally include either a spray system to douse the hot fuel, or reserve tanks to flood it with water. An accident such as a pipe or vessel rupture or other malfunction leading to loss of the reactor cooling water (a loss-of-coolant accident, or LOCA) would be expected to lead to rapid heat-up of fuel elements from radioactive waste heat-up even though power generation was halted. Uncontrolled heat-up would soon result in core meltdown, breaching of the reactor's pressure vessel and containment structures, and the release of radioactivity in large quantities. The controversy swirls around the effectiveness of the emergency core cooling systems that reactor manufacturers are required to install to provide replacement cooling promptly after an accident so as to blunt the heat-up and prevent the accident's developing to core meltdown.

Concern within the AEC over the effectiveness of emergency cooling systems began in 1965 with the sale of increasingly large reactors, many in the two or three thousand megawatt (thermal) range. (The electrical output of a nuclear reactor is approximately one-third its total power output. The remainder is waste heat. Thus, about two-thirds of the heat output of a nuclear reactor is wasted, adding generally to environmental problems.) At that time the AEC appointed a task force under the late W. K. Ergen of Oak Ridge National Laboratory to study the problem. Also the AEC Division of Reactor Development and Technology requested Oak Ridge National Laboratory and Battelle Memorial Institute to prepare assessments of the status of specific aspects of nuclear safety. These early reviews converged on one important conclusion: emergency core cooling effectiveness urgently needed experimental confirmation.

The Ergen task force recommended additional tests and studies be carried out on the effectiveness of emergency core cooling systems. Pursuant to its recommendations, a testing program was instituted at the Oak Ridge National Laboratory to study reactor fuel element failure modes and certain heat transfer problems relevant to an accident. A study of the cooling capability of emergency core cooling systems using full-length, electrically heated fuel rods (referred to as FLECHT tests) was instituted under a subcontract from the AEC to its safety contractor, Aerojet Nuclear Company (then the Phillips Petroleum Company) at the National Reactor Testing Station in Idaho Falls. The actual implementation of the tests was carried out by General Electric and Westinghouse, the major reactor vendors, under further subcontracts monitored by Aerojet. Increased attention was given to other series of tests being implemented by Aerojet in Idaho.

An effort was also made to formulate a water-cooled reactor safety program plan that would identify specific priorities and form the basis for a timely and efficient resolution of basic problems. The Water Reactor Safety Program Plan appeared in February 1970 after circulating for two years in draft form. It is noteworthy that this plan represented a consensus among industry and AEC groups involved as to priorities in reactor safety research. The plan identified "all of the factors affecting ECCS effectiveness" as "the most urgent problem areas in the safety program today." The plan identified several areas connected with emergency core cooling systems—for example, behavior of normal cooling water following a pipe rupture, fuel rod failure and effects, effectiveness of boiling-water reactor spray cooling systems and other effects as "very urgent, key problem areas, the solution of which would clearly have impact, either directly or indirectly, on a major critical aspect of reactor safety."

According to the UCS report submitted as testimony in the present hearings[5] the manufacturers' reports on the FLECHT tests that they carried out show numerous crippling failures in the test program including narrow scope, inadequate design of the test components, faulty conduct of the tests, and finally, most generous interpretations of the test results that in some cases were grossly inconsistent with the accumulated data. The Aerojet engineer in charge of the General Electric subcontract said in cross-examination at the hearings that the test conclusions were "tremendously slanted." Post-test analysis by Aerojet of one particularly important FLECHT run relied on by a vendor showed that 50 percent of the electrical energy delivered to the test bundle could not be accounted for and that in consequence "apparently significant errors exist in one or more of the measured test parameters." In 1969 and 1970, after many of the FLECHT tests had been run, Aerojet voiced its concern that the mechanical failures that occurred during such critical tests invalidated the test results.

Because of the expense and difficulty of testing the effects of accidents on fullscale nuclear plants, mathematical calculations are made to predict the effects of such accidents. The calculations require a computer, and are sometimes referred to as a mathematical "model" of a real reactor accident.

In the fall of 1970, Aerojet ran a series of tests, using a nine-inch-diameter model reactor core, to test the accuracy of mathematical models designed to evaluate the effects of loss of cooling water in a pressurized-water reactor. The model emergency system failed completely in these tests to deliver water to the core and the computer models were basically unable to predict the test results. The results sent a minor shock wave through the nuclear community. On February 11, 1971, a meeting was held by the representatives of the AEC, Aerojet, Oak Ridge National Laboratory, and the reactor manufacturers. The purpose of the meeting was to discuss criteria for the performance of emergency systems in a loss-of-coolant accident. There was a consensus reported from the February 11 meeting that the physical phenomena which might occur during a major accident were not adequately understood and were still open to serious question. During previous months, results from revised computer codes for accident analysis had been received by the Regulatory Staff, results which differed substantially from previous predictions.

Unresolved Questions

In the spring of 1971, therefore, as a result of the failure of the semiscale tests and the other disturbing developments concerning emergency cooling system adequacy, the AEC Regulatory Staff formed a task force under Dr. Stephen Hanauer to review the state-of-the-art of emergency core cooling systems. The AEC Regulatory Staff did not have a sufficient number of technically qualified people in-house to perform this review. Instead, in keeping with the Regulatory Staff's usual practice, expert consultants were asked to provide technical backup. The staff's chief consultant in this area is the Aerojet Nuclear Company. Thus, the emergency core cooling system task force commissioned Aerojet to prepare a detailed review of the state-of-the-art of the evaluation models used, in the absence of large experiments, to predict what would happen during a loss-of-coolant accident. The Aerojet Nuclear Company's two-volume report, which was produced following the staff's request, was in fact ignored by the Regulatory Staff during its preparation of criteria to be used for evaluating ECCS. The disregarded document stated that in several critical areas the ability to predict what would happen during a loss-of-coolant accident was simply "beyond the present capability of engineering science." The Regulatory Staff disregarded the detailed reports that contained those conclusions. Task force head Dr. Stephen Hanauer testified that he hadn't read the Aerojet report, despite the fact that he had in part commissioned its preparation.

The 60 internal memoranda released by the AEC in February of 1972 allowed a public view of the workings of the task force, and showed that there were other opinions available to it, which were similarly disregarded. Consider, for example, a memo written by the man in the AEC Regulatory Staff who had been in charge of emergency core cooling system analysis since its inception. This man, Dr. Morris Rosen, then chief of the Systems Performance Branch of AEC's Division of Reactor Standards, and his deputy, Robert J. Colmar, in a memo of June 1, 1971, stated:

> We believe that the consummate message in the accumulated code outputs is that the system performance cannot be designed with sufficient assurance to provide a clear basis for licensing.

Apparently the AEC's primary technical contractors and several of the most knowledgeable members of the task force seriously questioned whether any adequate criteria and evaluation models could be devised until further experimental testing had been carried out. Nevertheless, the AEC adopted on June 29, 1971 Interim Acceptance Criteria which merely codified existing informal criteria which were then in use within the AEC and made minor changes in the accepted evaluation models. Nowhere in the Interim Policy Statement, of which the Criteria were a part, did the AEC acknowledge the existence of unresolved questions or the fact that its own Regulatory Staff was split over these questions. However, the AEC was sufficiently concerned that it issued the Interim Acceptance Criteria to take effect immediately upon publication in the Federal Register. In its Interim Policy Statement, the AEC stated:

> In view of the large amount of new information available, the AEC has again conducted a review of the present state of ECCS technology and has reevaluated the basis previously used for adopting system designs for current types of light water reactors . . . In view of the public's health and safety considerations discussed above, the Commission has found that the Interim Acceptance Criteria contained herein should be promulgated without delay . . . and that good cause exists for making the statement of policy effective upon publication in the *Federal Register*.

The Interim Acceptance Criteria did not meet the substantive criticisms which had been raised before June 29 by the National Laboratories and dissenters within the Regulatory Staff. They did allow reactor licensing to continue without reductions in reactor power, since the designs of most reactors then operating or about to be licensed met the new criteria—a most interesting coincidence. Although alteration in the design of reactors already built or under construction might not have been possible, should the criteria have been made more stringent, reductions in the power levels at which the plants were allowed to operate could have been made to provide added safety margins. Such reductions, which would have resulted in substantial loss of revenue to the affected power companies, were not required under the Interim Criteria.

Between the issuance of the Interim Criteria in June 1971 and the start of the hearing in January 1972 there were three significant developments.

The first concerned a report from Aerojet, which is primarily responsible to the AEC's Division of Reactor Development and Technology and is only allowed by that division to do limited consulting work for the AEC licensing group. In this regard, the hearing brought to light the disturbing fact that the Reactor Development and Technology division was withholding information from the AEC licensing group. Thus, a report which Aerojet prepared in August of 1971 (subsequent to the two-volume review prepared for and disregarded by the Regulatory Staff) on the areas where additional safety research was urgently needed was not transmitted by the Division of Reactor Development and Technology to the Regulatory Staff. The report in question was devastating in its criticism of the emergency core cooling system in pressurized water reactors (about one-half of all reactors). Among other evidence, the report included an extended discussion of a table of 28 areas of knowledge or capability critical to effective emergency core cooling operation. In 7 areas techniques for dealing with the problems were identified as completely missing. The status of the remaining 23 was described as incomplete, unverified, inadequate, imprecise, or preliminary. Eight of the areas were identified as of very high urgency, 15 of high urgency. The AEC Regulatory Staff saw this evaluation of needed safety research for the first time in March 1972 when it was handed to them by the Intervenors, who had received an unsolicited copy.

The second significant development was in November 1971: The Division of Reactor Development and Technology issued its Water Reactor Safety Program Augmentation Plan. In a cover letter for the report, Milton Shaw, director of the division, stated that the "recommended new analytical and experimental efforts (were) considered urgent. . . ." The report stated in pertinent part:

> A major loss of coolant accident is extremely unlikely and present ECC systems as designed are expected to mitigate the consequences of such an accident should one occur. However, present *experimental data and analysis techniques are not now sufficient to provide the degree of ECC assurance deemed necessary by the AEC.* It is for this reason that a major emphasis in both the basic program and the augmented program is on the performance of ECC systems. . . . [Emphasis added.]
>
> As reactor designs and their operating characteristics changed, the analysis methods were 'patched up,' rather than redeveloped, with the net result that overall, existing methods are inefficient, inflexible, and do not adequately represent the physical phenomena intended.

The third prehearing development took place in December 1971. William Cottrell, director of Oak Ridge National Laboratory's Nuclear Safety Program wrote to the AEC's director of regulation, L. M. Muntzing:

> To summarize what follows herein, we are not certain that the Interim Criteria for ECCS adopted by the AEC will, as stated in the Federal Register "provide reasonable assurance that such systems will be effective in the unlikely event of a loss-of-coolant accident."

One of Cottrell's superiors called Muntzing and had the letter returned to Oak Ridge National Laboratory on the basis that it was a "draft" and that it did not represent the views of the laboratory's safety personnel. Cottrell's subsequent testimony has shown that the letter did in fact represent the views of the relevant Oak Ridge experts and that it was not a draft. Indeed, the sequence of events regarding the Cottrell letter is in hindsight a bellwether of the attitude the AEC was to take at the hearings themselves.

On December 28, 1971, the AEC Regulatory Staff filed its written testimony in the present emergency core cooling system hearings. The central conclusion of the staff testimony was "the conservatism in today's criteria and today's evaluation models suffice to provide reasonable assurance of the safety of today's reactors." The position of the Regulatory Staff was presented without indicating the nature or basis for any disagreements within the AEC. In fact, the staff testimony gave the impression that disagreements within the staff, if any, were of a minor and inconsequential nature. The clear impression was conveyed that no unresolved issues remained within the AEC Regulatory Staff. Yet, as we have seen, there were serious disagreements on important technical matters.

The AEC Criticized

Another divergent view came to light among the material released by the AEC. The director of the Oak Ridge National Laboratory, Alvin Weinberg, in a February 1972 letter to AEC Chairman James Schlesigner stated his "basic distrust" of reliance on the computer codes that the assurance of safety systems performance must depend on. Weinberg also went on to lament the fact that the Oak Ridge National Laboratory was not involved in the formulation of the Interim Criteria and that this "reflects a deficiency in the relation between the Laboratory and the Commission that troubles me."

When the Intervenors had forced release of the internal memoranda which documented publicly for the first time the extent and depth of the split of opinion within the AEC, Morris Rosen and Robert Colmar were called as witnesses at the hearing. Rosen termed the Hanauer task force review of emergency core cooling systems "superficial." He was disturbed and discouraged "to continue to see the advice of what I believe can be considered a significant portion of, more likely a majority of, the knowledgeable people available to the Regulatory Staff, still being basically disregarded." In Rosen's opinion, the Regulatory Staff does not have "knowledge sufficiently adequate to make licensing decisions for the approximately 100 reactors

UNITED STATES
ATOMIC ENERGY COMMISSION
WASHINGTON, D.C. 20545

July 28, 1972

Mr. Julian McCaull
Managing Editor
Environment Magazine
438 North Skinker Boulevard
St. Louis, Missouri 63130

Dear Mr. McCaull:

Thank you for the opportunity afforded us by your letter of July 14 to comment on the article by Daniel Ford and Henry Kendall which is being reviewed for publication in Environment magazine.

As the article notes in one of its few non-argumentative assertions, a public hearing on emergency core cooling systems—which has as its objective resolution of many questions raised in the article—is currently in progress. Therefore, we do not believe it would be appropriate for the Commission to comment at this time on the merit, or lack of it, of the many charges set forth in the article.

It should be noted that both Mr. Ford and Dr. Kendall are involved in this hearing, and their article sets forth the position they have taken. As such, it is an article advocating a contested point of view, which the article itself does not adequately put in context. It is fair to say that there is a very substantial amount of information in the record of the hearing which takes the opposite view from many of the statements made by the authors.

The Commission does not feel it would be proper, prior to the completion of the evidence-taking phase of the hearing, to attempt to summarize what the bulk of the testimony and documentation tends to show thus far.

For these reasons, we would only observe that the open examination which the Commission has provided for in this innovative rulemaking proceeding, and the Commission's commitment to make its ultimate decision based on the entire record, should serve to enhance public confidence in the integrity of the proceeding and of the final Commission determination in this matter.

Sincerely,

John A. Harris, Director
Office of Information Services

operating or under construction." Finally, Rosen stated that present AEC methods for analyzing emergency cooling systems are "based on numerous unrealistic models, suffer from a number of restrictive assumptions and lack applicable experimental verification. Under these conditions . . . the effectiveness of emergency core cooling systems cannot be established."

Hanauer in cross-examination said that three members of the Advisory Committee on Reactor Safety, an important committee which functions as the AEC's "watchdog," had "serious reservations about the effectiveness of ECCS." He identified these members as the most knowledgeable in the emergency core cooling system area. In a document released by the AEC, Hanauer noted that the advisory committee had recommended seeking design changes in the emergency cooling systems.

As expected by the Intervenors, the AEC and the reactor vendors filed written testimony in support of the technical foundation for the Interim Acceptance Criteria. However, Colmar expressed the sense of uneasiness felt by many knowledgeable people concerning the vendors' technical justification for the Interim Criteria when he stated that "if we have no one but the private sectors, the vendor, or the utilities on which to rely in terms of evaluating the safety of the emergency core cooling systems . . . I would find that somewhat disturbing."

Rosen supported that position, saying, "It is obvious to me the vendors have a point of view and a product to sell. And that's one of their inputs to us. And it's based on that. The Regulatory organization is going to go outside independently to review the inputs from the vendors." Rosen's reference to "outside" independent experts was to Oak Ridge, the other national laboratories, and Aerojet, whose participation in the formulation of emergency cooling system criteria was, as we have seen, either not solicited or ignored.

Following Rosen and Colmar's testimony and cross-examination a succession of reactor engineers from these laboratories and from within the AEC testified that they were in general agreement with the Rosen-Colmar position.

C. George Lawson, a heat transfer expert from Oak Ridge said that "any conclusion with respect to the effectiveness of ECCS is speculative," and went on to state that: "The assertion is that conservative assumptions are made [in the safety analysis] where possible and that is true. But there are some areas where, in my opinion, we don't know whether the assumptions we are making are conservative or not because we don't know what is occurring physically." Phillip Rittenhouse, in charge of the Oak Ridge fuel-failure studies program, outlined his own sense of insecurity and pointed out a number of acute technical deficiencies. He said, for example, that no reasonable scientist can tell if the FLECHT program demonstrates that a core is amenable to cooling, a claim which is in contrast to reactor vendors' claims. He went on to say that his views of emergency cooling reliability were widely shared by his colleagues. Under cross-examination, he identified many of these colleagues by reading into the record the names of 28 persons including Cottrell and his assistant, ten top officials of Aerojet, and Norman Lauben of the AEC Regulatory Staff. Lauben, a member of the group who had prepared the AEC's official hearing testimony, said that "There were certainly portions of the [AEC] testimony that I would have to consider personally as not being sufficient,"[6] and then testified that a minor change in one of the assumptions in the AEC's prediction models for accident analysis would make the difference between a controlled loss-of-coolant accident and uncontrolled core meltdown.

Weaknesses in the computer codes (that is, calculations used to predict the course of an accident) on which the Interim Criteria depend has been an important subject in the hearings.

The Water Reactor Safety Program Augmentation Plan of November 1971 made numerous references to these codes, among them:

> To date the evolution of the codes has not kept pace with the development of ECC systems.
>
> Additionally, the codes (used in LOCA analysis) are unable to describe important physical phenomena and therefore *unable to confidently define safety margins*, their treatment of common phenomena is inconsistent, and they overemphasize the use of empirical correlations. [Emphasis added.]

Code expert, Amir N. Nahavandi, the author of one of the important vendor codes, wrote in a document released by the AEC:

> ... under the present conditions, the core fluid flow and heat-removal capability and effectiveness of the ECCS in maintaining the fuel cladding temperature within allowable limits cannot be established.

In his report he proposed the development of a much improved computer code to rectify the situation, a code that would have to be ten times larger than any now running. Nahavandi spoke of the present codes as inadequate, lacking verification, and unrealistic.

In a memorandum in August 1971 to Milton Shaw, Edson Case, director of the AEC Reactor Standards Division, wrote:

> Current efforts on the development of a more sophisticated thermal-hydraulic LOCA code should be substantially increased both in priority and funding. What we have in mind is not 'patching' or adding to existing codes, but development from basic principles of a new code, better able to handle the complex physical problems related to a LOCA and to ECCS performance. . . . This is a difficult task, at the frontier of presently available knowledge but it is urgently needed . . .

Witnesses from Aerojet and from within the AEC outlined additional weaknesses in the codes. The AEC witnesses identified by name other experts at Oak Ridge, Aerojet, and at Battelle Memorial Institute who substantively supported the critics' views.

Much testimony from both Oak Ridge and Aerojet personnel hit at the quality of the AEC-supported safety research carried out by the vendors and the conclusions the vendors had drawn from it and from the research work at Oak Ridge National Laboratory and Aerojet. A telegram from C. M. Rice, president and general manager of Aerojet Nuclear to AEC Chairman Schlesinger, to AEC Director of Regulation Muntzing, and to the emergency core cooling system hearing board was an appeal to the AEC to allow Aerojet specialists to cross-examine the vendors' witnesses to offset the situation in which the reactor vendors "produced a very narrow and distorted picture of the significance and accomplishments of the AEC's reactor safety research and development program." The appeal was rejected.

The whole technical situation with regard to the emergency core cooling system problem was summarized in the UCS testimony prepared for the Intervenors and submitted at the hearing:

> The vendors' and AEC Regulatory Staff's claim that operating power reactors and those presently under construction will present an adequately low hazard to the general public is based on an experimental and analytical structure whose reputed integrity does not withstand a close, nor even a cursory, examination. It is not simply that one or two vital links are demonstrably missing or indeed weak. It is that the whole structure is unsound. It is based on engineering results whose quality and scope are broadly and seriously deficient and whose interpretation is in serious dispute by a host of persons whose views cannot be disregarded. The structure depends critically on computational predictions that are critically weak and unreliable and which are themselves in serious dispute. It is unimaginable to us that the present level of discord should have persisted through the design, construction and operation of nearly two dozen reactors without resulting in any serious reexamination as a prelude to construction of dozens more. The discord amongst professional and qualified persons is the clearest sign one can have that the assurances of reactor safety now offered to the American public are a fiction.

Backlash

One of the reactions by the AEC to the revelations of emergency core cooling system weaknesses has been a program of termination of many of the key reactor safety research programs at the independent laboratories. In the authors' view the reactor safety program has never had the resources or the priorities that it required nor has it been pursued with anything like adequate vigor. The AEC's recent moves have damaged it badly. The Oak Ridge National Laboratory studies of fuel rod failure modes under P. L. Rittenhouse were terminated abruptly in June 1971, with two months notice, just as the program was starting to develop important information on how severely fuel rod swelling can aggravate an accident. The support of reactor safety studies at Battelle Institute in Columbus, Ohio, was halted completely in 1971; important tests of the FLECHT series planned initially were never run; and a series of very informative in-reactor safety studies (one of which was described in an Oak Ridge report in March 1971 as "conducted under the most realistic loss-of-coolant accident conditions of any experiment to date") was terminated one-third short of completion. It was all of this, along with the lagging pace of the safety studies being conducted at Aerojet, that evidently stimulated a strongly worded letter from the Advisory Committee on Reactor Safety to AEC Chairman Schlesinger on February 10, 1972. The letter said "The ACRS recommends that a substantial increase in funds be made available for regulatory support of these activities and for reactor safety experiments which can be initiated in prompt response to items identified in regulatory review." The research areas identified in the letter relate to the majority of phenomena associated with a major LOCA.

Cottrell's December 1971 letter to Muntzing touched on the same lack: "Final criteria [for reactor safety] cannot be developed until more experimental evidence becomes available on these and other aspects of the ECCS situation including, in addition to the work identified above [fuel rod failure modes and fuel rod embrittlement, on which research had been terminated earlier], information on blowdown heat transfer and the development and verification of improved computational programs."

Along with terminating or curtailing safety research the AEC has acted to hinder the free flow of safety-related results developed in on-going programs. We mentioned earlier the Aerojet report withheld from the AEC Regulatory Staff. In addition, in the hearings the AEC's director of reactor development and technology, Milton Shaw, admitted that his division "censors" monthly and other reports from Aerojet. After this revelation, J. Curtis Haire, the head of Aerojet's nuclear safety program, was asked whether in fact censoring of reports has anything to do with technical judgment. Mr. Haire answered that the censoring did *not* have to do with technical judgment but rather was a deliberate inhibition of free and open discussion about nuclear safety by Aerojet Nuclear. Mr. Haire added: "I believe that RDT [the Division of Reactor Development and Technology] is trying to avoid the problem or burden, if you will, of having to spend a lot of time answering public inquiries that are addressed to Congress and referred to them . . . on general questions of nuclear safety."

The Maximum Accident

The AEC has selected a double-ended break (or guillotine rupture) of the largest reactor water recirculating line as the so-called design basis, or maximum credible accident. The AEC will not admit that if a water line can rupture, so can the primary reactor vessel itself. The UCS was not the first to point out the equivalent rupture potential of these two components, nor is it unique in so doing. The Water Reactor Safety Program Plan that appeared in February 1970 represented a consensus among industry and AEC groups involved as to priorities in reactor safety research. That report noted:

> . . . initiation of emergency core cooling (ECC) in the unlikely event of a loss-of-coolant accident presents one of the most extreme thermal conditions for the primary system, particularly large pressure vessels. The resulting thermal shock could conceivably have an adverse effect on primary system integrity, and, subsequently, on ECC effectiveness if the system should fail.

At the hearing, David Rose, a senior nuclear engineer at the Massachusetts Institute of Technology, was quoted:

> On the other hand, the main reactor pressure vessel is much larger [than the reactor water pipes]; though it is made with the very best technology, more uncertainty exists

about its quality, because of the nature of the thing. But also no protections could be envisaged for certain hypothetical failure modes of the main pressure vessel. Such failures have by edict been declared not credible.

Rose in his quote described this as an "intellectually unhealthy and irrational rationalization." Other experts, Rosen among them, have also suggested that this "incredible" failure should be reexamined, the AEC notwithstanding.

There are other more complex circumstances that can bring on a loss-of-coolant accident that involve postulated valve failures, failure of reactor shut-down mechanisms, and certain other classes of pipe ruptures. There is reason to believe that some among them are far more probable than the design basis accident. None, however, have been analyzed by the AEC, nor may they be discussed at the hearing. The probability of an accident is also excluded from discussion.

This refusal to consider the twin issues of what kinds of malfunctions could initiate a loss-of-coolant accident and what in fact is the probability of having an accident is, from a technical viewpoint, a quite serious error. Broadly stated, the object of the hearings is to determine whether the Interim Safety Criteria are adequate to ensure that a loss-of-coolant accident will be controlled by the emergency systems. Without an exploration of the possible events which can give rise to such an accident there can be no certain knowledge that the emergency systems can *in principle* have any effect. Our example of vessel rupture is a real and serious example of an event for whose mitigation all presently designed emergency cooling systems are most likely worthless. The AEC's arbitrary selection of a single kind of accident coupled with its refusal in the present hearings to discuss the root basis of accident control, an analysis of the kinds of events to be controlled, is capricious and technically unjustifiable. The AEC position that a loss-of-coolant accident is "extremely unlikely" is not based on detailed study of the probability of their design basis accident, for none have been carried out, nor does it include estimates of other events occurring. No arguments have been advanced to support their actions.

That not all events occurring in *normally* operating reactors important to loss-of-coolant accident analysis have necessarily been identified was illustrated in cross-examination of Westinghouse safety engineers by the Intervenors. They testified that in a recent power reactor refueling it was discovered that uranium dioxide pellet migration had occurred within the fuel cladding of approximately 5 percent of the fuel rods that were inspected. The pellets had moved up to 3 inches from their correct positions leaving gaps somewhat randomly positioned along the twelve-foot rods. The fuel cladding had collapsed at the larger gaps and rod bowing occurred. Analysis suggested that injured rods might reach temperatures in an accident perhaps 600 degrees F. greater than those of undamaged rods. It was admitted that the injured rods were more susceptible to accident-induced damage which would thus aggravate the severity of an accident. The explosive loss of water after a pipe rupture would make missiles of broken rod fragments, which could be expected to cause substantial damage to the remainder of the reactor core. The computed loss-of-coolant accident fuel rod temperatures for rods not injured by prior collapse is generally no more than from 30 degrees F. to 400 degrees F. below the maximum allowed by the Interim Criteria. The implications of these recent findings, taken alone, are that very substantial reductions in allowable reactor power may be required before certain reactors even meet the Interim Criteria. Cross-examination elicited the fact that all or part of the damaged fuel was reinstalled in the reactor from which it had been taken.

Implications

Many, if not a majority, of the reactor safety engineers outside of the vendor organizations, including among them important AEC personnel, have grave doubts about the quality of the safety assurances that are employed to support reactor licensing. For certain kinds of credible accidents there appears to be no assurance that the emergency cooling systems will prevent a major discharge of radioactive material to the environment. Such a discharge could cause a catastrophe of unparalleled scale and the discord among the professionals is a clear sign that the country does not have adequate assurance that such an accident will not occur.

AEC Chairman Schlesinger in a recent interview[7] said that "there have been questions, and I think erroneous questions, raised about safety" and went on to state that there "would be no threat whatever to public health or safety." In spite of the substantial and serious critism developed in the ECCS Hearings, the AEC has recently announced that it intends to proceed with licensing of seventeen nuclear power plants by February 1973 because of presumed power shortages. The proposal to license for full-power operation all those reactors now ready to go establishes that the message implicit in the ECCS Hearing results has not been heard. The AEC continues to rely upon the low probability of the occurrence of an accident and to stress the dependability of its quality assurance program and the conservatism of its safety criteria.

Notes

1. Hearings before a specially set Atomic Safety and Licensing Board, AEC Docket No. RM-50-1.
2. Novick, Sheldon, "A Mile from Times Square," Environment, Vol. 14, No. 1, Jan./Feb. 1969.
3. Lapp, Ralph, New Republic, Jan. 23, 1971.
4. The reports are summarized in Ian A. Forbes, Daniel F. Ford, Henry W. Kendall, and James J. MacKenzie, "Cooling Water," Environment, Vol. 14, No. 1, Jan./Feb. 1972.
5. Direct testimony of participant Consolidated National Intervenors, March 23, 1972.
6. It is interesting to note that the regulatory staff failed to point out in advance Mr. Lauben's reluctance as to the testimony he was asked to sponsor. Indeed, the hearings revealed that a document entitled "Hints to AEC Witnesses" had expressly directed AEC witnesses: "Never disagree with established policy."
7. Appearing in National Wildlife Magazine, Aug./Sept. 1972.

The Failure of Nuclear Power

I. Introduction

This paper is concerned with one of the largest-scale technological failures that has ever occurred in a major nation: the failure of nuclear power. Actively promoted in the period following World War II, nuclear power appeared to have unlimited prospects as a source of electricity. Today this promise is unfulfilled, the nuclear dream blighted. The public is no longer in favor of the technology, fearing catastrophic accident, and the electric utilities have long since abandoned new plant construction, owing to unreliable plant operation, uncontrolled costs, and public opposition. It has caused severe financial troubles for numerous utilities and has left the country with far too many reactors that produce uneconomical electricity. The future seems to hold only a slow decline for nuclear power, possibly punctuated by a disruptive accident.

During the wartime effort in which the atomic bomb was developed, it became clear that nuclear fission could provide benign as well as destructive power. Moreover, it was believed that fission power could be produced at very low cost and with minimal safety risks. Nuclear power seemed a very attractive alternative to fossil fuels, with their health and environmental costs. The enthusiasm of many experts from the wartime program and their allies in the Congress was enough to launch a major national program. By the start of the 1970s there were dozens of nuclear plants under construction and many more planned. It was expected that between 1000 and 1500 power reactors would supply electricity to the country by the end of the century.

But the troubles of nuclear power have had great impact on these plans. No nuclear plant ordered since 1974 remains under construction—all, some 114, have been canceled. No plants have been ordered since 1978 nor will any be ordered in the foreseeable future. Nearly half of the public believes that the operating plants

---

Excerpts from report prepared by Henry W. Kendall in connection with the Seminar on Risk and Organization, Yale School of Organization and Management. Physics Department, Massachusetts Institute of Technology, Cambridge, Massachusetts, February 16, 1990. Reprinted with permission of Kluwer Academic Publishers, Boston, 1991.

should be shut down and nuclear power phased out. The nuclear capacity at the end of the century will be no more than 10% of the original dream.

The United States will have spent something like a quarter of a trillion dollars on a technology that seems to be a dead end. The size and scale of the failure makes this unique and particularly unsettling in a country wedded to technology. And if there is a major accident, something that is a real possibility, the US may suffer a wrenching dislocation, as calls are made for the shutdown of existing plants.

C. Contributory Factors

There were a number of circumstances, some happenstance, some contrived, that paved the way for difficulty in the buildup of the program and made more likely its final decline.

One is the nature of nuclear technology. The requirements forced by the need to keep radioactive materials out of harm's way were far more demanding than the optimists at first realized. The task has remained technically highly complex and challenging. Some early bad choices were made. One was the selection of large water-cooled reactors, patterned after the small and successful submarine propulsion units. They were unforgiving in operation, and so hard to make safe.

> The existing generation of nuclear power plants has not turned out to be well matched to U.S. institutions. . . . The large [light-water reactor] is no better suited to U.S. electric power production than the hydrogen airship was to transatlantic passenger travel.
>
> —*MIT Nuclear Engineer L. L. Lidsky*[1]

The complexity of the nuclear technology helped keep laymen at a distance. Nuclear physics and nuclear engineering are difficult subjects, and, at the outset of the program, few outside the fields had even a nodding acquaintance with them, especially among the critics. This was coupled with an intense industry and AEC proclivity for secrecy by which many documents and insider concerns never saw the light of day. This delayed the arrival of well-prepared critics a decade or more.

The courts found that they, too, were unable to arrive at balanced assessments of the technical issues, and, lacking independent sources of reliable information, they tended to defer to the NRC when it came to matters of scientific expertise in challenges by critics. This was put explicitly by the Supreme Court in a 1983 decision:

> [T]he Commission is making predictions, within its area of expertise, at the frontiers of science. When examining this kind of scientific determination, as opposed to simple findings of fact, a reviewing court must generally be at its most deferential.[2]

Such deference may not always be warranted or appropriate when an independent judgment is sought through judicial review.

Another factor was that the important safety concern was major accidents that might release large quantities of radioactivity and irradiate many people. Such accidents were, by their nature, not going to occur often. As operating experience grew there were large numbers of reactor malfunctions but none that released substantial toxic material. So there was no experience of the sort we obtain with such technologies as air transportation, processing and transportation of hazardous chemicals or fuel, or the other multitude of human activities where many mishaps help establish, by trial and error, an acceptable level of risk. With respect to nuclear power, in contrast, there have been only predictions which could be manipulated, critics whose views could be dismissed, and long strings of "events" which hurt few if any. And this has "proved," at least to the satisfaction of industry, that the technology is under control. The dramatic mishaps come rarely. With no feedback, industry and its regulators grow complacent. Society's controls do not work well when faced with hazards of this kind.

Then there were the many successes the nuclear proponents achieved in eliminating the checks and balances that normally inhibit reckless or imprudent activities.* Congressional control of the program was concentrated in a single powerful committee of true believers. So there were no Congressional skeptics. The regulations were made by a single powerful agency, the AEC. The Atomic Energy Act of 1954 ruled out any role for the states in matters nuclear, reserving all safety concerns and plant licensing to the Federal government. This eliminated at the start any critical oversight the states might have provided. The obstacles to utility acceptance of reactors, liability, and open-ended claims in the event of catastrophe, were disposed of by the Price-Anderson Act.

The utilities operate in a rigid and controlled environment, one controlled by utility commissions and with captive customers. It was assumed, rightly for a long time, that people had to have electricity and would pay what was required for it. The utilities had little need for market studies. There was no skepticism from the financial community. It was believed that the utility commissions would go along, would come through with the money in the end. They did, also for many years. There was an assumption that no one would ever let a major utility go bankrupt.[3]

And finally there has been the array of subsidies, still continuing, that preferentially support the nuclear option. The Department of Energy estimated in 1981 that the cost of nuclear electricity would have been higher by a factor of between 1.66 and 2.0, except for subsidies running, in 1979, at $3 billion a year.[4] By 1984 this was up to $15.6 billion a year, about one-half the total subsidy to all forms of electricity, although nuclear power provided less than 14% of the country's electricity.[5] Without these subsidies the adverse economics of nuclear power would have been felt much earlier and acted as a brake on the program.

D. Prospects

The last of the reactors now under construction, if plans hold, will be completed in 1991.[6] Utilities will have to cope with the high costs of nuclear electricity. Their

---

*I am grateful to former NRC Commissioner Peter A. Bradford for a helpful discussion of this matter.

planning now centers on improved efficiency of electricity usage and reliance on smaller, more efficient plants, based on fossil fuels, to accommodate the low growth expected for the foreseeable future. Industry forecasts no nuclear additions up through 2005.[7] Provided there is no serious accident, the inventory of operating nuclear plants will be slowly depleted as older units are, one by one, decommissioned. Nuclear power "will gradually play a less important role in the nation's economy, and in 25 or 30 years it will be completely negligible."[8]

But the risk of serious mishap in the nuclear program hangs over the present, and any prospective nuclear program like a sword of Damocles. A nuclear plant malfunction in which a substantial release of radiation occurred would almost certainly have major public consequences very much beyond the direct damage it caused. A major accident, though smaller than Chernobyl, could lead to an abrupt, nationwide halt to reactor operation, causing a severe dislocation in the U.S. electric generating system, especially in those regions of the country highly dependent on nuclear power. A Chernobyl-scale event might cause a reaction of immense proportions. Rasmussen believes this risk is acceptable, that even an accident of the largest size would not lead to nuclear shutdown.[9] However, such an event would be the largest peacetime catastrophe in the nation's history. As noted earlier, over three-fourths of the public already oppose construction of more nuclear power plants and nearly half support phasing out existing plants. The stage seems set for public reaction to force severe operating restrictions, if not immediate shutdown, of operating plants in the event of a major accident. The reaction would surely include further deepening of distrust of technology, already fueled by the nuclear accidents at TMI and Chernobyl, as well as by the lethal chemical release at Bhopal, India.

E. Conclusions

That such a large-scale misfortune as the nuclear power failure should have occurred in the United States, the leader in adapting complex technology to national purposes, is puzzling. It has required decades for society's controls to come into effect, and they are even now not adequate. The failure developed over a period of thirty five years and no significant efforts were made in that period to reverse the decline and better the industry's prospects. Any one of the major participants could have acted: the reactor vendors, the electric utilities, the regulatory agency, the Congress, successive administrations. None did.

Whether this society *can* implement and control a technology having the characteristics of nuclear power is not yet known. If, in future years, nuclear energy were to become a necessity in order to deal with the grave problem posed by carbon dioxide from fossil fuels, one thing is certain: the present experience will make the transition much more difficult.

References and Notes

1. L. L. Lidsky, statement for U.S. House Committee on Internal and Insular Affairs, Subcommittee on Energy and the Environment, June 10, 1986.

2. *Baltimore Gas & Electric Co.* versus *Natural Resources Defense Council*, 462 U.S. 87, 103, 103 S.Ct. 2246, 2255, 76 L.Ed.2d 437 (1983).
3. *Boston Globe*, January 31, 1988, p. 29.
4. "Federal Subsidies to Nuclear Power: Reactor Design and the Fuel Cycle," Analysis Report of the U.S. Department of Energy, Energy Information Administration, March 1980.
5. "The Hidden Costs of Energy," Center for Renewable Resources, Washington, October 1985, and references cited therein. *Monthly Energy Review*, Energy Information Agency, Washington, 1986.
6. *Electrical World*, September 1986, p. 51.
7. *Electrical World*, September 1987, p. 37.
8. Lidsky, p. 20.
9. Interview with Norman Rasmussen, April 4, 1987.

PART TWO
SCIENCE

The two papers included here are descriptions of a discovery that I and some 17 other physics collaborators made as a result of a research program carried out at the Stanford Linear Accelerator Center (SLAC) in the late 1960s and early 1970s. For me as well as for most of the others, the program was a continuation of a long-term interest in the study of the internal structures of the proton and neutron, collectively referred to as the "nucleon," with high-energy electrons.

As a young graduate student at MIT, I had, in 1952, been introduced to electromagnetic interactions by my research supervisor, Martin Deutsch, the discoverer of the short-lived electron-positron atom, positronium. Following two postdoctoral years at MIT and Brookhaven National Laboratory, I had gone to Stanford University to join the research group run by Robert Hofstadter, who had been carrying out pioneering studies of the nucleon with high-energy electron beams from the then largest (300 feet) linear electron accelerator at the W. W. Hansen Laboratory at Stanford. At that time, studies of nucleon properties using electrons was not a popular branch of physics; the interests and priorities of most particle physicists centered on the large proton synchrotrons that were providing the bulk of the excitement in the field. Electrons were believed to be too intractable to deal with and important physics too hard to extract for it to become the focus of much attention. This was primarily the consequence of the electromagnetic radiation that electrons emit copiously when they interact with matter and that clouds the reactions of interest to investigators. Stanford was far in the lead in a field of few competitors in exploiting electron studies. It was here that Hofstadter and his collaborators were first able to discern something of the internal magnetic and electric structure of the nucleon, work that won Hofstadter the 1983 Nobel prize in physics.

Returning to MIT in 1961, I rejoined a friend from Stanford, Jerome I. Friedman, who had also been in Hofstadter's group, in a collaborative research program at the Harvard-MIT electron synchrotron. A very much larger linear electron accelerator, some two miles in length, had been proposed and was under construction at Stanford at SLAC. We joined the effort as collaborators and soon as coheads, along with another friend from earlier Stanford days, Richard E. Taylor, of a major experimental program at the machine under construction.

The details of the discoveries that emerged from our nearly ten-year program are set out in the articles that follow. The first is a semipopular article I wrote with the then director of the SLAC laboratory. The second is my Nobel lecture, setting out a scientific description of the experimental program and its results. That our discoveries proved to constitute an important contribution to knowledge of nucleon structure was confirmed by the award of the 1990 Nobel prize in physics to Jerry, Dick, and me. (For broad pictures of particle physics, see Andrew Pickering, *Constructing Quarks: A Sociological History of Particle Physics*, University of Chicago Press, Chicago, 1984, and Abraham Pais, *Inward Bound—Of Matter and Forces in the Physical World*, Clarendon Press, Oxford, and Oxford University Press, New York, 1986.)

Structure of the Proton and the Neutron

Sixty-five years ago Ernest Rutherford observed how alpha particles are scattered by thin metal foils and concluded that the atom is not a homogeneous body but consists of negatively charged electrons surrounding a small, massive, positively charged nucleus. Since that time physicists in many laboratories have conducted scattering experiments with particles of ever increasing energy in an effort to probe the structure first of the atom, then of the nucleus and now of the basic constituents of the nucleus: the proton and the neutron. Are these "elementary" nuclear particles homogeneous? Recent investigations with electrons brought to an energy of 21 billion electron volts by the two-mile accelerator at the Stanford Linear Accelerator Center (SLAC) [see Figure 1] strongly suggest that history may be repeating itself on a scale 100,000 times smaller than that of the atom. It turns out that ultrahigh-energy electrons are scattered by protons and neutrons in ways that no one had predicted. The tentative conclusion is that the nuclear particles have a complex internal structure consisting of pointlike entities now called partons. And there is evidence that partons share some of the properties assigned earlier to those hypothetical particles named quarks.

Knowledge of the internal structures of the proton and the neutron may provide the key to understanding the "strong" force that holds the atomic nucleus together and endows the universe with its stability. The strong force makes its presence known in the nuclear reactions that fuel the stars and that, on a more modest scale, provide the energy for nuclear power and nuclear explosives. Although the exploitation of the strong force has become a commonplace in technology, the nature and origin of the force is still poorly understood.

In addition to exhibiting the strong force, protons and neutrons also respond to the electromagnetic force, which is some 100 times weaker. Both nuclear particles behave like tiny magnets and both comprise electric charges (although the neutron's

---

By Henry W. Kendall and Wolfgang K. H. Panofsky. Reprinted with permission from *Scientific American* **224** (6), 59–78 (June 1971).

Figure 1. Two-mile-long electron accelerator at the Stanford Linear Accelerator Center (SLAC) was used to obtain the experimental results reported in this article. The electron beam is raised to a maximum energy of 21 billion electron volts (GeV) as it travels down a vacuum pipe lined with klystron tubes and focusing magnets. Near the end of its trip the electron beam passes through a "beam switchyard" before reaching the target areas, which are located inside the two large buildings in the foreground.

net charge is zero). Whereas the strong force operates only when the interacting particles are very close together (a distance roughly equivalent to their own diameter: about 10^{-13} centimeter), the electromagnetic force has an infinite range, falling off in strength with the square of the separation. Since the neutron and the proton respond to the electromagnetic force, they scatter electrons aimed at them. It is the pattern of the scattering that provides clues to their structure.

Since the Stanford experiments are fundamentally the same as Rutherford's it will be useful to briefly review his techniques and results. He placed a natural emitter of alpha particles (particles with a charge of +2, later identified as helium nuclei) in an evacuated box equipped with a collimator so that a well-defined beam of particles would strike a target consisting of a metal foil [Figure 2]. The box was also provided with a zinc sulfide screen that would scintillate when it was struck by an alpha particle. The screen could be moved to intercept particles scattered at any angle, and the scintillations were counted one at a time with the aid of a low-power microscope. Two of Rutherford's collaborators, Hans Geiger and Ernest Marsden, soon noticed that alpha particles were being scattered at large angles far more often than one would have predicted on the basis of the then current ideas of atomic structure. The electric charge in atoms was believed to be diffusely distributed and hence should not have exhibited the concentrated electric fields needed to produce such large particle deflections.

Rutherford concluded that "the positive charge associated with an atom is concentrated into a minute center or nucleus, and that the compensating negative charge is distributed over a sphere of radius comparable with the radius of the atom." He also worked out the mathematical law describing how one point of electric charge would be scattered by another point charge [Figure 3]. The force between two charged particles was assumed to be given by Coulomb's law. Knowing the charge and mass of the interacting particles, Rutherford combined Coulomb's law with Newton's laws of motion to relate the probability of scattering through a given angle to the energy of the incident particle. The probability of scattering by a single target atom is the "scattering cross section," defined as the area of the incident beam within which the influence of the target atom gives rise to the process observed—in this case scattering. The cross section is not necessarily related to the "true" physical size of the target particle but rather represents a measure of the force exerted on the incident particle by the target particle.

The cross section is experimentally determined for different angles (measured from the axis of the incident beam), and the results can be compared with theoretical predictions. Rutherford's formula predicts the scattering cross section from the mass m and charge of the incident particle, the mass and charge of the target particle, the velocity v of the incident particle, and the scattering angle θ. The formula depends directly on the particular combination of these variables that describes the vector difference, q, between the initial momentum and the final momentum of the scattered particle: $q = 2mv (\sin \theta/2)$. Another term for q is "momentum transfer" [Figure 4]. The formula assumes that the interacting particles are mathematical points, having neither size nor shape. In general, however, a scattering cross section will depend not only on the details of the forces (for example exactly how their

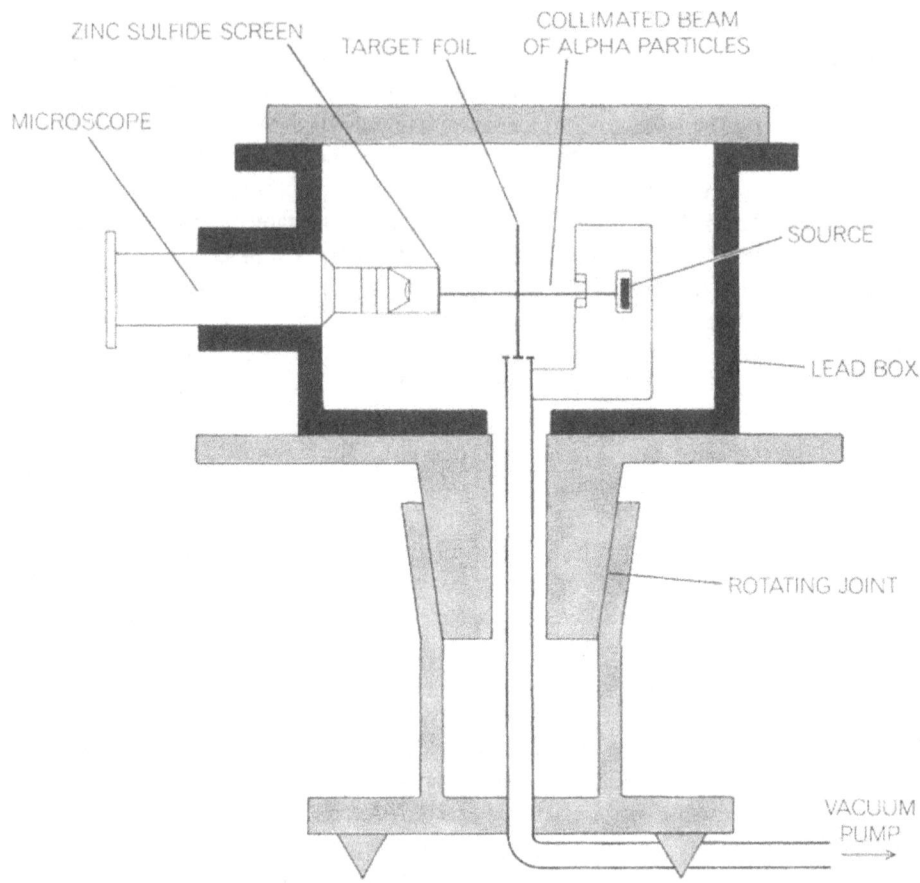

Figure 2. Original apparatus used by Ernest Rutherford and his co-workers to study how alpha particles are scattered by thin metal foils is shown in this illustration, which was adapted from a diagram published in *Philosophical Magazine* in 1913. A natural emitter of alpha particles was placed in an evacuated box equipped with a collimator so that a well-defined beam of particles would strike the target foil. A zinc sulfide screen that would scintillate when struck by an alpha particle was moved to intercept particles scattered at any angle and the scintillations were counted with the aid of a low-power microscope. It was on the basis of observations made with this device that Rutherford concluded that the atom consists of a massive, positively charged nucleus surrounded by negatively charged electrons. All later scattering experiments are essentially variations of this basic technique.

strength varies with distance) and on the laws of motion of the particles (which may involve non-Newtonian, or relativistic, considerations) but also on whatever internal structure the particles may have.

In scattering processes described by quantum mechanics the momentum transfer plays a central role, because it determines the scale of what is being studied. In quantum mechanics a particle that has a certain momentum p also has associated

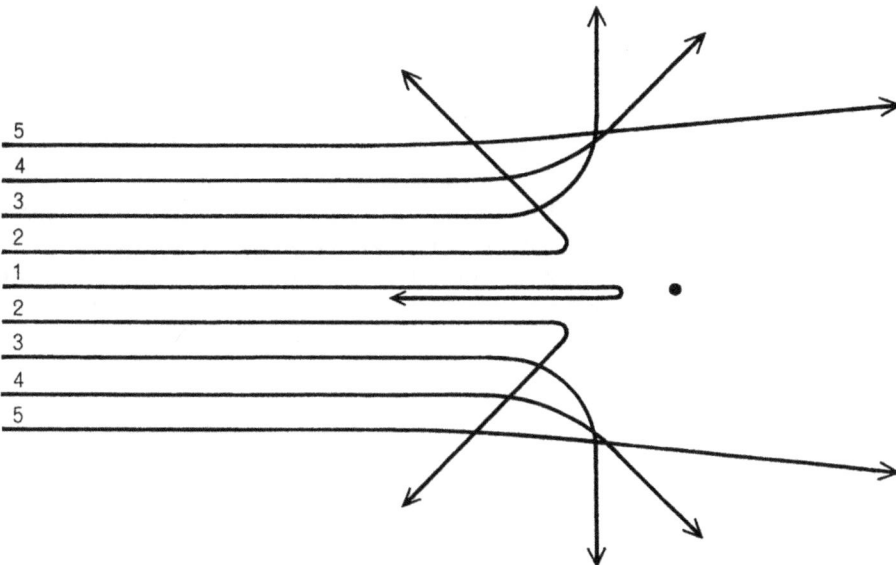

Figure 3. According to Rutherford, the scattering of one point of electric charge by another point charge could be described by a mathematical law that combined Coulomb's law (for the force of attraction or repulsion between two charged particles) with Newton's laws of motion to relate the probability of scattering through a given angle to the energy of the incident particles. In this diagram of the Rutherford scattering process the amount of scattering can be seen to depend also on the position of the incident particle's trajectory.

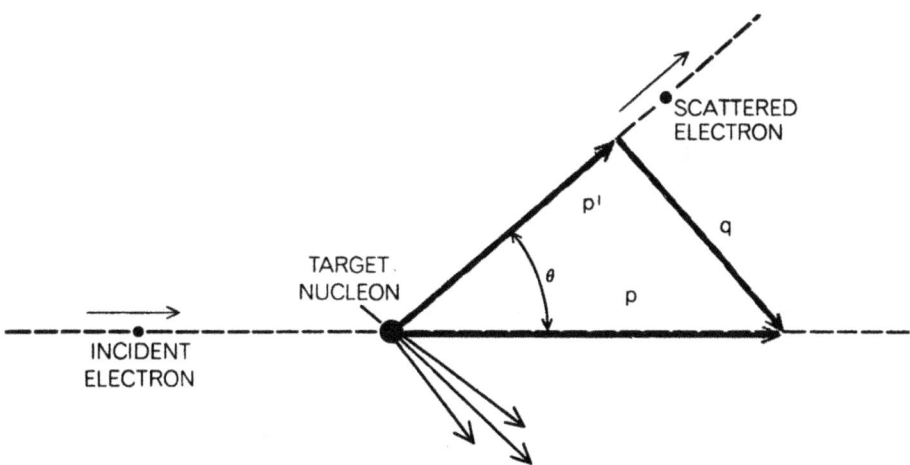

Figure 4. Momentum transfer, an important concept in the theoretical treatment of the scattering process, is defined as the vector difference (q) between the initial momentum (p) and the final momentum (p') of the scattered particle. The formula that expresses this relation is $q = 2mv (\sin \theta/2)$, where m is the mass of the incident particle, v is its velocity and θ is the scattering angle. In elastic scattering the target nucleon simply recoils; in inelastic scattering it either disintegrates to form other particles or it is left in an excited state.

with it a certain wavelength λ. The formula that relates these properties is $\lambda = h/p$, where h is the extremely small number (6.6×10^{-27} erg-second) known as Planck's constant. The accuracy to which a particle can be located is limited by the associated wave; the probability of finding the particle at a given point is governed by the behavior of the "wave packet" describing the particle's motion. To locate one particle with another, the two have to interact (that is, the experimenter must scatter one from the other), and this involves a transfer of momentum between the two. Thus it is reasonable that the accuracy Δx to which the details of an unknown structure can be examined is governed by the momentum transfer q experienced in the collision; the resulting relation is $\Delta x = h/q$ [Figure 5]. This formula implies that our ability to distinguish fine detail in the target particle depends on making q as large

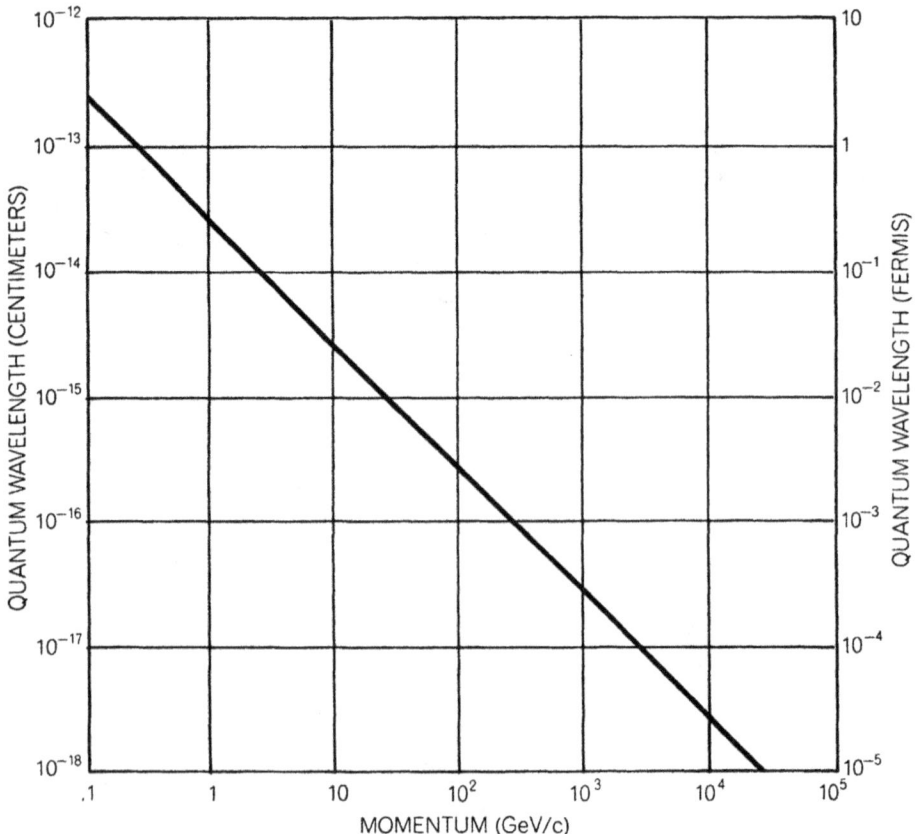

Figure 5. The possibility of locating a particle of momentum p is governed by its associated wavelength λ according to the relation $\lambda = h/p$, where h is Planck's constant (6.6×10^{-27} erg-second). In this graph of the relation momentum is measured in units of GeV/c, where "GeV" stands for giga (10^9) electron volts and c is the velocity of light. The quantum wavelength of the incident particle is given in both centimeters (*left*) and fermis (*right*).

as possible in order to make the wavelength λ as small as possible. (Momentum is the product of mass times velocity; at the energies of interest to physicists engaged in high-energy electron scattering the mass increases with increasing energy while the velocity remains essentially constant at the velocity of light.)

In scattering experiments of the type performed at Stanford momentum is measured in units of GeV/c, where "GeV" stands for giga (10^9, or one billion) electron volts and c is the velocity of light. An electron of 20 GeV lacks only one part in three billion of traveling at the velocity of light. Under these conditions the particle energy expressed in GeV and its associated momentum expressed in GeV/c are essentially equal.

Two Kinds of Scattering

The scattering of electrons can be either "elastic" or "inelastic." In elastic scattering the target particle recoils much as if it were a billiard ball, remaining in the same internal state it was in before the collision. In inelastic scattering the target particle either disintegrates or is left in an excited state, a state different from its original condition. There is a trade-off between the two processes: one robs the other. Both processes tell a good deal about the structure of the target particle.

Rutherford's formula does not adequately describe the elastic scattering of high-energy electrons for two reasons. First, the velocities are so great that one must use relativistic quantum theory to describe the wave nature and behavior of the incident and target particles. Second, electrons have "spin," that is, they have a unique angular momentum, as if they rotated around an internal axis. The more precise formula that must be used is known as the Mott cross section. Except for the term that accounts for the spin of the electron, the Mott equation can be reduced directly to Rutherford's equation in those cases where the velocity of the incident particle is much smaller than the speed of light, as was true in Rutherford's experiments [Figure 6]. Since Rutherford did not know that quantum mechanics governed his scattering experiments, it is only a happy accident that his formula correctly describes low-energy scattering. We now know that Newton's "classical" laws of motion can be successfully applied only when scattering is attributable primarily to those forces whose strength varies inversely with the square of the distance, as the electrical Coulomb force does.

The Mott formula itself must be modified if the electron is scattered not by another point charge but rather by an object of finite dimensions [Figure 7]. In that case each segment of the electron wave front is diffracted separately by each subunit of charge within the target particle. The individual wavelets scattered by the subunits then recombine to form an outgoing wave that describes the scattered electron. As one might expect, some of the wavelets add constructively and some interfere, thereby canceling one another. The elastic-scattering cross section from a charged particle of finite size is therefore generally less than the cross section from a point charge. The factor by which the scattering is decreased below that from a point charge is given by the square of a number called a form factor, designated F.

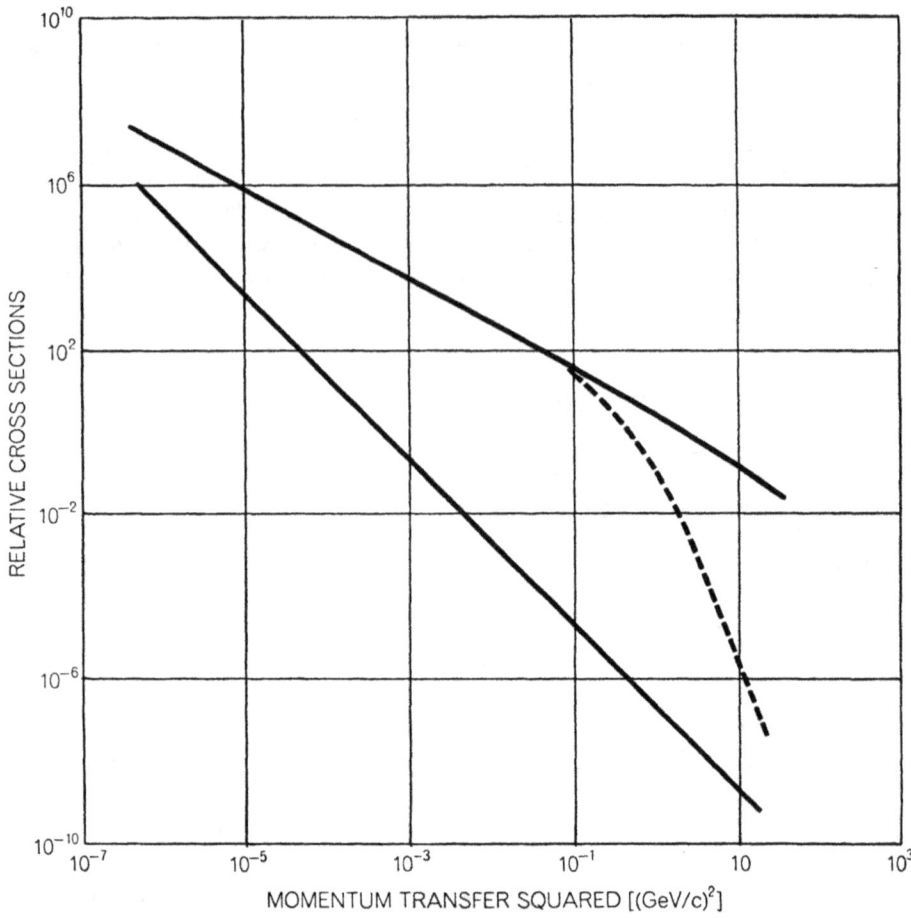

Figure 6. Scattering cross section, defined as the area of the incident beam within which the influence of a target atom gives rise to a certain kind of interaction, is given here for the scattering of an electron by a target nucleon according to the Rutherford formula (*bottom line*) and according to the Mott formula (*top line*). Except for the term that accounts for the spin of the electron, the Mott formula reduces directly to Rutherford's as the energy and the velocity of the incident electron become small. The broken curve shows Mott scattering from a finite proton. The curves are drawn for a scattering angle of 20 degrees.

The formula for the form factor is obtained by tracing the extra length each wavelet has to travel when it is scattered by charged subunits within the target particle. The formula depends solely on the momentum transfer, q, which is the vector difference in momentum between the ingoing and the outgoing electron. Given a sufficiently high value of q, the form factor will be sensitive to details of the target's structure; if q is too small, the experiment will reveal little.

If the target particle is a nucleon (a proton or a neutron), one would like to study its structure at distances smaller than its own radius, which is known to be about .8

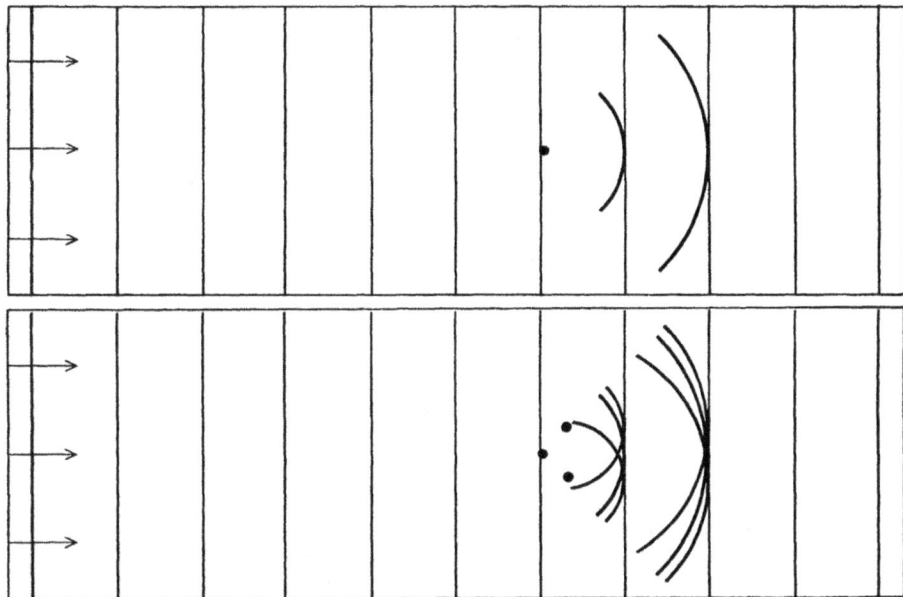

Figure 7. Modification of the scattering formula is required if the electron is assumed to scatter not from another point charge (*top*) but rather from an object of finite dimensions (*bottom*), represented here as composed of three point constituents. In the latter case each segment of the electron wave front is diffracted separately by each subunit of charge. The individual wavelets scattered by subunits then recombine to form an outgoing wave that represents the scattered electron. The amount by which the scattering cross section from a charged particle of finite size is reduced below that from a point charge is called the form factor (F).

fermi (one fermi is 10^{-13} centimeter). To have a resolution of, say, .1 fermi would require a momentum transfer of 2 GeV/c. In the present experiments the practical limit is about 5 GeV/c and is therefore small enough to provide substantial information about the proton. If the form factor were known for a wide range of values of the momentum transfer, the charge distribution in the target particle could be reconstructed.

The task of computing the distribution of charge within a particle such as a proton from electron-scattering data closely resembles the task of reconstructing the structure of a crystal from the complex diffraction pattern produced when it is bombarded by X rays. The electron-scattering problem is much more difficult, however, particularly when the velocity of the recoiling proton approaches the velocity of light. The effects of relativity on the motion of the proton introduce ambiguities that complicate our attempts to reconstruct the spatial distribution of the charge.

A further complication is introduced by the proton's spin, which produces a magnetic moment. As a result the incident electron can interact with the proton's magnetization as well as with its electric charge. Since the magnetization can also have a finite distribution in space, it gives rise to a second form factor, designated F_m to distinguish it from the electric form factor F_e. The effect of these complications is to modify Rutherford's original formula to take account of the following facts: Both the incident and the target particle carry spin, the target particle is extended in space, the collision velocities are so high that relativistic effects are introduced, and the motion of both particles is described by wave mechanics rather than by classical mechanics [Figure 8].

This somewhat elaborate discussion should not detract from the basic simplicity of the electron-scattering process. The process enables one to explore the unknown structure of subnuclear particles with the known forces of electromagnetism. This is in contrast to those experiments (interesting for other reasons) in which two particles of unknown structure collide, for example in proton-proton or pion-proton scattering. As far as is known to date, electrons behave like point charges and interact in scattering experiments only through the force of electromagnetism. (It is true, of course, that electrons also intereact through the "weak" force, which plays a role in radioactive decay processes, but since the weak force is roughly 10^{10} times smaller than the electromagnetic force it can be ignored in electron-scattering experiments.) The laws of electricity and magnetism as they are now embodied in the equations of quantum electrodynamics represent the one and only area in physics where a single quantitative description has proved valid over the entire range of ex-

RUTHERFORD CROSS SECTION $$\sigma_R = \frac{(2e^2\, m)^2}{q^4}$$	e = ELECTRON m = MASS OF ELECTRON $q = 2\sqrt{pp'}\sin\theta/2$ p = INITIAL MOMENTUM OF ELECTRON p' = FINAL MOMENTUM OF ELECTRON
MOTT CROSS SECTION $$\sigma_M = \frac{(2e^2\, E'/c^2)^2}{q^4} \cdot \frac{E'}{E}\left(\cos^2\frac{\theta}{2}\right)$$	θ = SCATTERING ANGLE OF ELECTRONS E = INITIAL ENERGY OF ELECTRON E' = FINAL ENERGY OF ELECTRON c = VELOCITY OF LIGHT F_e = ELECTRIC FORM FACTOR F_m = MAGNETIC FORM FACTOR
ROSENBLUTH CROSS SECTION $$\sigma = \sigma_M\left(\frac{F_e^2 + \tau F_m^2}{1+\tau} + 2\tau F_m^2 \tan^2\frac{\theta}{2}\right)$$	$\tau = q^2/4M^2c^2$ M = MASS OF NUCLEON

Figure 8. Increasing complexity is introduced to the scattering equation as one proceeds from the Rutherford formula to the Mott formula to the Rosenbluth formula (*left*). The final equation takes into account the following facts: both the incident and the target particle carry spin, the target particle is extended in space, the collision velocities are so high that relativistic effects are introduced, and the motion of both particles is described by wave mechanics rather than by classical mechanics. Symbols are defined in key at right.

periments for which it has been tested, from cosmic dimensions down to 10^{-15} centimeter. Thus the assumption that these particular forces are understood seems well justified.

The Two-Mile Accelerator

Before discussing the results of elastic and inelastic scattering experiments obtained with the Stanford electron accelerator, we shall briefly describe the facility and the techniques involved. The electron beam is raised to a maximum energy of 21 GeV as it travels down a two-mile evacuated pipe lined with 245 klystron tubes that pour electromagnetic energy into the beam. During its two-mile trip the beam is kept tightly focused by magnetic "lenses" spaced every 100 meters. At the end of its trip the beam passes through a final "purgatory" of magnets and slits that closely define the width and energy range of the electron beam that reaches the target. A typical scattering experiment requires a target containing hydrogen or deuterium and a means for selecting and identifying electrons scattered at different angles and measuring their momenta in the presence of many other particles produced by the collisions of electrons and nuclei.

A vessel containing liquid hydrogen provides the target protons; the nucleus of ordinary hydrogen consists of a single proton. Using liquid deuterium or heavy hydrogen is the next best thing to having a target of free neutrons; the deuterium nucleus consists of a proton and a neutron. To a good approximation the scattering from deuterium nuclei is simply the sum of the scattering from neutrons and protons. Because the beam striking these liquefied gases is very intense they must be cooled continuously by means of a heat exchanger, not simply to prevent boiling but to minimize changes in density that would throw off the results.

To separate and classify the electrons emerging from the target the Stanford installation is equipped with three magnetic spectrometers, which funnel the electrons into a system of detectors. They were designed and constructed as a collaborative effort by physicists from the California Institute of Technology, the Massachusetts Institute of Technology, and the group at SLAC. Very high resolution in both energy and angle is required, since we must be able to distinguish between elastically and inelastically scattered electrons and to resolve the detailed structure in the spectra of electron energies produced by inelastic scattering.

In the inelastic scattering one or more pions can be produced in the scattering collision. Since the energy required to create a pion is 139 MeV (million electron volts) the resolution needed must be considerably better than the ratio of 139 MeV to the incident energy, which can exceed 20 GeV. A resolution of better than .7 percent in energy is therefore needed. A similar analysis of the collision kinematics indicates that the resolution in angle should be a fraction of a milliradian, which is about three minutes of arc. What counts is the precision in *relative* angle and in *relative* energy between the incident and the scattered electrons; therefore these requirements for the resolution of both angle and energy apply equally to the incident beam and to the spectrometers analyzing the scattered beam.

The spectrometers are large and complicated machines [Figure 9]. They consist of magnetic lenses and bending magnets that deflect the scattered electrons vertically and then bring them to a focus. The amount of vertical deflection is a measure of the electron's momentum; the horizontal position is a measure of the scattering angle. Hundreds of counters, the equivalent of the zinc sulfide scintillation screen used in Rutherford's experiments, identify the momentum and angle of each electron. The counters are narrow bars of specially prepared transparent plastic that scintillate briefly when they are struck by a high-energy particle. Each bar is viewed by a photomultiplier tube that signals each tiny light flash.

The signals from the counters and other particle-identification devices are processed and passed on to a large computer. The computer is run "on line," storing data for later detailed analysis at the same time it is performing a simplified partial analysis. In addition to displaying such results the computer provides status information on the equipment and performs many routine "housekeeping" chores, such as adjusting currents in the spectrometer magnets and logging beam currents and other quantities of interest.

Figure 9. Large magnetic spectrometers in one of the experimental areas at the SLAC site are used to separate and classify the scattered electrons emerging from the target and to funnel them into a system of detectors. Three spectrometers, each consisting of a complicated array of magnetic lenses and bending magnets, are installed around a common pivot point in this area; two are visible in this view. The scale of the instruments can be appreciated by noting the two men standing near the "middle-sized" device.

Nucleon Form Factors

The elastic-scattering experiments carried out by Cal Tech, M.I.T. and SLAC physicists have yielded measurements of the four elastic form factors that describe the structure of the proton and the neutron. The quality and quantity of these data, however, are quite variable. The most accurate measurements are those that give the magnetic form factor of the proton [Figure 10]. The magnetic form factor of the neutron, obtained by subtracting from the deuterium scattering the scattering attributable to the proton, looks similar to the proton curve except that the errors are larger. The electric form factor of the proton resembles its magnetic form factor, but the electric curve has been determined for only a much smaller range of vari-

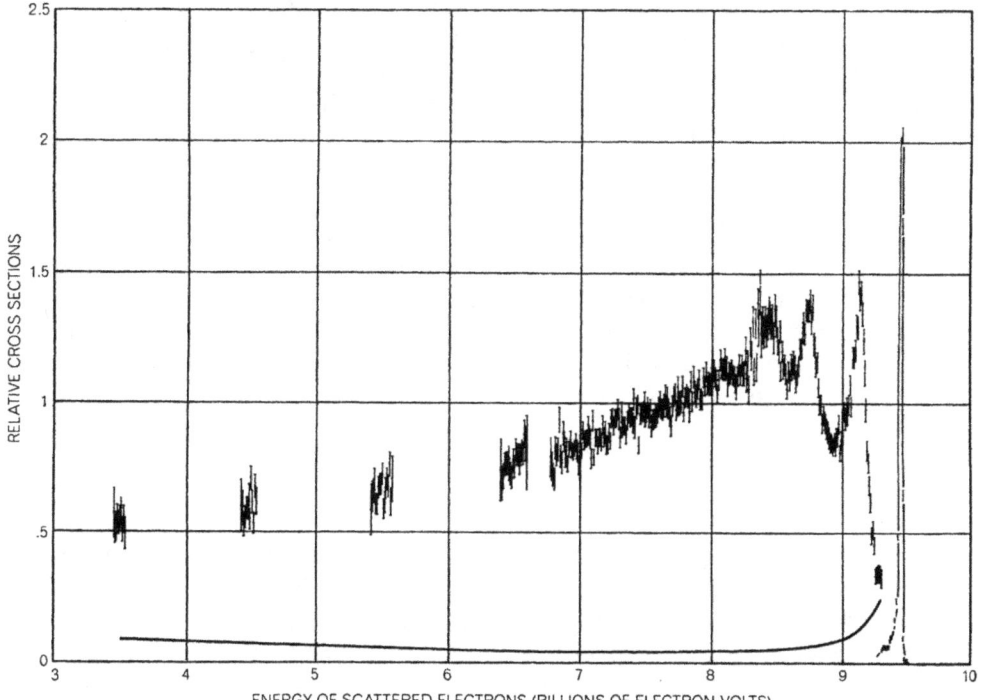

Figure 10. Typical scattering spectrum produced by an electron beam with an energy of 10 GeV colliding with stationary protons includes both elastic events (smooth lines, bottom and far right) and inelastic events. The elastic peak at right has been reduced in height by a factor of five; the asymmetry of its tail arises because the electrons can emit "soft" X rays that "rob" various amounts of energy and thus blur out the elastic peak on the low-energy side. The smaller peaks or bumps in the inelastic spectrum correspond to excited states of the proton; they are called resonance excitations, or simply resonances. To the left of these bumps is a smoother continuous spectrum called the continuum. As one goes to higher incident energies the resonances tend to disappear but the continuum remains.

ables. The electric form factor of the neutron is known to be practically zero; the errors in the existing measurements, however, are large.

One might ask: Why are electrons scattered by the neutron at all, since the neutron has no electric charge? The answer has two parts. First, the neutron's spin produces a magnetic moment; this alone would show up in the scattering described by the magnetic form factor. Second, the electric current that gives rise to the neutron's magnetism can produce localized accumulations of charge within the particle even though the particle's net charge is zero. Such accumulations give rise to electric scattering whenever the values of momentum transfer exceed zero. Thus elastic electron scattering not only responds to the overall charge and magnetic moment of the neutron but also reveals what is going on inside.

The experiments indicate that the magnetic structures of the neutron and the proton are almost identical but that the magnitude of the scattering from each is proportional to the magnetic properties of each particle as found from static experiments. In other words, the magnetic form-factor curves of the two particles are identical in shape as far as we can tell from experiment. It is probably also significant that over the limited range accessible to experiment the electric scattering of the proton is proportional to the magnetic scattering. This suggests that the distribution of electric charge within the proton is directly related to the magnetic structure.

The scattered wavelets create a diffraction pattern similar to the shadow pattern formed when parallel rays of light strike the edge of an object. If the object has a sharp edge, the pattern will consist of alternate dark and light bands. Similarly, if the proton were an object with a sharply defined surface, one would see much more structure in the form-factor curve than is in fact seen. Evidently, therefore, the proton has a fuzzy boundary. Details of the curve give the proton's average radius: about .8 fermi, or $.8 \times 10^{-13}$ centimeter.

Particles Real and Virtual

One of the most surprising findings to physicists is the fact that the curve representing the magnetic form factor of the proton, shown [in Figure 11], is smooth over an enormous range of experimental variables. The observed scattering cross section, which varies as the square of the form factor multiplied by the Mott formula for point scattering, falls off by 10^{12} over the range of variables for which measurements have been made. The cross sections associated with the lowest part of the curve are extremely small: the smallest cross section measured was about 2×10^{-39} square centimeter per steradian, which under the conditions of the experiment means that only one out of every 10^{18} electrons was scattered into the detector. The scattering decreases as the fourth power of the momentum transfer. This rapid falling off is one of the current puzzles in high-energy physics. To understand how the puzzle arises and how it may be explained it is necessary to dwell briefly on the concept of "virtual particles."

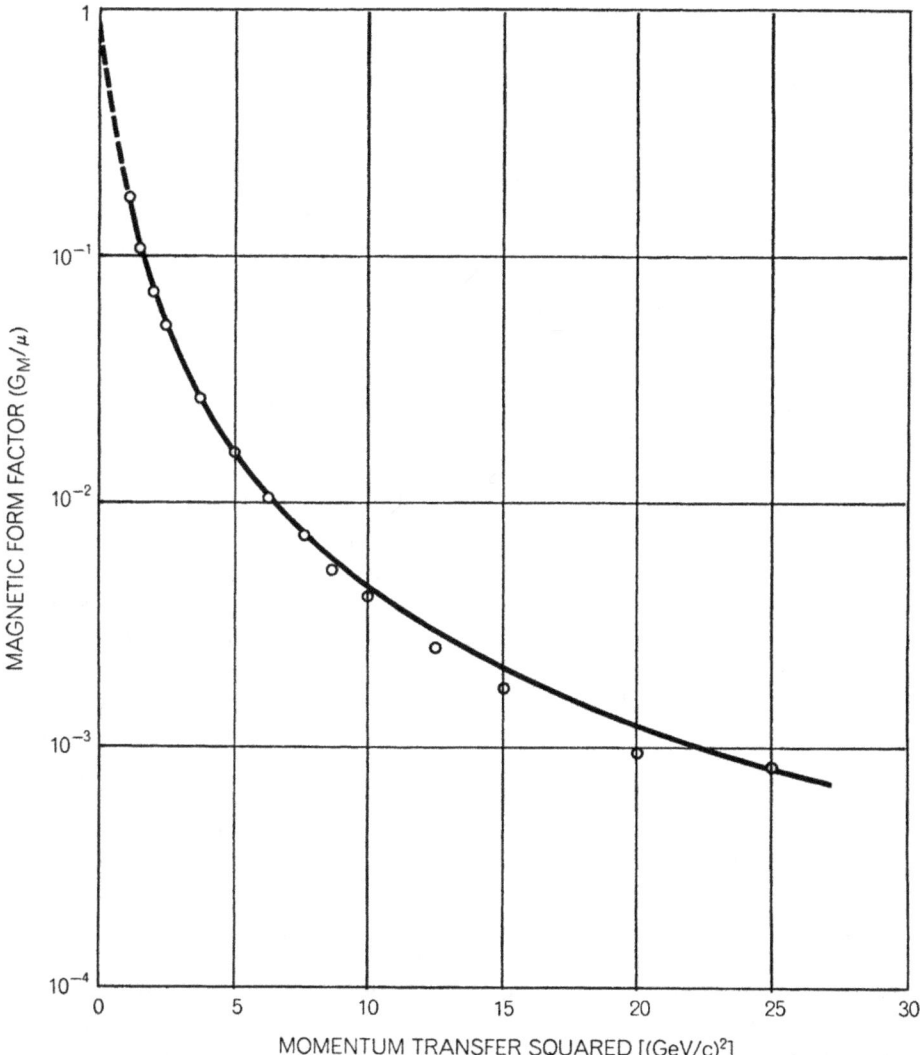

Figure 11. Magnetic form factor of proton was found by the Cal Tech-M.I.T.-SLAC group to be unexpectedly smooth over an enormous range of experimental variables. (The square of the magnetic form factor is the amount by which the scattering cross section attributable to the magnetization of a charged particle of finite size is less than that of a point charge.) The fact that the form factor decreases as fourth power of the momentum transfer, which is faster than theorists had predicted, is a current puzzle of high-energy physics.

The concept of virtual particles is related to the uncertainty principle enunciated by Werner Heisenberg more than 40 years ago. In the wave description of matter it is impossible to determine simultaneously a particle's wavelength and its momentum. Heisenberg's principle relates the uncertainty in the measurement of the particle's wavelength, Δx, with the uncertainty in the particle's momentum, Δp. The product of the two uncertainties is proportional to Planck's constant h ($\Delta p \cdot \Delta x \cong h$). Equivalently one can relate the uncertainty in the particle's measured energy, ΔE, to the uncertainty in the time, Δt, within which the measurement was made, in which case $\Delta E \cdot \Delta t \cong h$.

Now, in relativity theory mass and energy are equivalent, as expressed by Einstein's relation $E = mc^2$. One can imagine, therefore, that for a very short time Δt any given amount of energy ΔE can be converted into a mass m equivalent to the rest mass of some particle, provided that the product of ΔE and Δt does not exceed h. In other words, without violating the uncertainty principle one or more particles can appear in a system and exist for immeasurably brief periods. In a sense their existence is "hidden" by an irreducible uncertainty in our knowledge of the system. Particles that appear in this way are called virtual particles; they cannot be observed directly as real particles can.

Most models that describe the interaction between the electron and the proton visualize the photon (the quantum of light) as the carrier of the electromagnetic force. It too can be real or virtual. Real photons are the packets of waves that carry energy from a radiating source (such as a star) to an absorber (such as the pigments in the eye). In quantum electrodynamics the electromagnetic forces that act between two (or more) moving charges are attributed to the emission and absorption of virtual photons. Hence in electron scattering a virtual photon emitted by the electron interacts with and is absorbed by the electric charge and magnetism within the proton. Virtual photons can carry energy and momentum in any proportion, unlike real photons, whose energy and momentum are uniquely related.

Although it may seem that virtual particles violate fundamental conservation laws, the violation is closely delimited to those areas where the uncertainty principle applies. It does not apply, for example, to the conservation of electric charge. Thus it is not possible for a single virtual electron to appear in a vacuum; it must always be accompanied by a particle of opposite charge, the positron.

There is a class of unstable particles, the neutral vector mesons, whose members resemble photons in many ways, with two important exceptions: they have mass and they exhibit the strong force. The most prominent is the rho meson, which has a mass equivalent to about 750 MeV. (The mass of the proton is equivalent to 939 MeV.) Rho mesons can be created as real particles in the laboratory, and their decay products can be detected. Neutral vector mesons can also be created as single virtual particles by photons propagating in a vacuum—and the photons that create them can be either real or virtual. In a sense the photon is a vector meson a tiny fraction of the time.

Because vector mesons are massive they become a significant factor in modifying photon processes only in experiments at very high energies, such as those we are describing. In addition, as carriers of the strong force, the vector mesons play an important role when real photons of very high energy interact with nucleons.

Before the recent scattering experiments were conducted theorists thought they could predict how the vector mesons would participate in both elastic and inelastic scattering at high energies. In particular they predicted that if elastic scattering is dominated by vector mesons, the form-factor curve should fall off as the inverse square of the momentum transfer. Instead the curve decreases as the inverse fourth power. Clearly the simple model does not work.

Inelastic v. Elastic Scattering

In a collaborative program of measurements carried out by workers at M.I.T. and SLAC very large cross sections were discovered for the inelastic-scattering processes. When one looks at a typical scattering spectrum produced by electrons of 10 GeV colliding with protons, one sees first of all a broad peak with an asymmetric tail [see Figure 10]. The peak represents elastic scattering; the asymmetry of the tail results from the fact that the electrons can emit "soft" photons (X rays) that steal various amounts of energy and so blur the elastic peak on the low-energy side.

In addition, the scattered electron spectrum contains two features produced by inelastic processes. First one sees a number of bumps that correspond to excited states of the proton. They are often called resonance excitations, or simply resonances. The position of the bumps corresponds to now well known excited states of the proton, identified in many high-energy experiments. Four specific resonances have been identified in inelastic electron scattering; the size of the associated bumps depends strongly on the magnitude of the momentum transfer to the proton. The bumps shrink rapidly in size as the momentum transfer increases. The shrinkage occurs just about as fast as the shrinkage of the elastic-scattering peak itself. From this we conclude that the radial dimensions of the excited states represented by the bumps are comparable to the dimensions of the proton itself in its unexcited condition. This implies that in some way most of the nucleon structure is involved when it is in a resonance, or excited, state.

The second feature of the scattered-electron spectrum produced by inelastic processes is called the continuum: the smooth distribution in the energies of those scattered electrons that do not fall in the resonance peaks. Physicists regard the continuum as perhaps the most exciting and puzzling part of all the recent Stanford results. As we go to larger scattering angles or to higher incident energies the resonances tend to disappear but the continuum remains.

When the inelastic-scattering program was formulated, theorists had believed that the continuum cross sections would decrease nearly as rapidly as the elastic cross sections when the momentum transfer was raised. Instead the results show that for incident electron energies ranging from 4.5 to 19 GeV the inelastic scattering cross sections more closely resemble those that would be produced by point targets [Figure 12]. In one comparison the best predictions available before the experiments turned out to be low by as much as a factor of 40 [Figure 13]. The factor of error is even higher in other spectra. The tentative conclusion is that the internal structures from which inelastic scattering takes place are much smaller than the nucleons either in their ground state or in their excited state.

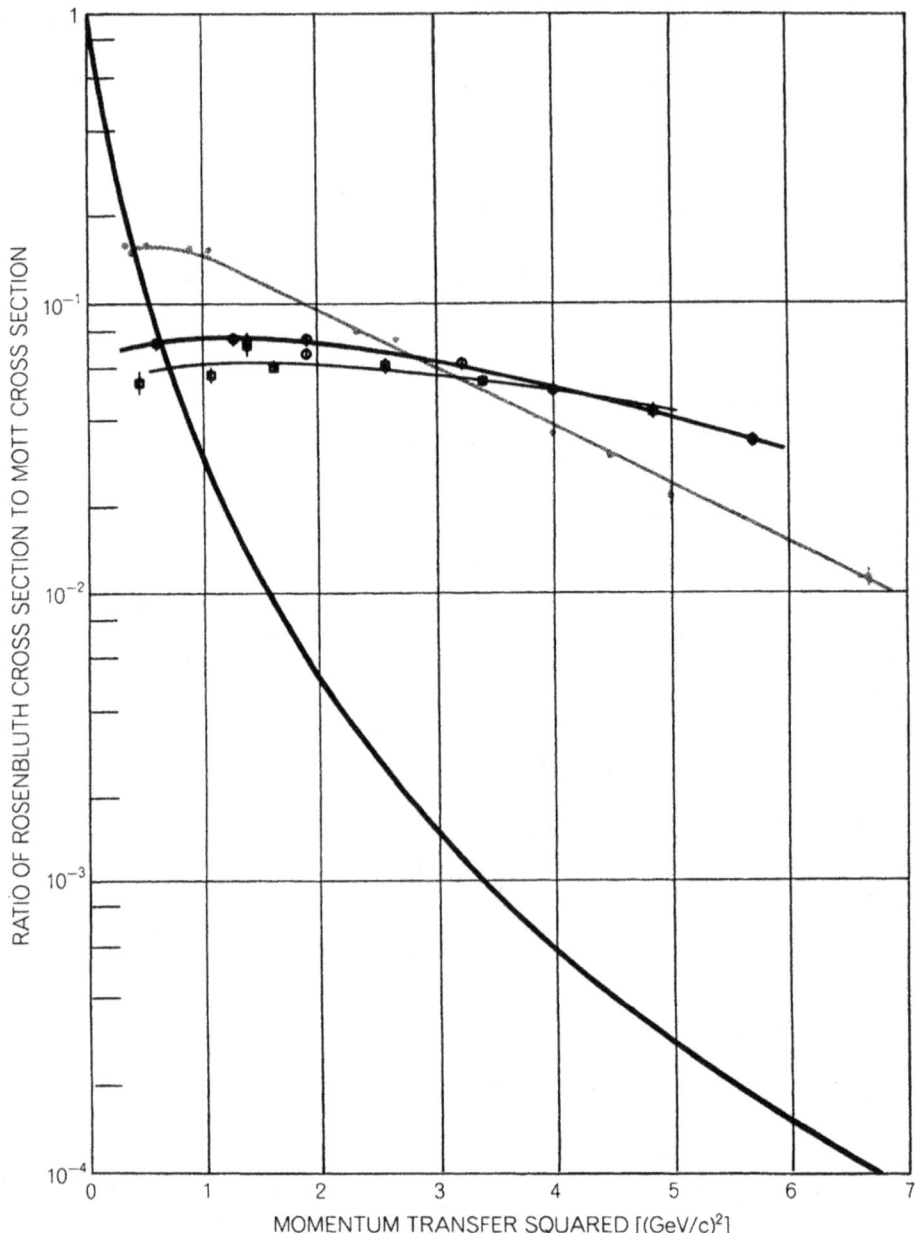

Figure 12. Further evidence that the observed inelastic-scattering cross sections may be produced by point targets is presented in this graph, in which the ratio of Rosenbluth scattering to Mott scattering is given for elastic electron scattering [longest curve] and for three different portions of the inelastic-scattering spectrum [other curves]. Before these results were obtained it had been assumed that the inelastic-continuum cross sections would decrease as rapidly as the elastic cross sections when the momentum transfer was raised.

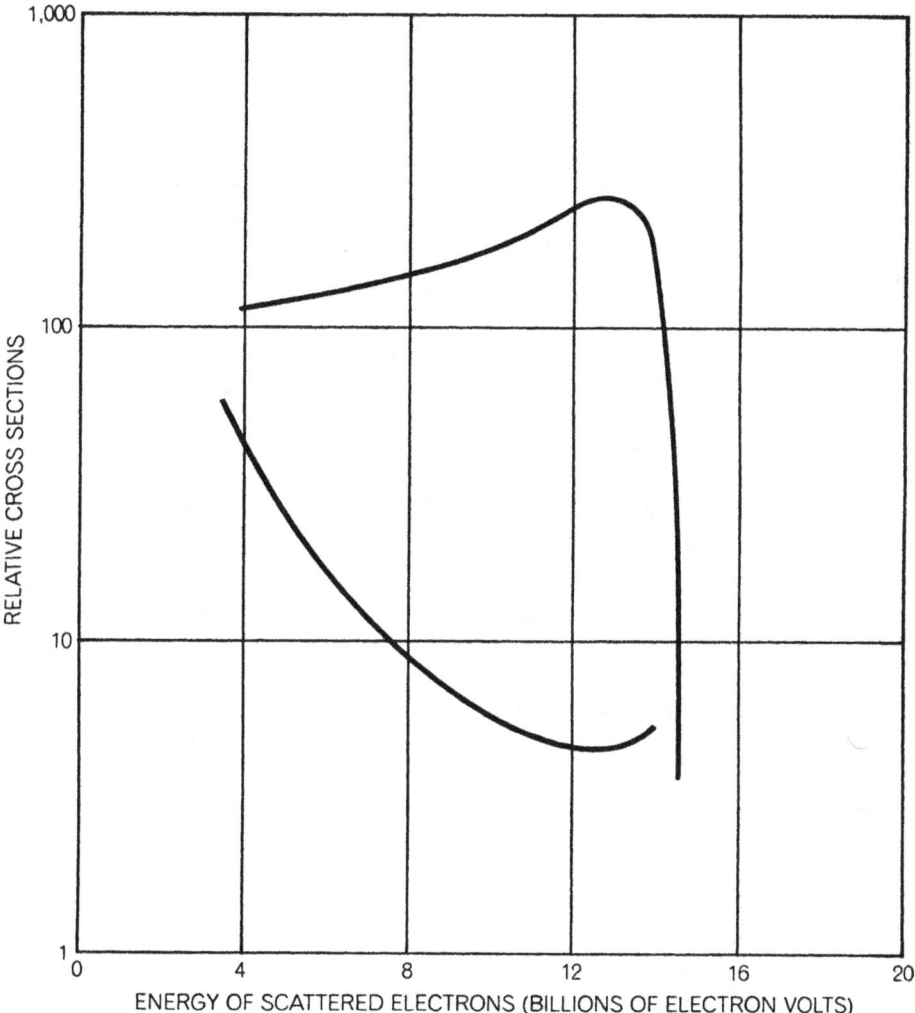

Figure 13. Evidence that the internal structures of the proton and the neutron from which inelastic scattering takes place are much smaller than the nucleons either in their ground state or in their excited state is summarized in this graph, which covers a portion of the spectrum recorded by the M.I.T.-SLAC group in which the predicted scattering cross section (*bottom curve*) is lower by a factor of 40 than the observed cross section (*top curve*). The data were obtained at a scattering angle of six degrees; the energy of the incident electrons was 16 GeV.

The Parton Model

Richard P. Feynman of Cal Tech has been developing a theoretical model of the nucleon that may explain the inelastic-scattering results. He has given the name "parton" to the unknown constituents of the proton and the neutron that inelastically scatter high-energy electrons. Feynman assumes that partons are point particles. He and others have examined the possibility that partons may be one or another of the great array of previously identified subnuclear particles. The mesons that contribute to the "clouds" of nucleonic charge are obvious candidates, but there is strong experimental evidence that partons, if they exist at all, do not exhibit the known properties of mesons.

It has also been suggested that partons may be identical with the hypothetical entities known as quarks, the curious particles proposed independently in 1964 by Murray Gell-Mann and George Zweig of Cal Tech. Quarks are unlike all known particles in having a fractional electric charge: either $+2/3$ or $-1/3$ ($-2/3$ or $+1/3$ for antiquarks). Gell-Mann and Zweig suggested that mesons could be assembled from a quark and an antiquark. Nucleons and other particles with similar properties (that is, the baryons) would have to be assembled from three quarks. No real particles with fractional charge have yet been observed, in spite of long and continuing searches. Nevertheless, a fairly detailed picture of the nucleon's properties, as exhibited in inelastic scattering, can be constructed mathematically by arbitrarily assuming that the hypothetical partons have the properties formerly assigned to the equally hypothetical quarks.

Conceptual models such as the parton model represent the theorist's effort to describe the nucleon's internal structure in accordance with the most advanced information provided by high-energy experiments. The theorist tries to solve the mathematical problems that arise when the model is used to "predict" the properties observed in experiments that have already been completed; he also suggests further measurements to test the validity of the model. Models fail either because the mathematical difficulties cannot be overcome or because their predictions do not agree with experiment. The verification of a model, such as occurred with Rutherford's nuclear atom, can greatly extend the range and scope of the physicist's understanding. It is through the interplay of observation, prediction and comparison that the laws of nature are slowly clarified.

Another unexpected result is that inelastic scattering from the proton is distinctly different from inelastic scattering from the neutron [Figure 14]. It turns out, however, that the electron-scattering results can be greatly simplified if one introduces a variable representing the ratio of the square of the momentum transfer to the difference in energy of the electron before and after scattering. If the various observations are plotted as a function of this simple ratio, the data recorded over a large range of scattering angles and initial and final energies coalesce into a single curve for the proton and a single curve for the neutron [Figure 15]. This unexpected coalescence has a simple explanation if one assumes that the scattering is produced by individual partons, since a "scaling" relation involving the square of the momentum transfer arises naturally in the kinematics of scattering from point particles. In addition the difference between neutron scattering and proton scattering can be accounted for qualitatively by the different configurations of the three quarks needed to produce protons and neutrons.

Because the partons, whatever they may be, are so intertwined with one another their individual properties are difficult to determine. Paradoxically the problem becomes simpler if one conceives of a cloud of partons moving in a frame of refer-

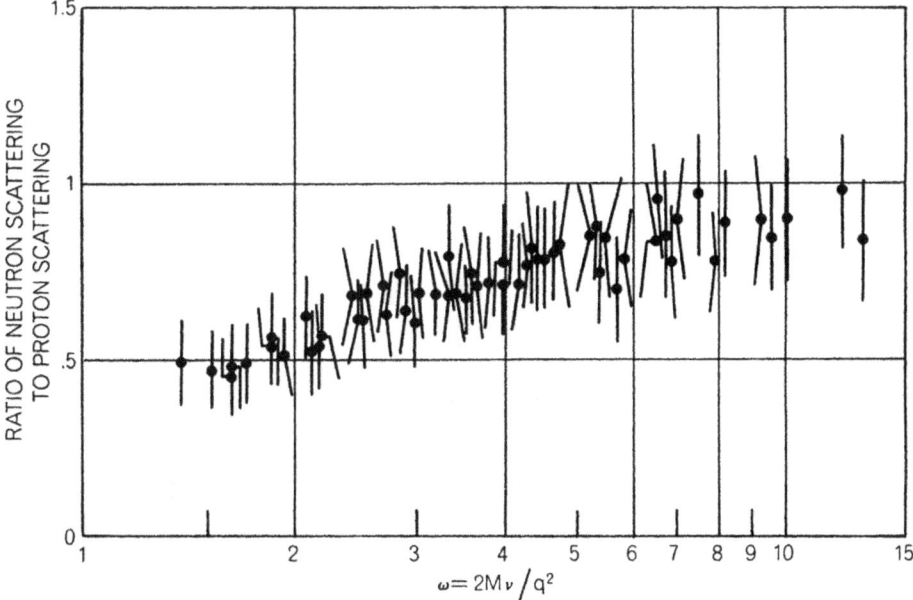

Figure 14. Another unexpected result of the scattering experiments is that inelastic scattering from the proton is distinctly different from inelastic scattering from the neutron. In this graph the ratio of the inelastic-scattering cross sections of the two types of nucleon is plotted as a function of a new variable, ω, which is defined as the ratio of the square of the momentum transfer (q) to the difference in energy of the electron before and after scattering.

ence at nearly the velocity of light, so that the entire nucleon is relativistically contracted into a flat disk. Here the virtual photon that carries the electromagnetic force exerted by the scattered electron interacts with only one of the partons; the parton (owing to the relativistic dilation of time) exists as a free object long enough for it to retain its individual character. Therefore the theoretical analysis of events in the rapidly moving frame can be made with some degree of confidence and transformed back to the laboratory frame. In this way theory can be compared with experiment. Although the parton model is qualitatively quite successful in explaining the scattering results, its quantitative predictions are not uniformly reliable. There is evidently a need both for more experimental information and for more theoretical studies.

Even though the parton model is incomplete, it has already been used to interpret experimental results from other particle reactions, and it has supplied the motivation for several experiments now in the planning stage. At the Italian nuclear research center in Frascati an intense beam of high-energy electrons circulating in a storage ring has been made to cross a counterflowing beam of positrons. A certain fraction of the positrons and electrons interact and annihilate each other, frequently giving rise to two or more pions. The cross sections for annihilation and pion production turn out to be much larger than was expected. Electron-positron an-

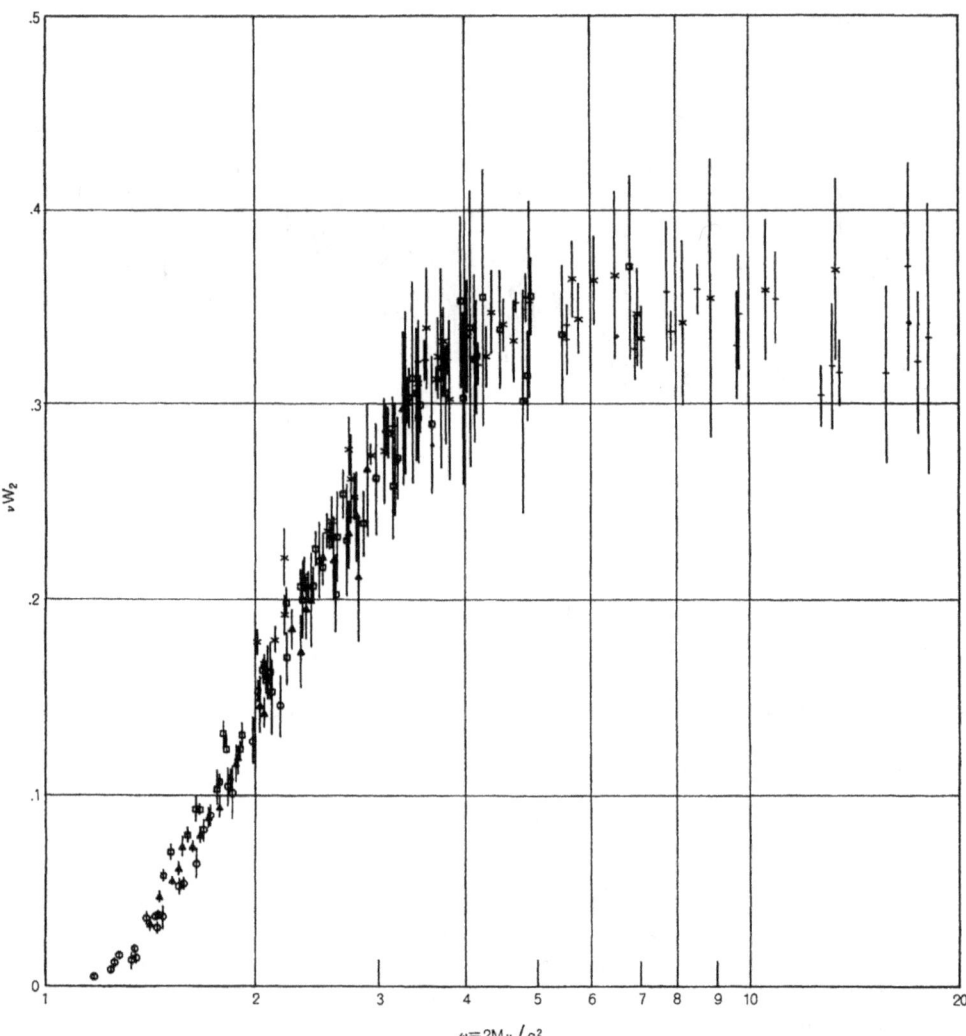

Figure 15. A "universal curve" results when the inelastic-scattering data taken over a large range of scattering angles and initial and final energies are plotted as a function of the new variable ω introduced in the bottom illustration on the preceding page. This coalescence into a single curve (one for the proton and one for the neutron) is consistent with the idea that the scattering of the high-energy electrons actually takes place from pointlike objects within the individual nucleons. The physical nature of these objects, which have been called "partons," remains uncertain. The coalescence illustrated by the curve has been given the name "scaling." This kind of relation, involving the square of the momentum transfer, occurs naturally in the kinematics of scattering from pointlike particles.

nihilation and the "deep" inelastic scattering of electrons observed at Stanford are directly related phenomena; in a fundamental way they can be regarded as inverse reactions of each other. Hence the large cross sections at Frascati support and confirm the large scattering cross sections at Stanford. A further related result is that neutrino beams from the huge accelerator at CERN (the European Organization for Nuclear Research) have initiated inelastic reactions whose cross sections too are unexpectedly large. Again the parton model provides the best available explanation for the observations.

Several related experiments are now being planned. One calls for a comparison of neutrino and antineutrino scattering (which one expects to have equal cross sections). Another involves a search for positron-electron annihilation at high energies to yield a proton and an antiproton in addition to pions (a reaction that may also exhibit a pointlike cross section). A third experiment is being designed to measure the highly inelastic scattering of real photons (which one expects to show large cross sections similar to those observed in electron scattering).

The unpredicted electron-scattering results obtained with the two-mile linear accelerator at Stanford have stimulated a fresh wave of theoretical speculation and experimental study. It is still too early to say whether the parton model will lead to an understanding of the nucleon's structure or whether entirely new ideas may be required. Whatever the case, it seems likely that a full explanation of the electron-scattering studies will clarify not only the nature of the nucleon's constituents but also the nature of the strong interaction and the families of particles that are governed by it.

Deep Inelastic Scattering: Experiments on the Proton and the Observation of Scaling

I. Introduction

A. Overview of the Electron-scattering Program

In late 1967 the first of a long series of experiments on highly inelastic electron scattering was started at the two-mile accelerator at the Stanford Linear Accelerator Center (SLAC) using liquid hydrogen and, later, liquid deuterium targets. Carried out by a collaboration from the Massachusetts Institute of Technology (MIT) and SLAC, the object was to look at large-energy-loss scattering of electrons from the nucleon (the generic name for the proton and neutron), a process soon to be dubbed deep inelastic scattering. Beam energies up to 21 GeV, the highest electron energies then available, and large electron fluxes made it possible to study the nucleon to very much smaller distances than had previously been possible. Because quantum electrodynamics provides an explicit and well-understood description of the interaction of electrons with charges and magnetic moments, electron scattering had, by 1968, already been shown to be a very powerful probe of the structures of complex nuclei and individual nucleons.

Hofstadter and his collaborators had discovered, by the mid-1960s, that as the momentum transfer in the scattering increased, the scattering cross section dropped sharply relative to that from a point charge. The results showed that nucleons were roughly 10^{-13} cm in size, implying a distributed structure. The earliest MIT-SLAC studies, in which California Institute of Technology physicists also collaborated, looked at elastic electron-proton scattering, later ones at electro-production of nucleon resonances with excitation energies up to less than 2 GeV. Starting in 1967,

---

Lecture delivered by Henry W. Kendall on December 8, 1990, on the occasion of the presentation of the 1990 Nobel Prize in Physics. Reprinted with permission from *Review of Modern Physics*, 63(3), 597–613 (July 1991). Copyright © 1991 The American Physical Society.

the MIT-SLAC collaboration employed the higher electron energies made available by the newly completed SLAC accelerator to continue such measurements before beginning the deep inelastic program.

Results from the inelastic studies arrived swiftly: the momentum-transfer dependence of the deep inelastic cross sections was found to be weak, and the deep inelastic form factors—which embodied the information about the proton structure—depended unexpectedly only on a single variable rather than the two allowed by kinematics alone. These results were inconsistent with the current expectations of most physicists at the time. The general belief had been that the nucleon was the extended object found in elastic electron scattering but with the diffuse internal structure seen in pion and proton scattering. The new experimental results suggested point-like constituents but were puzzling because such constituents seemed to contradict well-established beliefs. Intense interest in these results developed in the theoretical community, and, in a program of linked experimental and theoretical advances extending over a number of years, the internal constituents were ultimately identified as *quarks*, which had previously been devised in 1964 as an underlying, quasi-abstract scheme to justify a highly successful classification of the then-known hadrons. This identification opened the door to development of a comprehensive field theory of hadrons (the strongly interacting particles), called Quantum Chromodynamics (OCD), that replaced entirely the earlier picture of the nucleons and mesons. QCD in conjunction with electroweak theory, which describes the interactions of leptons and quarks under the influence of the combined weak and electromagnetic fields, constitutes the Standard Model, all of whose predictions, at this writing, are in satisfactory agreement with experiment. The contributions of the MIT-SLAC inelastic experiments program were recognized by the award of the 1990 Nobel Prize in Physics.

B. Organization of Papers

The three Nobel lectures, taken together, describe the MIT-SLAC experiments. The first, written by R. E. Taylor, sets out the early history of the construction of the two-mile accelerator, the proposals made for the construction of the electron scattering facility, the antecedent physics experiments at other laboratories, and the first scattering experiments which determined the elastic proton structure form factors. The second, this paper, describes the knowledge and beliefs about the nucleon's internal structure in 1968, including the conflicting views on the validity of the quark model and the "bootstrap" models of the nucleon. This is followed by a review of the inelastic scattering program and the series of experiments that were carried out, and the formalism and variables. Radiative corrections are described and then the results of the inelastic electron-proton scattering measurements and the physics picture—the naive parton model—that emerged. The last paper, by J. I. Friedman, is concerned with the later measurements of inelastic electron-neutron and electron-proton measurements and the details of the physical theory—the constituent quark model—which the experimental scattering results stimulated and subsequently, in conjunction with neutrino studies, confirmed.

II. Nucleon and Hadronic Structure in 1968

At the time the MIT-SLAC experiments started in 1968, there was no detailed model of the internal structures of the hadrons. Indeed, the very notion of "internal structure" was foreign to much of the then-current theory. Theory attempted to explain the soft scattering—that is, rapidly decreasing cross sections as the momentum transfer increased—which was the predominant characteristic of the high-energy hadron-hadron scattering data of the time, as well as the hadron resonances, the bulk of which were discovered in the late 1950s and 1960s. Quarks had been introduced, quite successfully, to explain the static properties of the array of hadrons. Nevertheless, the available information suggested that hadrons were "soft" inside and would yield primarily distributions of scattered electrons reflecting diffuse charge and magnetic moment distributions with no underlying point-like constituents. Quark constituent models were gleams in the eyes of a small handful of theorists, but had serious problems, then unsolved, which made them widely unpopular as models for the high-energy interactions of hadrons.

The need to carry out calculations with forces that were known to be very strong introduced intractable difficulties: perturbation theory, in particular, was totally unjustified. This stimulated renewed attention to S-matrix theory,[1] an attempt to deal with these problems by consideration of the properties of a matrix that embodied the array of strong-interaction transition amplitudes from all possible initial states to all possible final states.

A. Theory: Nuclear Democracy

An approach to understanding hadronic interactions and the large array of hadronic resonances was the bootstrap theory,[2] one of several elaborations of S-matrix theory. It assumed that there were no "fundamental" particles: each was a composite of the others. Sometimes referred to as "nuclear democracy," the theory was at the opposite pole from constituent theories.

Regge theory,[3] a very successful phenomenology, was one elaboration of S-matrix theory which was widely practiced.[4] Based initially on a new approach to non-relativistic scattering, it was extended to the relativistic S matrix applicable to high-energy scattering.[5] The known hadrons were classified according to which of several "trajectories" they lay on. It provided unexpected connections between reactions at high energies to resonances in the crossed channels, that is, in disconnected sets of states. For scattering, Regge theory predicted that at high energy, hadron-hadron scattering cross sections would depend smoothly on s, the square of the center-of-mass energy, as $A(s) \sim s^{(\alpha(0))}$, and would fall exponentially with t, the square of the space-like momentum transfer, as $A(t) \sim \exp(\alpha' t \ln(s/s_0))$. Regge theory led to duality, a special formulation of which was provided by Veneziano's dual resonance model.[6] These theories still provide the best description of soft, low-momentum-transfer scattering of pions and nucleons from nucleons, all that was known in the middle 1960s. There was a tendency, in this period, to extrapolate these low-momentum-transfer results so as to conclude there would be no hard scattering at all.

S-matrix concepts were extended to the electromagnetic processes involving hadrons by the Vector-Meson-Dominance (VMD) model.[7] According to VMD, when a real or virtual photon interacts with a hadron, the photon transforms, in effect, into one of the low-mass vector mesons that has the same quantum numbers as the photon (primarily the rho, omega, and phi mesons). In this way electromagnetic amplitudes were related to hadronic collision amplitudes, which could be treated by S-matrix methods. The VMD model was very successful in phenomena involving real photons and many therefore envisaged that VMD would also deal successfully with the virtual photons exchanged in inelastic electron scattering. Naturally, this also led to the expectation that electron scattering would not reveal any underlying structure.

All of these theories, aside from their applications to hadron-hadron scattering and the properties of resonances, had some bearing on nucleon structure as well and were tested against the early MIT-SLAC results.

B. Quark Theory of 1964

The quark was born in a 1964 paper by Murray Gell-Mann[8] and, independently, by George Zweig.[9] For both, the quark (a term Zweig did not use until later) was a means to generate the symmetries of SU(3), the "Eightfold Way," Gell-Mann and Ne'emann's highly successful 1961 scheme for classifying the hadrons.[10] Combinations of spin $-1/2$ quarks, with fractional electric charges and other appropriate quantum numbers, were found to reproduce the multiplet structures of all the observed hadrons. Fractional charges were not necessary but provided the most elegant and economical scheme. Three quarks were required by baryons, later referred to as "valence" quarks, and quark-antiquark pairs for mesons. Indeed the quark picture helped solve some difficulties with the earlier symmetry groupings.[11] The initial successes of the theory stimulated numerous free-quark searches. There were attempts to produce them with accelerator beams, studies to see if they were produced in cosmic rays, and searches for "primordial" quarks by Millikan oil drop techniques sensitive to fractional charges. None of these has ever been successful.[12]

C. Constituent Quark Picture

There were serious problems in having quarks as physical constituents of nucleons, and these problems either daunted or repelled the majority of the theoretical community, including some of its most respected members.[13] The idea was distasteful to the S-matrix proponents. The problems were, first, that the failure to produce quarks had no precedent in physicists' experience. Second, the lack of direct production required the quarks to be very massive, which, for the paired quark configurations of the mesons, meant that the binding had to be very great, a requirement that led to predictions inconsistent with hadron-hadron scattering results. Third, the ways in which they were combined to form the baryons meant that they could not obey the Pauli exclusion principle, as required for spin-one-half particles. Fourth, no fractionally charged objects had ever been unambiguously identified. Such charges were very difficult for many to accept, for the integer character of elementary charges was long established. Enterprising theorists did construct quark theories employing integrally charged quarks, and others contrived ways to circumvent the other objections. Nevertheless, the idea of constituent quarks was not accepted by the bulk of the physics community, while others sought to construct tests that the quark model was expected to fail.[14]

Some theorists persisted, nonetheless. Dalitz[15] carried out complex calculations to help explain not only splittings *between* hadron multiplets but the splittings *within* them also, using some of the theoretical machinery employed in nuclear spectroscopy calculations. Calculations were carried out on other aspects of hadron dynamics, for example, the successful prediction that Δ^+ decay would be predominantly magnetic dipole.[16] Owing to the theoretical difficulties just discussed, the acceptance of quarks as the basis of this successful phenomenology was not carried over to form a similar basis for high-energy scattering.

Gottfried studied electron-proton scattering with a model assuming point quarks, and argued that it would lead to a total cross section (elastic plus inelastic) at fixed momentum transfer, identical to that of a point charge, but he expressed great skepticism that this would be borne out by the forthcoming data.[17] With the exception of Gottfried's work and one by Bjorken stimulated by current algebra, discussed below, all of the published constituent quark calculations were concerned with low-energy processes or hadron characteristics rather than high-energy interactions. Zweig carried out calculations assuming that quarks were indeed hadron constituents, but his ideas were not widely accepted.[18]

Thus one sees that the tide ran against the constituent quark model in the 60s.[19] One reviewer's summary of the style of the 60s was that "quarks came in handy for coding information but should not be taken seriously as physical objects."[20] While quite helpful in low-energy resonance physics, it was for some "theoretically disreputable," and was felt to be largely peripheral to a description of high-energy soft scattering.[21]

D. Current Algebra

Following his introduction of quarks, Gell-Mann, and others, developed "current algebra," which deals with hadrons under the influence of weak and electromagnetic interactions. Starting with an assumption of free-quark fields, he was able to find relations between weak currents that reproduced the current commutators postulated in constructing his earlier hadronic symmetry groups. Current algebra had become very important by 1966. It exploited the concept of *local observables*—the current and charge densities of the weak and electromagnetic interactions. These are field theoretic in character and could only be incorporated into *S*-matrix *cum* bootstrap theory by assumptions like VMD. The latter are plausible for moderate-momentum transfer, but hardly for transfer large compared to hadron masses. As a consequence, an important and growing part of the theoretical community was thinking in field-theoretic terms.

Current algebra also gave rise to a small but vigorous "sum-rule" industry. Sum rules are relationships involving weighted integrals over various combinations of cross sections. The predictions of some of these rules were important in confirming the deep inelastic electron and neutrino scattering results, after these became available.[22]

Gell-Mann made clear that he was not suggesting that hadrons were made up of quarks,[23] although he kept open the possibility that they might exist.[24] Nevertheless current algebra reflected its constituent quark antecedents, and Bjorken used it to demonstrate that sum rules derived by him and others required large cross sections for these to be satisfied. He then showed that such cross sections arose naturally in a quark constituent model,[25] in analogy to models of nuclei composed of constituent protons and neutrons, and also employed it to predict the phenomena of scaling, discussed at length below. Yet Bjorken and others were at a loss to decide how the point-like properties that current algebra appeared to imply were to be accommodated.[26]

E. Theoretical Input to the Scattering Program

In view of the theoretical situation as set out above, there was no consideration that point-like substructure of the nucleon might be observable in electron scattering during the planning and design of the electron scattering facility. Deep inelastic processes were, however, assessed in preparing the proposal submitted to SLAC for construction of the facility.[27] Predictions of the cross sections employed a model assuming off-mass-shell photo-meson production, using photoproduction cross sections combined with elastic scattering structure functions, in what was believed to be the best guide to the yields expected. These were part of extensive calculations, carried out at MIT, designed to find the magnitude of distortions of inelastic spectra arising from photon radiation, necessary in planning the equipment and assessing the difficulty of making radiative corrections. It was found ultimately that these had underpredicted the actual yields by between one and two orders of magnitude.

III. The Scattering Program

The linear accelerator that provided the electron beam employed in the inelastic scattering experiments was, and remains to the date of this paper, a device unique among high-energy particle accelerators. See Figure 1. An outgrowth of the smaller, 1 GeV accelerator employed by Hofstadter in his studies of the charge and magnetic moment distributions of the nucleon, it relied on advanced klystron technology devised by Stanford scientists and engineers to provide the high levels of microwave power necessary for one-pass acceleration of electrons. Proposed in 1957, approved by the Congress in 1962, its construction was initiated in 1963. It went into operation in 1967, on schedule, having cost $114 million.

The experimental collaboration began in 1964. After 1965 R. E. Taylor was head of SLAC Group A, with J. I. Friedman and the present author sharing responsibil-

Figure 1. View of the Stanford Linear Accelerator. The electron injector is at the top, the experimental area in lower center. The deep inelastic scattering studies were carried out in End Station A, the largest of the buildings in the experimental area.

ity for the MIT component. A research group from California Institute of Technology joined in the construction cycle and the elastic studies but withdrew before the inelastic work started in order to pursue other interests.

The construction of the facility to be employed in electron scattering was nearly concurrent with accelerator's construction. This facility was large for its time. A 200 ft by 125 ft shielded building housed three magnetic spectrometers with an adjacent "counting house" containing the fast electronics and a computer, also large for its time, where experimenters controlled the equipment and conducted the measurements. See Figure 2(a) and 2(b). The largest spectrometer would focus electrons up to 20 GeV and was employed at scattering angles up to 10°. A second spectrometer, useful to 8 GeV, was used initially out to 34°, and a third, focusing to 1.6 GeV, constructed for other purposes, was employed in one set of large-angle measurements to help determine the uniformity in density of the liquified target gases. The detectors were designed to detect only scattered electrons. The very short duty cycle of the pulsed beam precluded studying the recoil systems in coincidence with the scattered electrons: it would have given rise to unacceptable chance coincidence rates, swamping the signal.

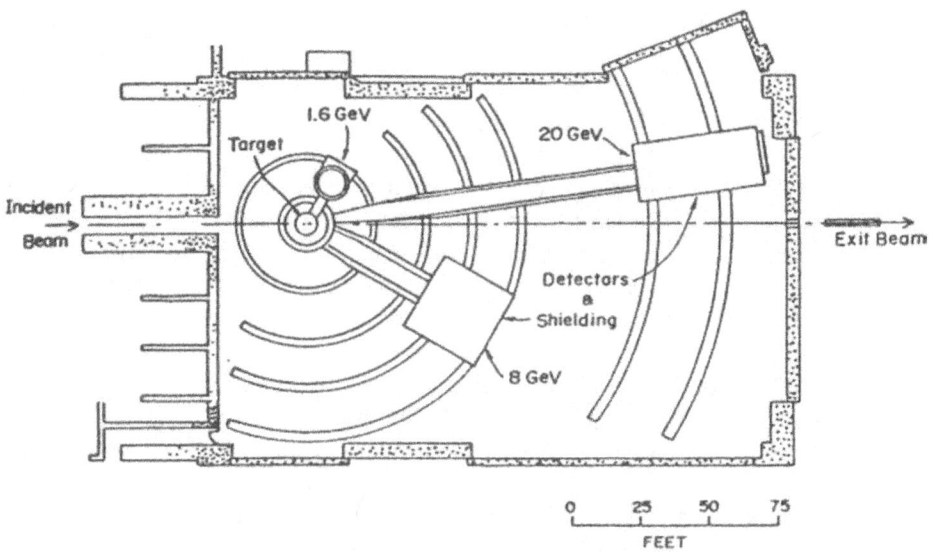
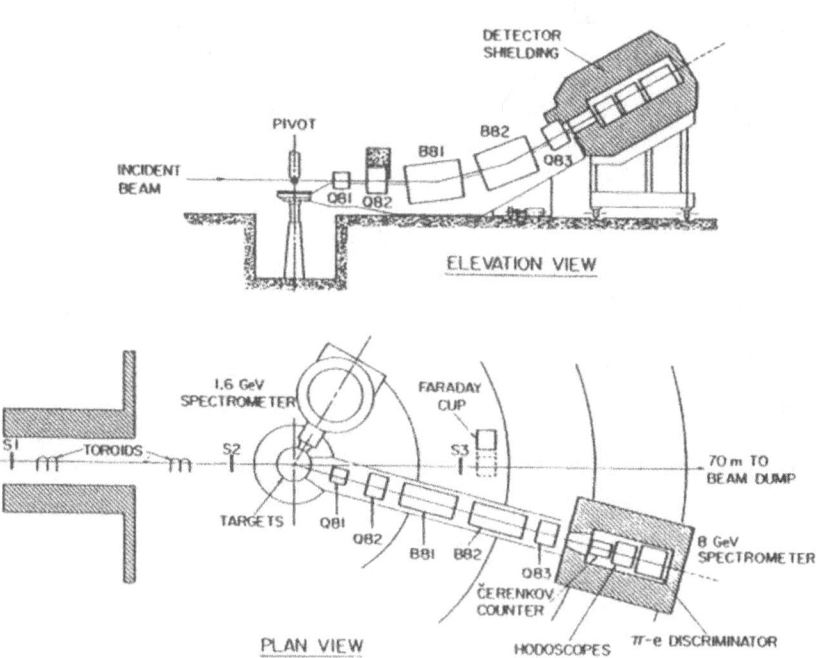

Figure 2. (a) Plan view of End Station A and the two principal magnetic spectrometers employed for analysis of scattered electrons. (b) Configuration of the 8 GeV spectrometer, employed at scattering angles greater than 12°.

The elastic studies started in early 1967 with the first look at inelastic processes from the proton late the same year. By the spring of 1968 the first inelastic results were at hand. The data were reported at a major scientific meeting in Vienna in August and published in 1969.[28] Thereafter a succession of experiments were carried out, most of them, from 1970 on, using both deuterium and hydrogen targets in matched sets of measurements to extract neutron scattering cross sections with a minimum of systematic error. These continued well into the 1970s. One set of measurements[29] studied the atomic-weight dependence of the inelastic scattering, primarily at low-momentum transfers, studies that were extended to higher-momentum transfers in the early 1980s, and involve extensive reanalysis of earlier MIT-SLAC data on hydrogen, deuterium, and other elements.[30]

The collaboration was aware from the outset of the program that there were no accelerators in operation, or planned, that would be able to confirm the entire range of results. The group carried out independent data analyses at MIT and at SLAC to minimize the chance of error. One consequence of the absence of comparable scattering facilities was that the collaboration was never pressed to conclude either data taking or analysis in competitive circumstances. It was possible throughout the program to take the time necessary to complete work thoroughly.

IV. Scattering Formalism and Radiative Corrections

A. Fundamental Processes

The relation between the kinematic variables in elastic scattering, as shown in Figure 3, is

$$\nu = E - E' = q^2/(2M) \quad q^2 = 2EE'(1 - \cos(\theta)), \tag{1}$$

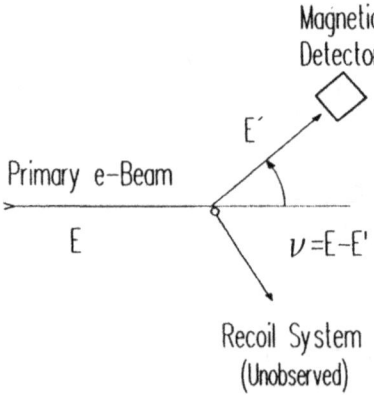

Figure 3. Scattering kinematics.

where E is the initial, and E' the final, electron energy; θ the laboratory angle of scattering; ν the electron energy loss; q the four-momentum transferred to the target nucleon; and M the proton mass.

The cross section for elastic electron-proton scattering has been calculated by Rosenbluth[31] in first Born approximation, that is, to leading order in $\alpha = \frac{1}{137}$:

$$\frac{d\sigma}{d\Omega}(E) = \sigma_M(E)\left[\frac{E'}{E}\right]\left[\frac{G_{Ep}^2(q^2) + \tau G_{Mp}^2(q^2)}{1+\tau}\right.$$
$$\left. + 2\tau G_{Mp}^2 \tan^2(\theta/2)\right] \quad (2)$$

where

$$\sigma_M = \frac{4\alpha^2 E'^2}{q^4}\cos^2(\theta/2)$$

is the Mott cross section for elastic scattering from a point proton, and

$$\tau = q^2/(4M^2).$$

In these equations, and in what follows, $\hbar = c = 1$, and the electron mass has been neglected. The functions $G_{Ep}(q^2)$ and $G_{Mp}(q^2)$, the electric and magnetic form factors, respectively, describe the time-averaged structure of the proton. In the nonrelativistic limit the squares of these functions are the Fourier transforms of the spatial distributions of charge and magnetic moment, respectively. As can be seen from Eq. (2), magnetic scattering is dominant at high q^2. Measurements[32] show that G_{Mp} is roughly described by the "dipole" approximation:

$$G_{Mp}/\mu = 1/(1 + q^2/0.71)^2,$$

where q^2 is measured in $(GeV)^2$ and $\mu = 2.79$ is the proton's magnetic moment. Thus at large q^2 an additional $1/q^8$ dependence beyond that of σ_M is imposed on the elastic scattering cross section as a consequence of the finite size of the proton. This is shown in Figure 4.

In inelastic scattering, energy is imparted to the hadronic system. The invariant or missing mass W is the mass of the final hadronic state. It is given by

$$W^2 = (2M\nu + M^2 - q^2).$$

When only the electron is observed the composition of the hadronic final state is unknown except for its invariant mass W. On the assumption of one photon exchange (Figure 5), the differential cross section for electron scattering from the nucleon target is related to two structure functions, W_1 and W_2 according to Drell and Walecka,[33]

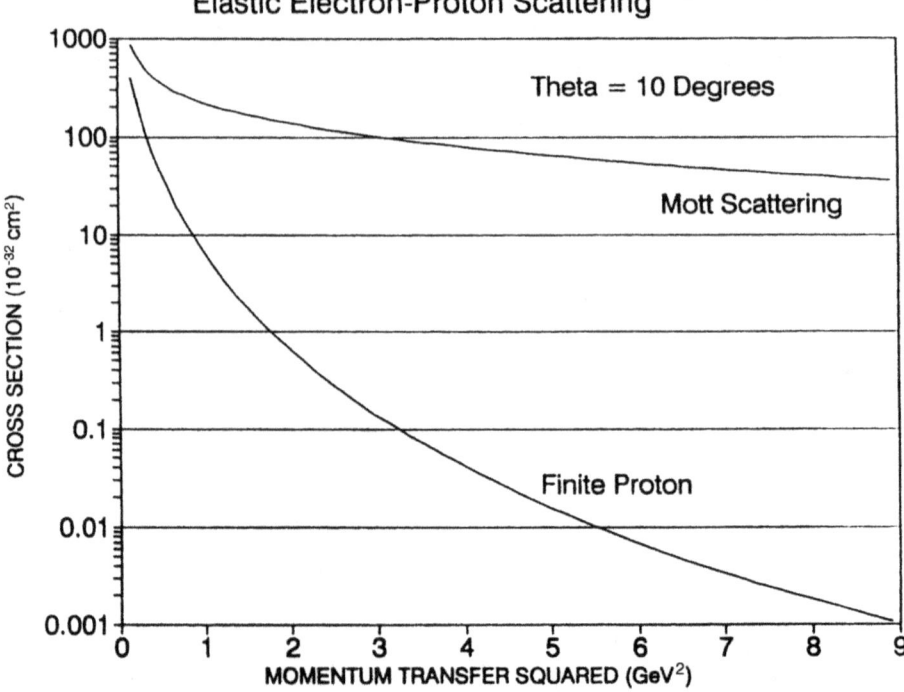

Figure 4. Elastic scattering cross sections for electrons from a "point" proton and for the actual proton. The differences are attributable to the finite size of the proton.

$$\frac{d^2\sigma}{d\Omega dE'}(E,E',\theta) = \sigma_M[W_2(\nu,q^2) + 2W_1(\nu,q^2)\tan^2(\theta/2)]. \qquad (3)$$

This expression is the analog of the Rosenbluth cross section given above. The structure functions W_1 and W_2 are similarly defined by Eq. (3) for the proton, deuteron,

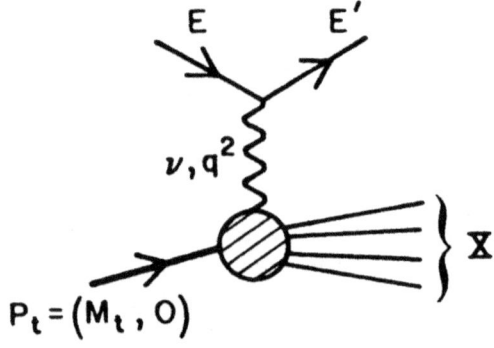

Figure 5. Feynman diagram for inelastic electron scattering.

or neutron; they summarize all the information about the structure of the target particles obtainable by scattering unpolarized electrons from an unpolarized target.

Within the single-photon-exchange approximation, one may view inelastic electron scattering as photoproduction by "virtual" photons. Here, as opposed to photoproduction by real photons, the photon mass q^2 is variable and the exchanged photon may have a longitudinal as well as a transverse polarization. If the final-state hadrons are not observed, the interference between these two components averages to zero, and the differential cross section for inelastic electron scattering is related to the total cross sections for absorption of transverse, σ_T, and longitudinal, σ_L, virtual photons according to Hand,[34]

$$\frac{d^2\sigma}{d\Omega dE'}(E,E'\theta) = \Gamma[\sigma_T(\nu,q^2) + \epsilon\sigma_L(\nu,q^2)], \quad (4)$$

where

$$\Gamma = \frac{\alpha}{4\pi^2}\frac{KE'}{q^2 E}\left[\frac{2}{1-\epsilon}\right],$$

$$\epsilon = [1 + 2(1 + \nu^2/q^2)\tan^2(\theta/2)]^{-1},$$

and

$$K = (W^2 - M^2)/(2M).$$

The quantity Γ is the flux of transverse virtual photons, and ϵ is the degree of longitudinal polarization. The cross sections σ_T and σ_L are related to the structure functions W_1 and W_2 by

$$W_1(\nu,q^2) = \frac{K}{4\pi^2\alpha}\sigma_T(\nu,q^2),$$

$$W_2(\nu,q^2) = \frac{K}{4\pi^2\alpha}\left[\frac{q^2}{q^2+\nu^2}\right][\sigma_T(\nu,q^2) + \sigma_L(\nu,q^2)]. \quad (5)$$

In the limit $q^2 \to 0$, gauge invariance requires that $\sigma_L \to 0$ and $\sigma_T \to \sigma_\gamma(\nu)$, where $\sigma_\gamma(\nu)$ is the photoproduction cross section for real photons. The quantity R, defined as the ratio σ_L/σ_T, is related to the structure functions by

$$R(\nu,q_2) \equiv \sigma_L/\sigma_T = (W_2/W_1)(1 + \nu^2/q^2) - 1. \quad (6)$$

A separate determination of the two inelastic structure functions W_1 and W_2 (or, equivalently, σ_L and σ_T) requires values of the differential cross section at several values of the angle θ for fixed ν and q^2. According to Eq. (4), σ_L is the slope and σ_T is the intercept of a linear fit to the quantity Σ where

$$\Sigma = \frac{1}{\Gamma}\frac{d^2\sigma}{d\Omega dE'}(\nu,q^2,\theta).$$

The structure functions W_1 and W_2 are then directly calculable from Eq. (5). Alternatively, one can extract W_1 and W_2 from a single differential cross-section measurement by inserting a particular functional form for R in the equations

$$W_1 = \frac{1}{\sigma_M} \frac{d^2\sigma}{d\Omega dE'} \left[(1+R)\left[\frac{q^2}{q^2+\nu^2}\right] + 2\tan^2(\theta/2) \right]^{-1},$$

$$W_2 = \frac{1}{\sigma_M} \frac{d^2\sigma}{d\Omega dE'} \left[1 + \left[\frac{2}{1+R}\right] \right. \qquad (7)$$

$$\left. \times \left[\frac{q^2+\nu^2}{q^2}\right] \tan^2(\theta/2) \right]^{-1}.$$

Equations (5) through (7) apply equally well for the proton, deuteron, or neutron.

In practice it was convenient to determine values of σ_L and σ_T from straight-line fits to differential cross sections as functions of ϵ. R was determined from the values of σ_L and σ_T, and W_1 and W_2 were, as shown above, determined from R.

B. Scale Invariance and Scaling Variables

By investigating models that satisfied current algebra, Bjorken (1969) had conjectured that in the limit of q^2 and ν approaching infinity, with the ratio $\omega = 2M\nu/q^2$ held fixed, the two quantities νW_2 and W_1 become functions of ω only.[35] That is,

$$2MW_1(\nu,q^2) = F_1(\omega),$$

$$\nu W_2(\nu,q^2) = F_2(\omega).$$

It is this property that is referred to as "scaling" in the variable ω in the "Bjorken limit." The variable $x = 1/\omega$ came into use soon after the first inelastic measurements; we will use both in this paper.

Since W_1 and W_2 are related by

$$\nu W_2/W_1 = (1+R)/(1/\nu + \omega/(2M)),$$

it can be seen that scaling in W_1 accompanies scaling in νW_2 only if R has the proper functional form to make the right-hand side of the equation a function of ω. In the Bjorken limit, it is evident that the ratio $\nu W_2/W_1$ will scale if R is constant or is a function of ω only.

C. Radiative Corrections

Radiative corrections must be applied to the measured cross sections to eliminate the effects of the radiation of photons by electrons which occurs during the nucleon scattering itself and during traversals of material before and after scattering. These corrections also remove higher-order electrodynamic contributions to the electron-photon vertex and the photon propagator. Radiative corrections as extensive as were required in the proposed scattering program had been little studied previously.[36] Friedman[37] in 1959 had calculated the elements of the required "triangle," discussed in more detail below, in carrying out corrections to the inelastic scattering of 175 MeV electrons from deuterium. Isabelle and Kendall (1964), studying the inelastic scattering of electrons of energy up to 245 MeV from Bi^{209} in 1962, had measured inelastic spectra over a number of triangles and had developed the computer procedures necessary to permit computation of the corrections. These studies provided confidence that the procedures were tractable and the resulting errors of acceptable magnitude.[38]

The largest correction has to be made for the radiation during scattering, described by diagrams (a) and (b) in Figure 6. A photon of energy k is emitted in (a) after the virtual photon is exchanged, and in (b) before the exchange. Diagram (c) is the cross section which is to be recovered after appropriate corrections for (a) and (b) have

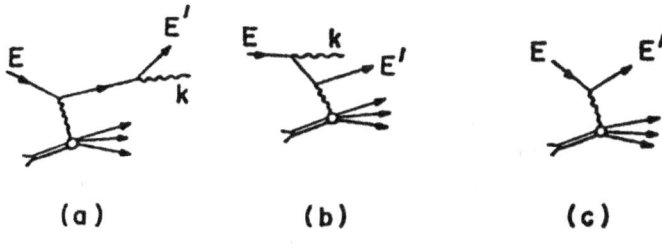

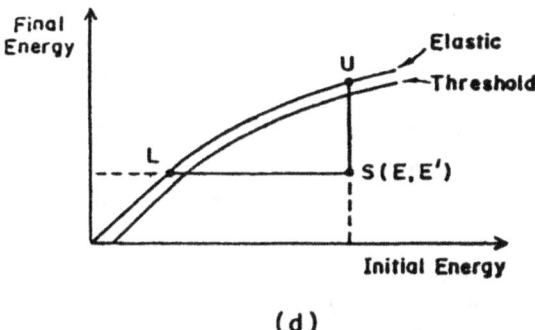

Figure 6. Diagrams showing radiation in electron scattering (a) after exchange of a virtual photon (b) before exchange of a virtual photon. Figure (c) is the diagram with radiative effects removed. Figure (d) is the kinematic plane relevant to the radiative corrections program. The text contains a further discussion of corrections procedures. A "triangle" as discussed in the text is formed by points L, U, and S.

been made. A measured cross section at fixed E, E', and θ will have contributions from (a) and (b) for all values of k that are kinematically allowed. The lowest value of k is zero, and the largest occurs in (b) for elastic scattering of the virtual electron from the target particle. Thus, to correct a measured cross section at given values of E and E', one must know the cross section over a range of incident and scattered energies.

To an excellent approximation, the information necessary to correct a cross section at an angle θ may all be gathered at the same value of θ. Diagram (d) of Figure 6 shows the kinematic range in E and E' of cross sections which can contribute by radiative processes to the fundamental cross section sought at point S for fixed θ. The range is the same for contributions from bremsstrahlung processes of the incident and scattered electrons. For single hard-photon emission, the cross section at point S will have contributions from elastic scattering at points U and L, and from inelastic scattering along the lines SL and SU, starting at inelastic threshold. If two or more photons are radiated, contributions can arise from line LU and the inelastic region bounded by lines SL and SU. The cross sections needed for these corrections must themselves have been corrected for radiative effects. However, if uncorrected cross sections are available over the whole of the "triangle" LUS, then a one-pass radiative correction procedure may be employed, assuming the peaking approximation,[39] which will produce the approximately corrected cross sections over the entire triangle, including the point S.

The application of radiative corrections required the solution of another difficulty, as it was generally not possible to take measurements sufficiently closely spaced in the E-E' plane to apply them directly. Typically, five to ten spectra, each for a different E, were taken to determine the cross sections over a "triangle." Interpolation methods had to be developed to supply the missing cross sections and had to be tested to show that they were not the source of unexpected error. Figure 7(a), (b), and (c) shows the triangles, and the locations of the spectra, for data taken in one of the experiments in the program.

In the procedures that were employed, the radiative tails from elastic electron-proton scattering were subtracted from the measured spectra before the interpolations were carried out. In the MIT-SLAC radiative correction procedures, the radiative tails from elastic scattering were calculated using the formula of Tsai,[40] which is exact to lowest order in α. The calculation of the tail included the effects of radiative energy degradation of the incident and final electrons, the contributions of multiple photon processes, and radiation from the recoiling proton. After the subtraction of the elastic peak's radiative tail, the inelastic radiative tails were removed in a one-pass unfolding procedure as outlined above. The particular form of the peaking approximation used was determined from a fit to an exact calculation of the inelastic tail to lowest order which incorporated a model that approximated the experimental cross sections. One set of formulas and procedures are described by Miller et al.[41] and were employed in the SLAC analysis. The measured cross sections were also corrected in a separate analysis, carried out at MIT, using a somewhat different set of approximations.[42] Comparisons of the two gave corrected cross

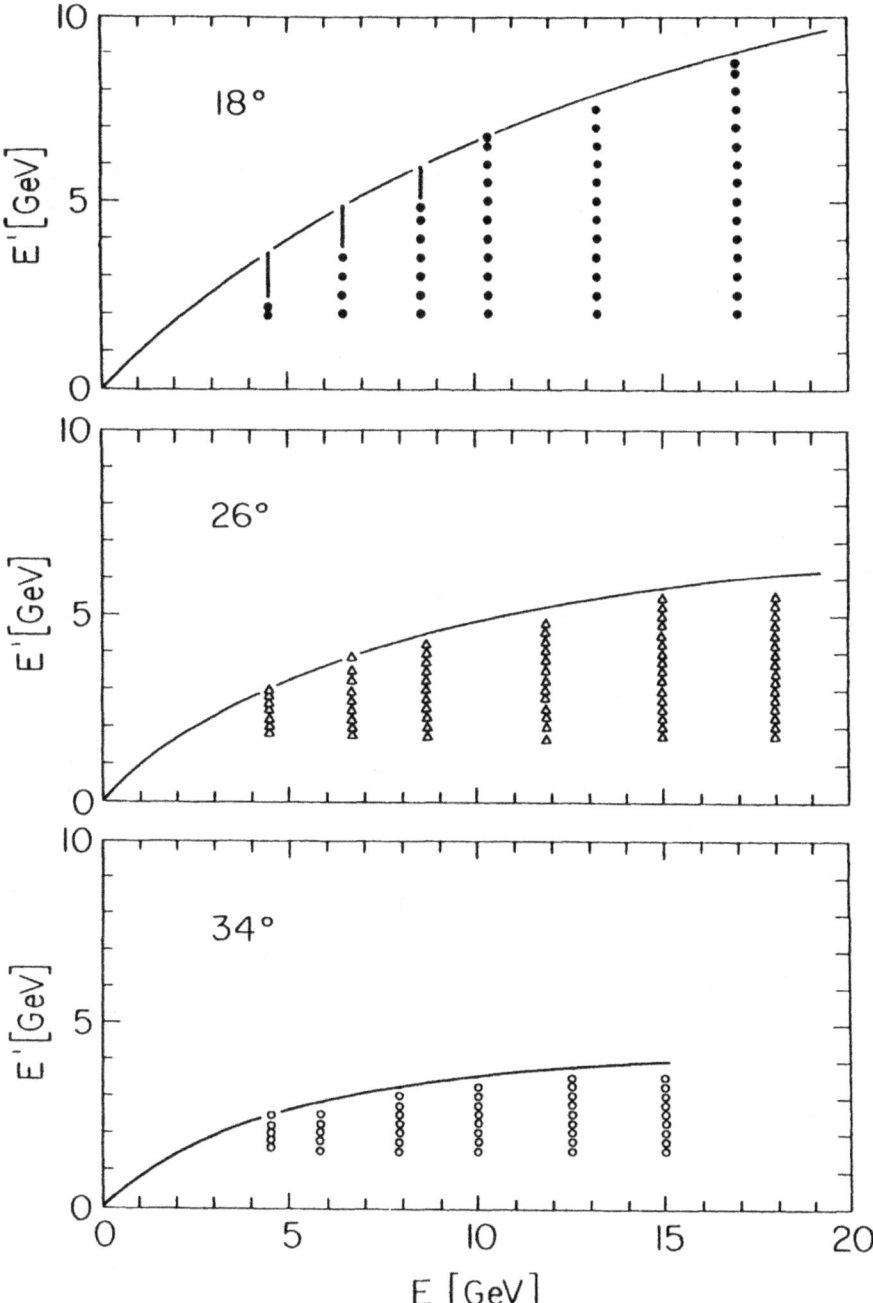

Figure 7. Inelastic measurements: where spectra were taken to determine "triangles" employed in making radiative corrections for three angles selected for some of the later experiments. The solid curves represent the kinematics of elastic electron-proton scattering.

sections which agreed to within a few percent. Bodek et al.[43] give a complete description of the MIT radiative correction procedures that were applied, the cross checks that were carried out, and the assessment of errors arising both from the radiative corrections and from other sources of uncertainty in the experiment. Figure 8 shows the relative magnitude of the radiative corrections as a function of W for a typical spectrum with a hydrogen target. While radiative corrections were the largest corrections to the data and involved a considerable amount of computation, they were understood to a confidence level of 5% to 10% and did not significantly increase the total error in the measurements.

V. Electron-Proton Scattering: Results

The scattered electron spectra observed in the experiment had a number of features whose prominence depended on the initial and final electron energies and the scattering angle. At low q^2 both the elastic peak and the resonance excitations were large, with little background from nonresonant continuum scattering either in the resonance region or at higher missing masses. As q^2 increased, the elastic and resonance cross sections decreased rapidly, with the continuum scattering becoming more and more dominant. Figure 9 shows four spectra of differing q^2. Data points taken at the elastic peak and in the resonance region were closely spaced in E' so as to allow fits to be made to the resonance yields, but much larger steps were employed for larger excitation energies.

Figures 10(a) and 10(b) show visual fits to spectra over a wide range in energy and scattering angle including one spectrum from the accelerator at the Deutsches Electronen Synchrotron (DESY), illustrating the points discussed above.

Two features of the nonresonant inelastic scattering that appeared in the first continuum measurements were unexpected. The first was a quite weak q^2 dependence of the scattering at constant W. Examples for $W = 2.0$ and $W = 3.0$ GeV, taken from data of the first experiment, are shown in Figure 11 as a function of q^2. For comparison the q^2 dependence of elastic scattering is shown also.

The second feature was the phenomenon of scaling. During the analysis of the inelastic data, J. D. Bjorken suggested a study to determine if νW_2 was a function of ω alone. Figure 12(a) shows the earliest data so studied: W_2, for six values of q^2, as a function of ν. Figure 12(b) shows $F_2 = \nu W_2$ for 10 values of q^2, plotted against ω. Because R was at that time unknown, F_2 was shown for the limiting assumptions $R = 0$ and $R = \infty$. It was immediately clear that the Bjorken scaling hypoth-

Figure 8. Spectra of 10 GeV electrons scattered from hydrogen at 6°, as a function of the final hadronic state energy W. Diagram (a) shows the spectrum before radiative corrections. The elastic peak has been reduced in scale by a factor of 8.5. The computed radiative "tail" from the elastic peak is shown. Diagram (b) shows the same spectrum with the elastic peak's tail subtracted and inelastic corrections applied. Diagram (c) shows the ratio of the inelastic spectrum before to the spectrum after radiative corrections.

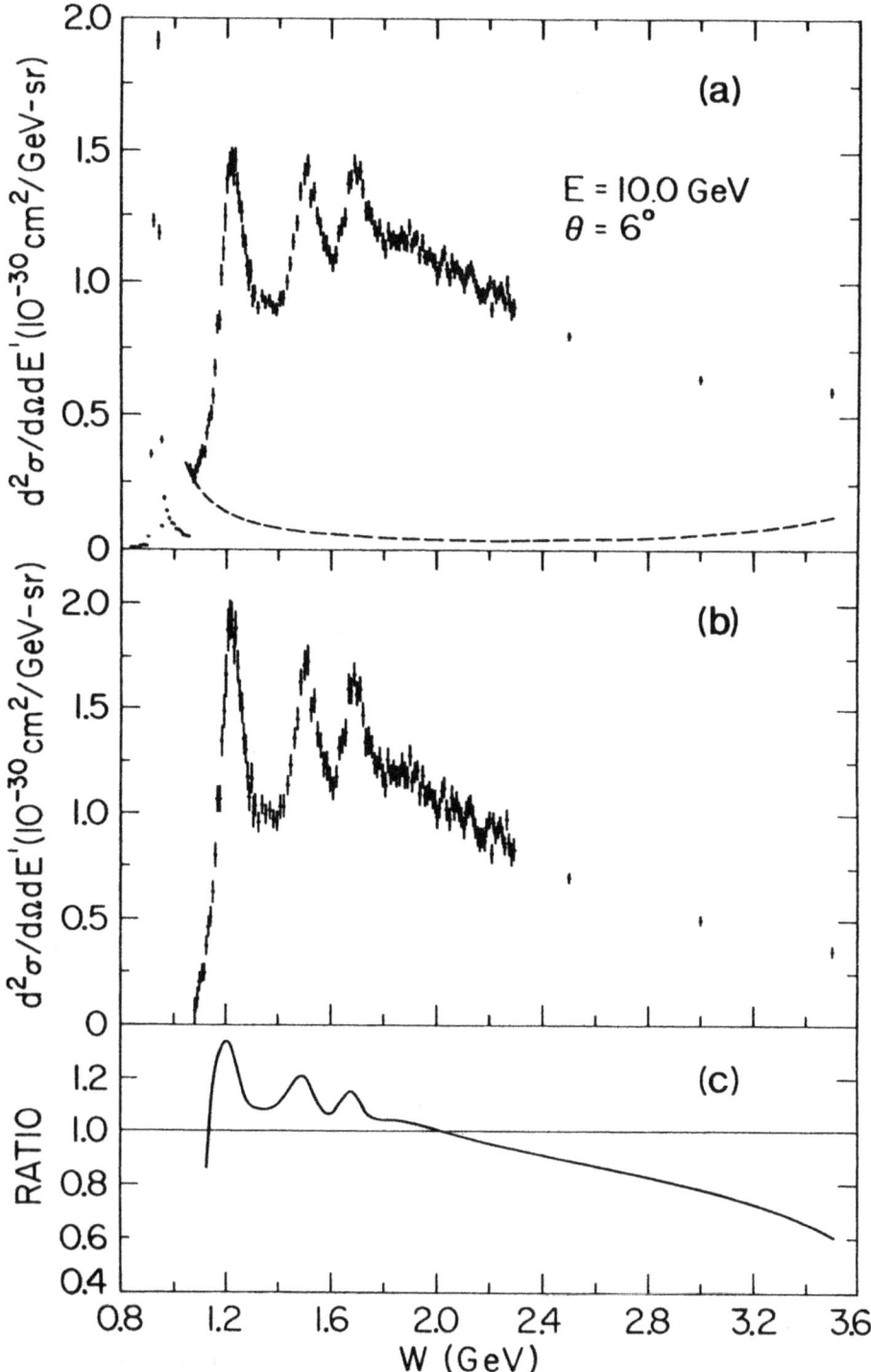

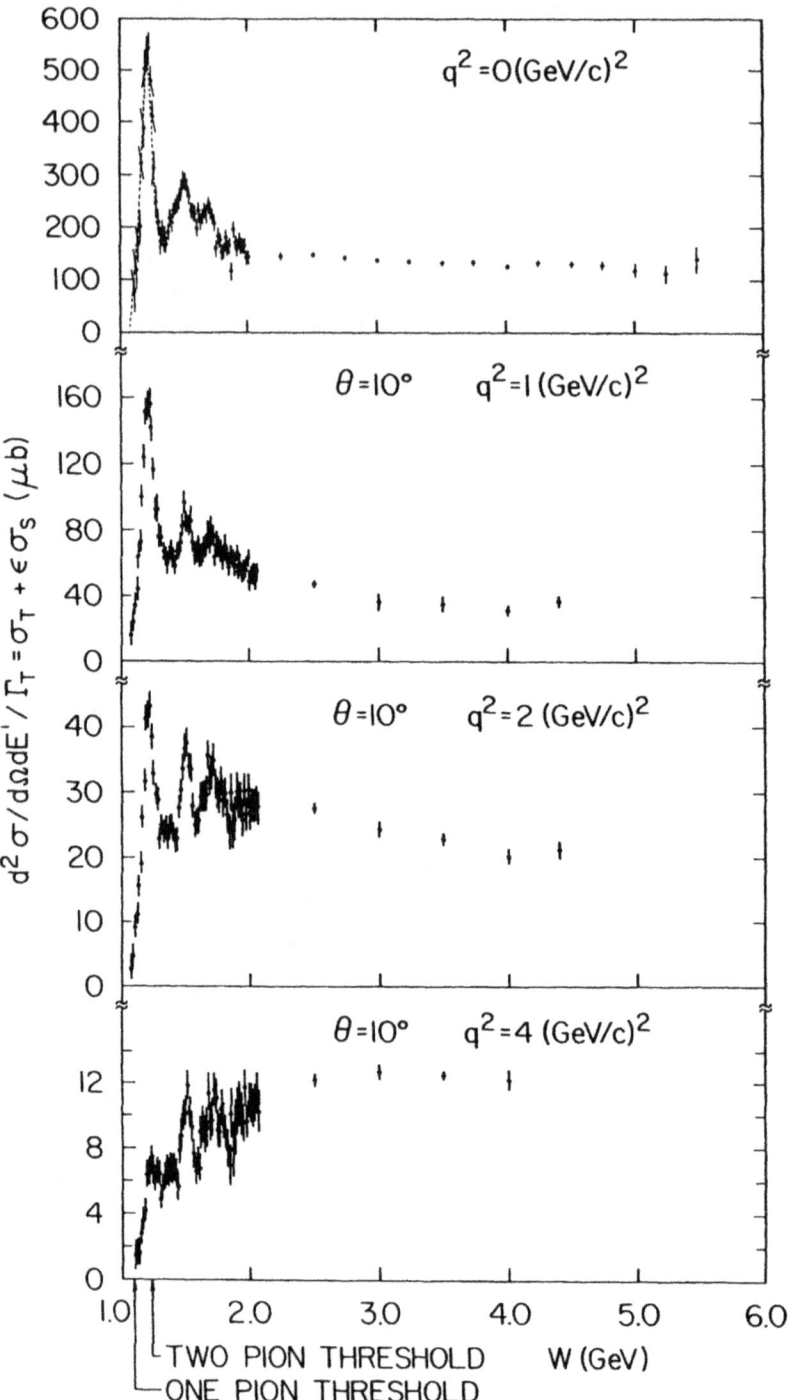

Figure 9. Spectra of electrons scattered from hydrogen at q^2 up to 4 (GeV/c)^2. The curve for $q^2 = 0$ represents an extrapolation to $q^2 = 0$ of electron scattering data acquired at $\theta = 1.5°$. Elastic peaks have been subtracted and radiative corrections have been applied.

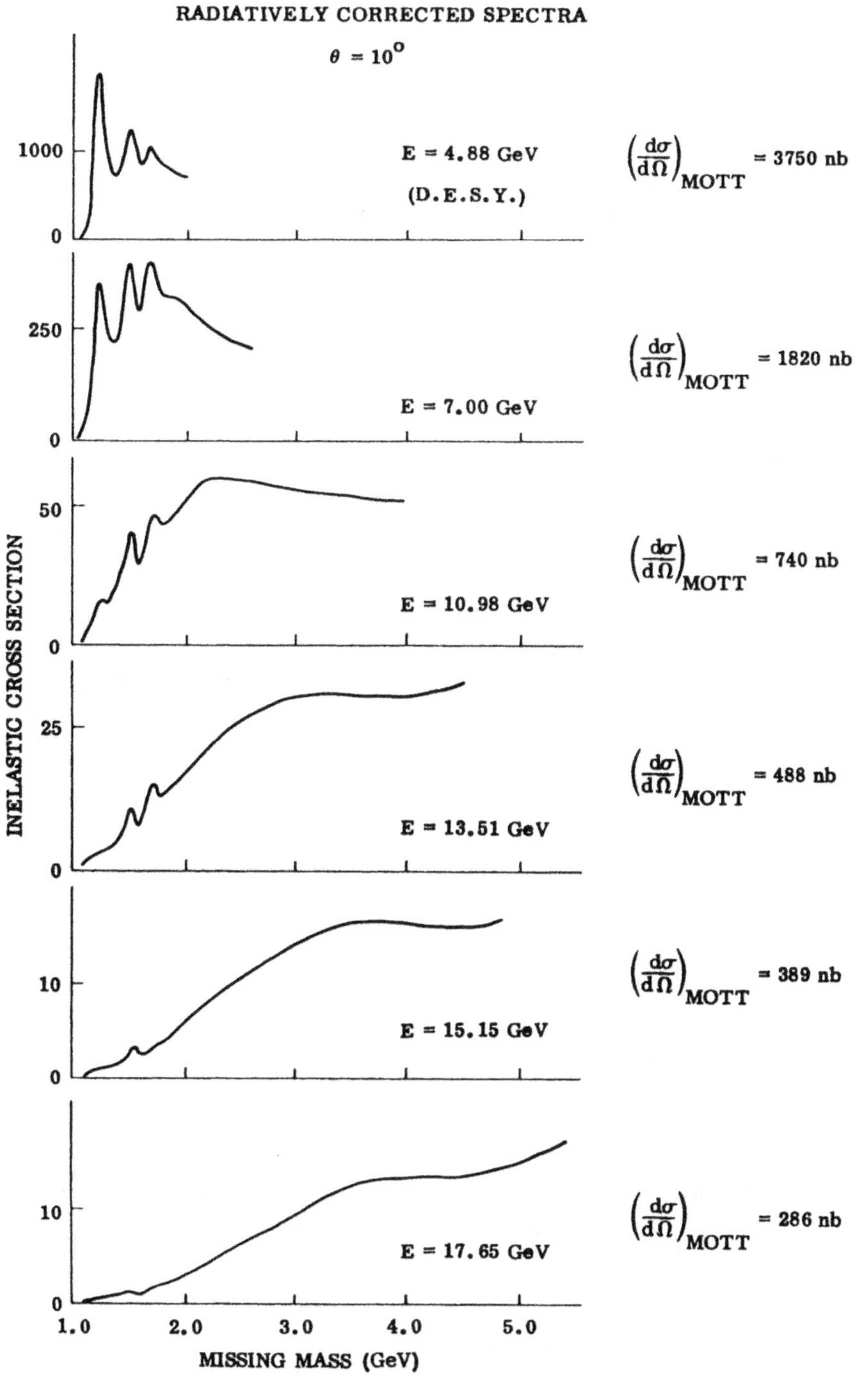

Figure 10. (a) Visual fits to spectra showing the scattering of electrons from hydrogen at 10° for primary energies, E, from 4.88 GeV to 17.5 GeV. The elastic peaks have been subtracted and radiative corrections applied. The cross sections are expressed in nanobarn per GeV per steradian. The spectrum for E = 4.88 GeV was taken at DESY.[44]

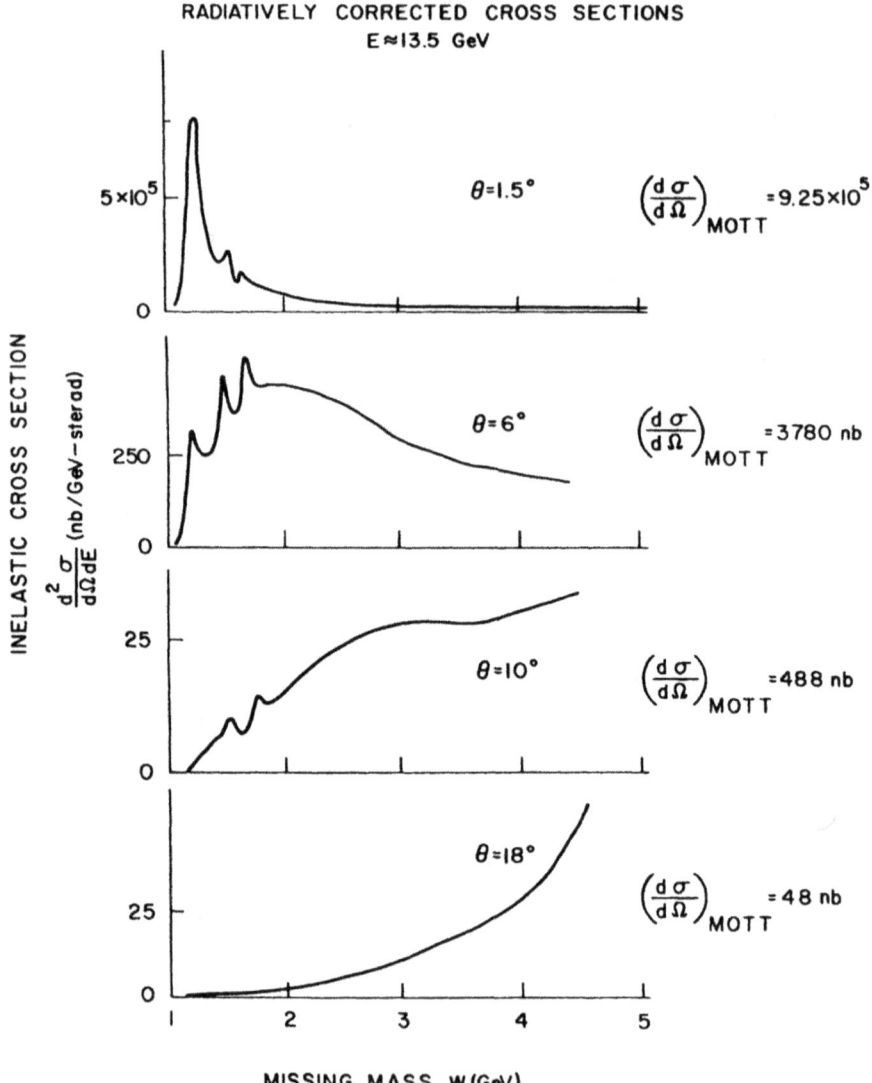

Figure 10. (continued) (b) Visual fits to spectra showing the scattering of electrons from hydrogen at a primary energy E of approximately 13.5 GeV, for scattering angles from 1.5° to 18. The 1.5° curve is taken from MIT-SLAC data used to obtain photoabsorption cross sections.

esis was, to a good approximation, correct. This author, who was carrying out this part of the analysis at the time, recalls wondering how Balmer may have felt when he saw, for the first time, the striking agreement of the formula that bears his name with the measured wavelengths of the atomic spectra of hydrogen.

More data showed that, at least in the first regions studied and within sometimes large errors, scaling held nearly quantitatively. As we shall see, scaling holds over

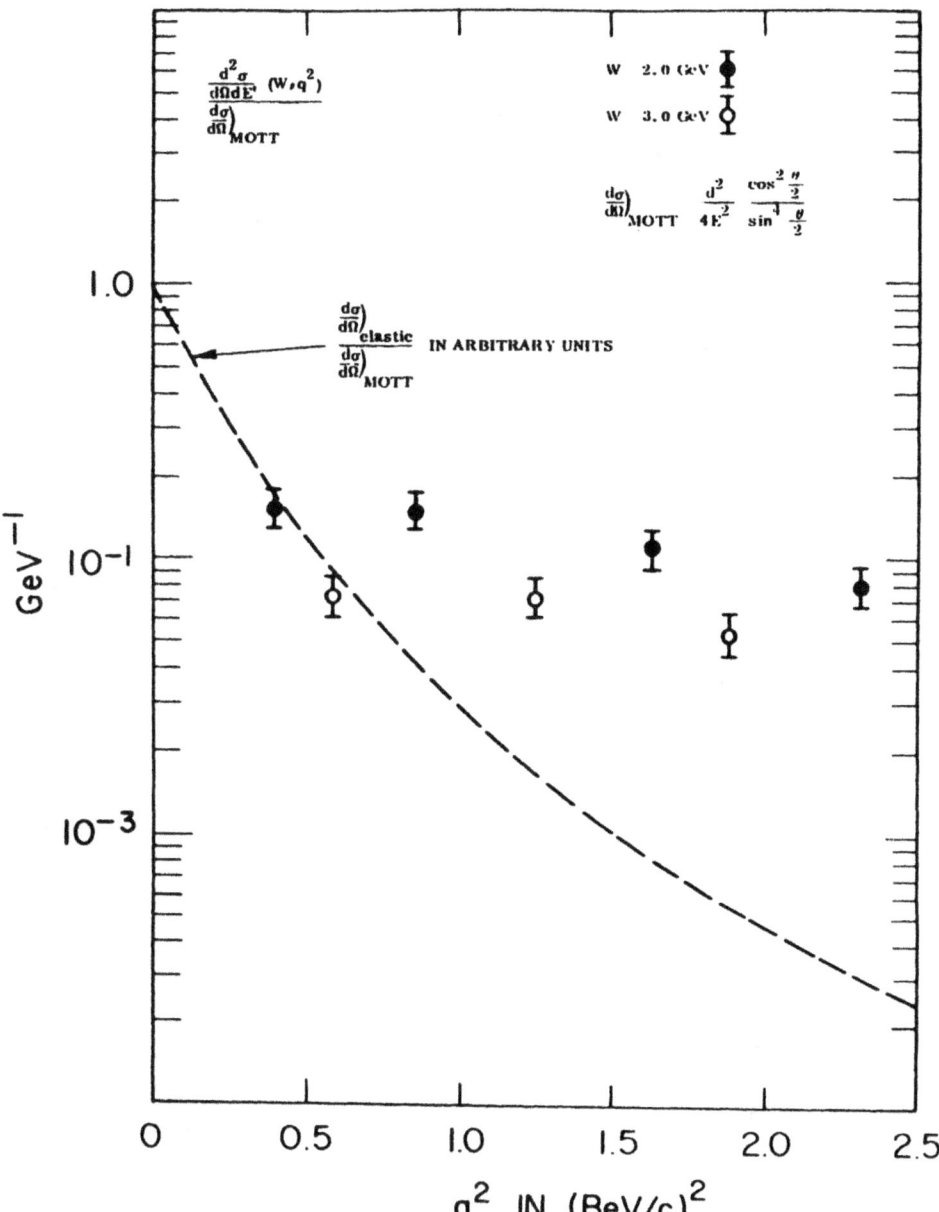

Figure 11. Inelastic data for $W = 2$ and 3 GeV as a function of q^2. This was one of the earliest examples of the relatively large cross sections and weak q^2 dependence that were later found to characterize the deep inelastic scattering and which suggested point-like nucleon constituents. The q^2 dependence of elastic scattering is shown also; these cross sections have been divided by σ_M.

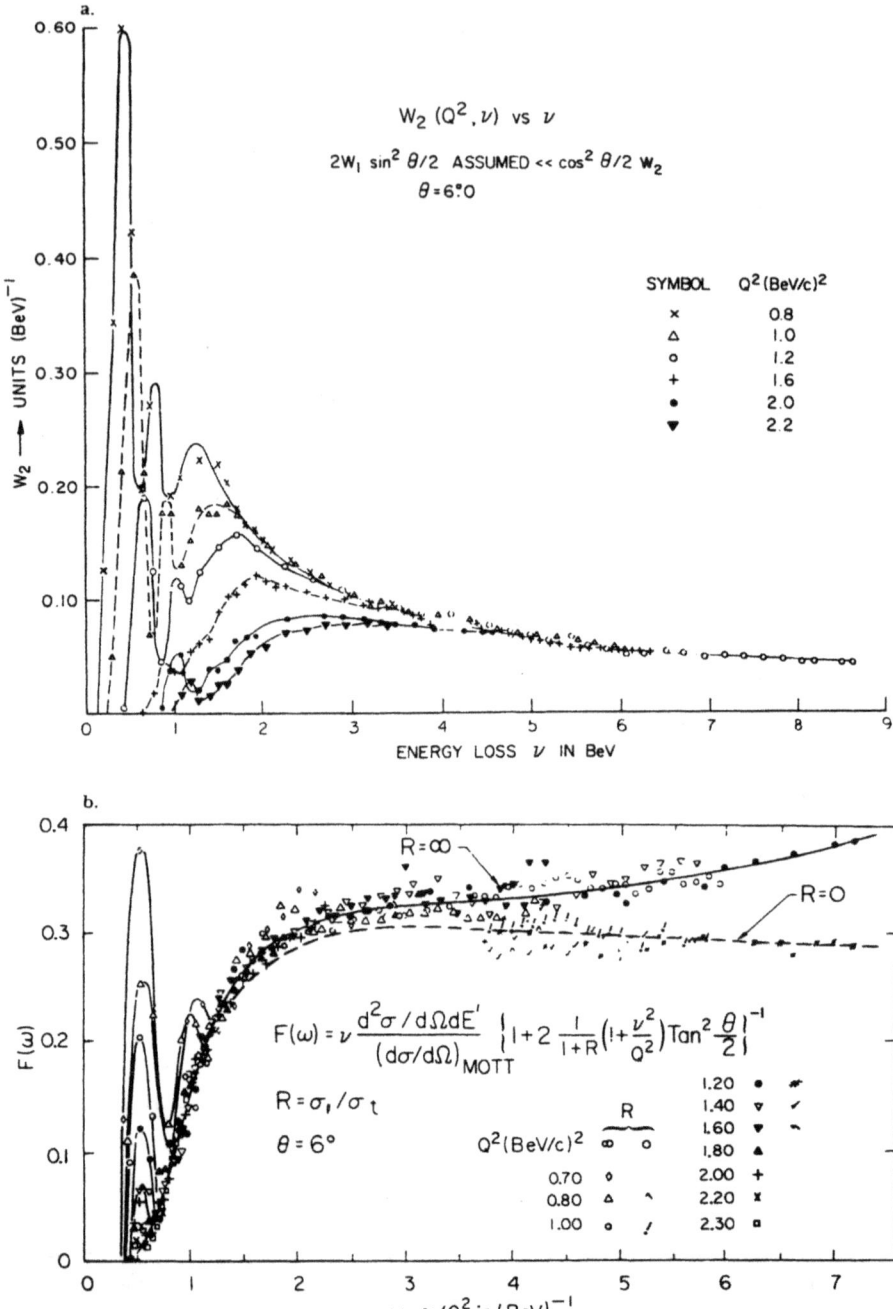

Figure 12. (a) The inelastic structure function $W_2(\nu, q^2)$ plotted against the electron energy loss ν. (b) The quantity $F_1 = \nu W_2(\omega)$. The "nesting" of the data observed here was the first evidence of scaling. The figure is discussed further in the text.

a substantial portion of the ranges of ν and q^2 that have been studied. Indeed the earliest inelastic e-p experiments[45] showed that approximate scaling behavior occurs already at surprisingly nonasymptotic values of $q^2 \geq 1.0$ GeV^2 and $W \geq 2.6$ GeV.

The question quickly arose as to whether there were other scaling variables that converged to ω in the Bjorken limit and that provided scaling behavior over a larger region in ν and q^2 than did the use of ω. Several were proposed[46] before the advent of QCD, but because this theory predicts small departures from scaling, the search for such variables was abandoned soon after.

Figure 13 shows early data on νW_2, for $\omega = 4$, as a function of q^2. Within the errors there was no q^2 dependence.

A more complex separation procedure was required to determine R and the structure functions, as discussed above. The kinematic region in q^2-W^2 space available for the separation is shown in Figure 14. This figure also shows the 75 kinematic points where, after the majority of the experiments were complete, separations had been made. Figure 15 displays sample least-squares fits to $\Sigma(\nu,q^2,\theta)$ vs $\epsilon(\nu,q^2,\theta)$, as defined earlier, in comparison with data, from which σ_L and σ_T and then R were found.

A rough evaluation of scaling is provided by, for example, inspecting a plot of the data taken by the collaboration on νW_2 against x as shown in Figure 16. These data, to a fair approximation, describe a single function of x. Some devi-

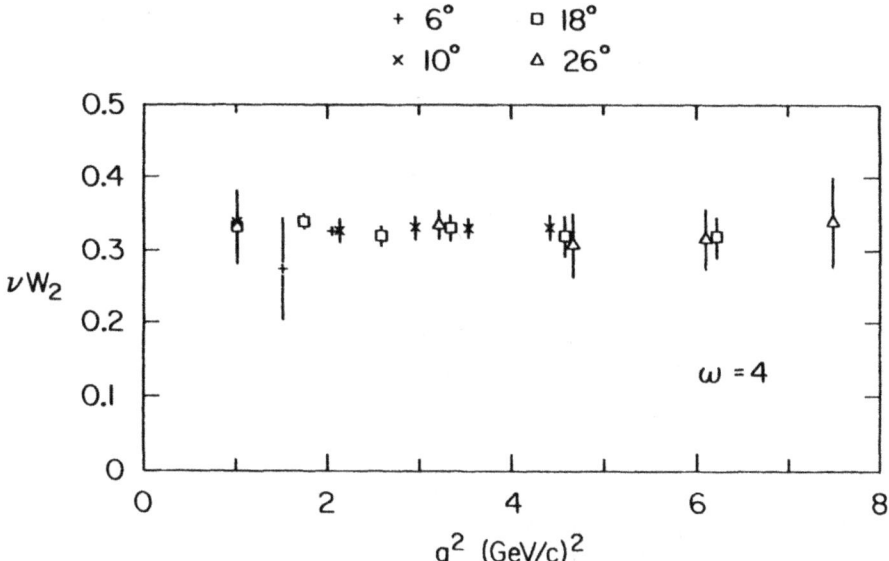

Figure 13. An early observation of scaling: νW_2 for the proton as a function of q^2 for $W > 2$ GeV, at $\omega = 4$.

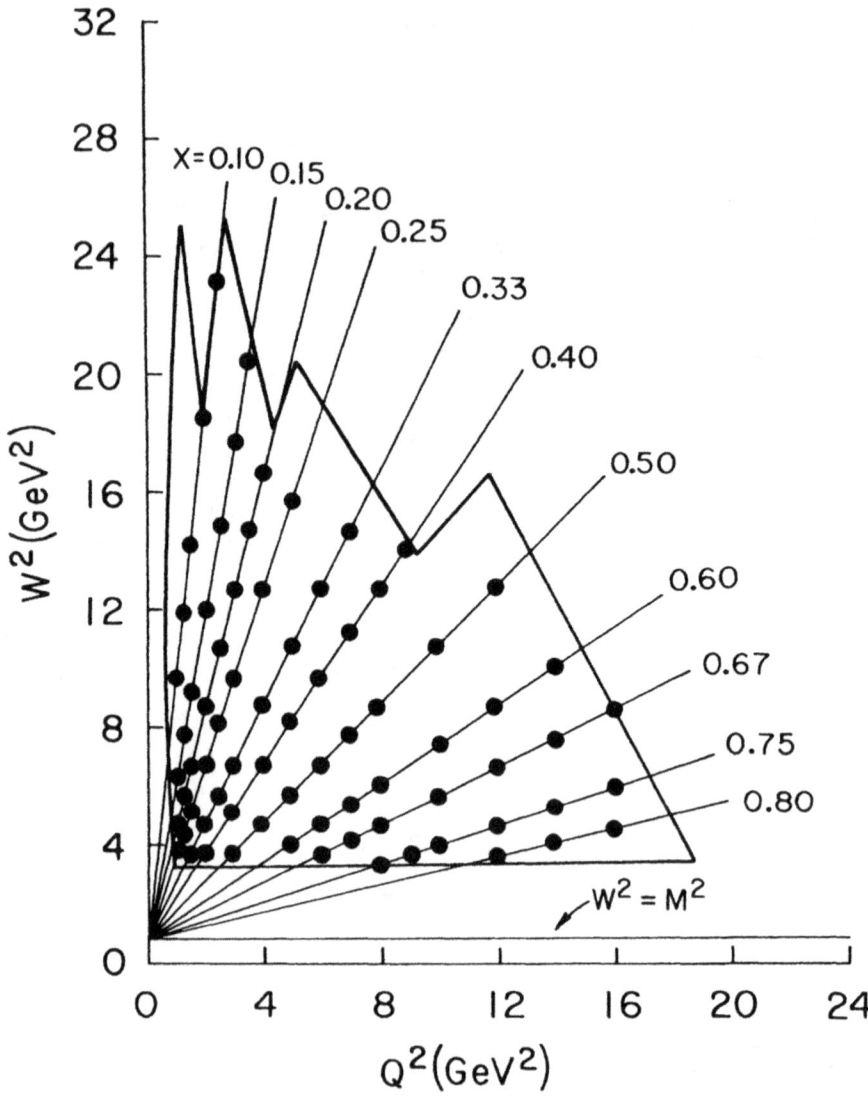

Figure 14. The kinematic region in $q^2 - W^2$ space available for the extraction of R and the structure functions. Separations were made at the 75 kinematic points (ν, q^2) shown.

ations, referred to as scale breaking, are observed. They are more easily inspected by displaying the q^2 dependence of the structure functions. Figure 17 shows separated values of $2MW_1$ and νW_2 from data taken late in the program, plotted against q^2 for a series of constant values of x. With extended kinematic coverage and with smaller experimental errors, sizable scale breaking was observable in the data.

VI. Theoretical Implications of the Electron-Proton Inelastic Scattering Data

As noted earlier, the discovery, during the first inelastic proton measurements, of the weak q^2 dependence of the structure function νW_2, coupled with the scaling concept inferred from current algebra and its roots in the quark theory, at once suggested new possibilities concerning nucleon structure. At the 1968 Vienna Meeting, where the results were made public for the first time, the rapporteur, W. K. H. Panofsky, summed up the conclusions: "Therefore theoretical speculations are focused on the possibility that these data might give evidence on the behavior of point-like, charged structures within the nucleon."[47]

Theoretical interest at SLAC in the implications of the inelastic scattering increased substantially after an August, 1968 visit by R. P. Feynman. He had been trying to understand hadron-hadron interactions at high energy assuming constituents he referred to as *partons*. On becoming aware of the inelastic electron scattering data, he immediately saw in partons an explanation both of scaling and of weak q^2 dependence. In his initial formulation,[48] now called the naive parton theory, he assumed that the proton was composed of point-like partons, from which the electrons scattered incoherently. The model assumed an infinite momentum frame of refer-

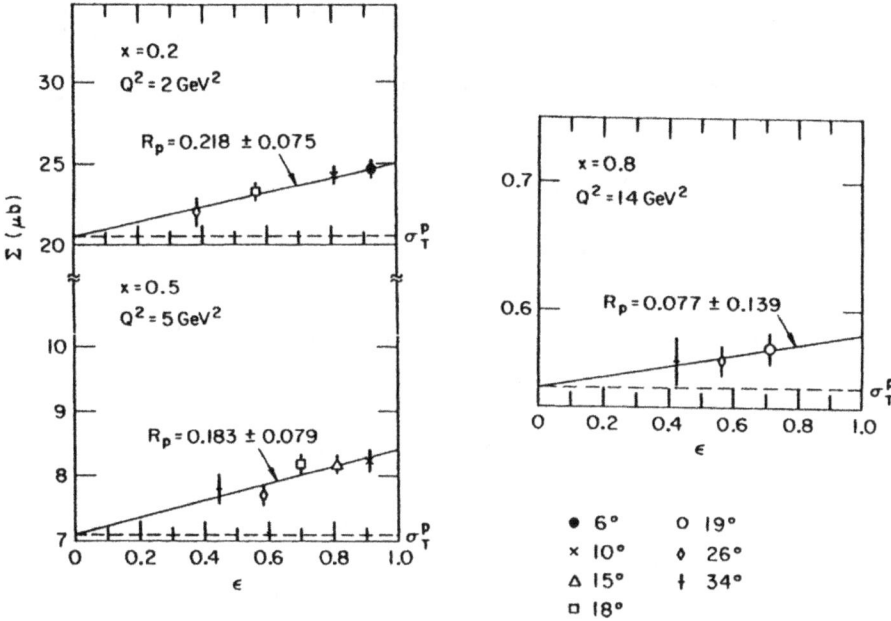

Figure 15. Sample least-squares fits to Σ vs ϵ in comparison with data from the proton. The quantities R and σ_T were available from the fitting parameters, and from them σ_L was determined.

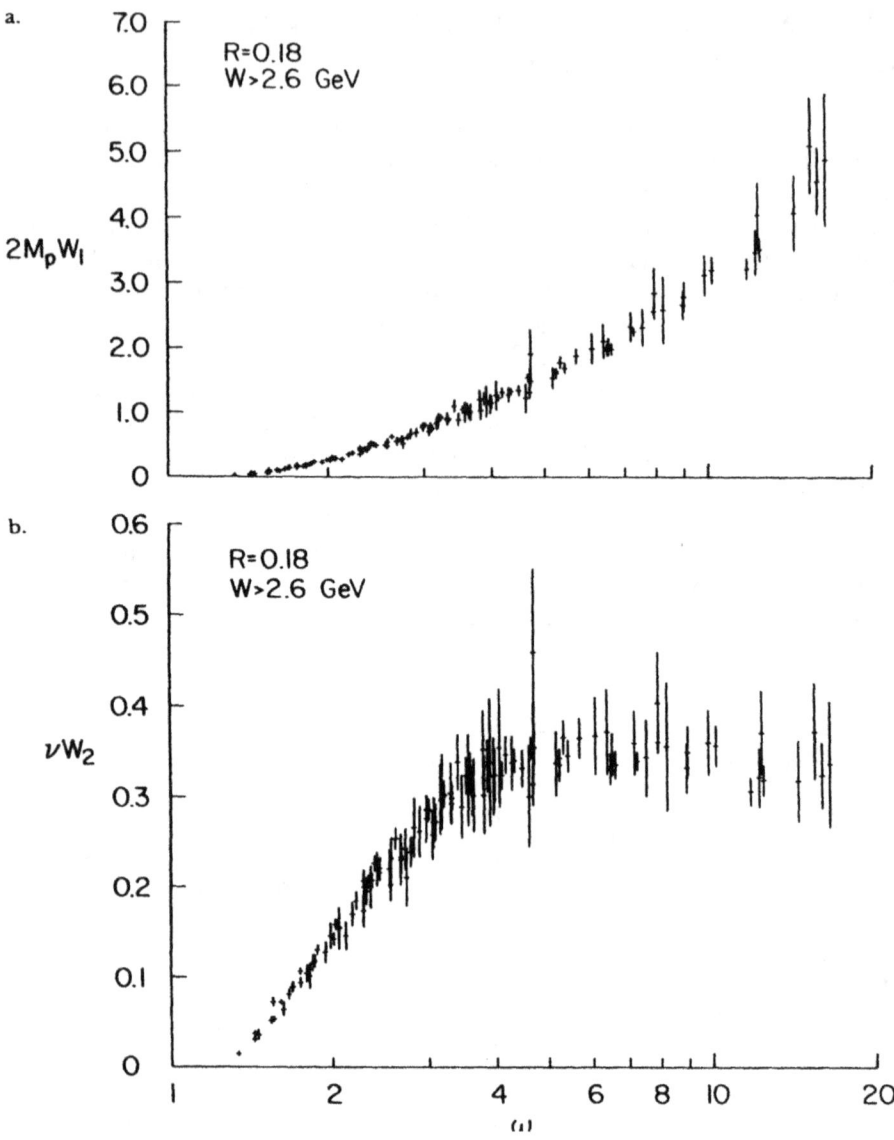

Figure 16. Scaling: $F_1 = 2MW_1(\omega)$ vs ω, and $F_2 = \nu W_2(\omega)$ vs ω.

ence, in which the relativistic time dilation slowed down the motion of the constituents. The transverse momentum was neglected, a simplification relaxed in later elaborations. The partons were assumed not to interact with one another while the virtual photon was exchanged: the impulse approximation of quantum mechanics. Thus, in this theory, electrons scattered from constituents that were "free," and therefore the scattering reflected the properties and motions of the constituents. This as-

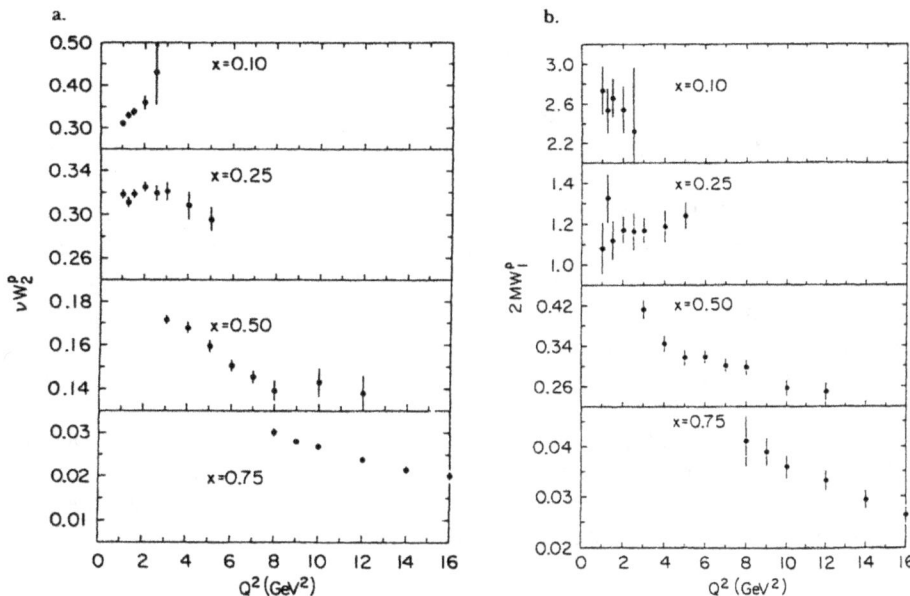

Figure 17. F_1 and F_2 as functions of q^2, for fixed values of x.

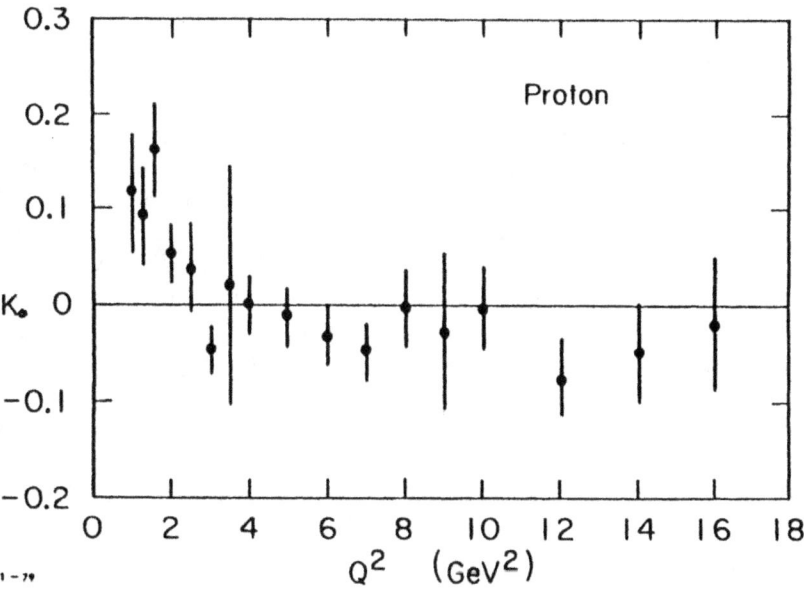

Figure 18. The Callan-Gross relation: K_0 vs q^2, where K_0 is defined in the text. These results established the spin of the partons as 1/2.

sumption of a near-vanishing of the parton-parton interaction during lepton scattering, in the Bjorken limit, was subsequently shown to be a consequence of QCD known as *asymptotic freedom*. Feynman came to Stanford again, in October, 1968, and gave the first public talk on his parton theory, stimulating much of the theoretical work that ultimately led to the identification of his partons with quarks.

In November 1968, Curt Callan and David Gross[49] showed that R, given in Eq. (6), depended on the spins of the constituents in a parton model and that its kinematic variation constituted an important test of such models. For spin-1/2, R was expected to be small, and, for the naive parton model, where the constituents are assumed unbound in the Bjorken limit, $R = q^2/\nu^2$ (i.e., $F_2 = xF_1$). More generally, for spin-1/2 partons, $R = g(x)(q^2/\nu^2)$. This is equivalent to the scaling of νR.

Spin-zero or spin-one partons led to the prediction $R = 0$ in the Bjorken limit and would indicate that the proton cloud contains elementary bosons. Small values of R were found in the experiment, and these were totally incompatible with the predictions of vector-meson dominance. Later theoretical studies[50] showed that deviations from the general Callan-Gross rule would be expected at low x and low q^2. A direct evaluation of the Callan-Gross relation for the naive parton model may be found from

$$K_0 = F_2/(xF_1) - 1,$$

which vanishes when the relation is satisfied. K_0 is shown in Figure 18, as a function of q^2. Aside from the expected deviations at low q^2, K_0 is consistent with zero, establishing the parton spin as 1/2.

VII. Epilogue

After the initial inelastic measurements were completed, deuteron studies were initiated to make neutron structure functions accessible. Experiments were made over a greater angular range and statistical, radiative, and systematic errors were reduced. The structure functions for the neutron were found to differ from the proton's. Vector-meson dominance was abandoned and by 1972 all diffractive models, and nuclear democracy, were found to be inconsistent with the experimental results. Increasingly detailed parton calculations and sum-rule comparisons, now focusing on quark constituents, required sea quarks—virtual quark-antiquark pairs—in the nucleon, and, later, gluons—neutral bosons that provided the inter-quark binding.

On the theoretical front, a special class of theories was found that could incorporate asymptomatic freedom and yet was compatible with the binding necessary to have stable nucleons. Neutrino measurements confirmed the spin-1/2 assignment for partons and that they had fractional, rather than integral electric charge. The number of "valence" quarks was found to be 3, consistent with the original 1964 assumptions.

By 1973 the picture of the nucleon had clarified to such an extent that it became possible to construct a comprehensive theory of quarks and gluons and their strong interactions: QCD. This theory was built on the concept of "color," whose introduction years before[51] made the nucleons' multi-quark wave functions compatible with the Pauli principle, and, on the assumption that only "color-neutral" states exist in nature, explained the absence of all unobserved multi-quark configurations (such as quark-quark and quark-quark-antiquark) in the known array of hadrons. Furthermore, as noted earlier, QCD was shown to be asymptotically free.[52]

By that year the quark-parton model, as it was usually called, satisfactorily explained electron-nucleon and neutrino-nucleon interactions and provided a rough explanation for the very-high-energy "hard" nucleon-nucleon scattering that had only recently been observed. The experimenters were seeing quark-quark collisions.

By the end of the decade, the fate of quarks recoiling within the nucleon in high-energy collisions had been understood; for example, after quark pair production in the electron-positron colliders, they materialized as back-to-back jets composed of ordinary hadrons (mainly pions), with the angular distributions characteristic of spin-1/2 objects. Gluon-jet enhancement of quark jets was predicted and then observed, having the appropriate angular distributions for the spin 1 they were assigned within QCD. Theorists had also begun to deal, with some success, with the problem of how quarks remained confined in stable hadrons.

Quantum chromodynamics describes the strong interactions of the hadrons and so can account, in principal at least, for their ground-state properties as well as hadron-hadron scattering. The hadronic weak and electromagnetic interactions are well described by electroweak theory, itself developed in the late 1960s. The picture of the nucleon, and the other hadrons, as diffuse, structureless objects was gone for good, replaced by a successful, nearly complete theory.

References and Notes

1. S. C. Frautschi, 1963, *Regge Poles and S-Matrix Theory* (Benjamin, New York).
2. G. F. Chew and S. C. Frautschi, 1961, Phys. Rev. Lett. 8, 394.
3. P. D. B. Collins and E. J. Squires, 1968, *Regge Poles in Particle Physics* (Springer, Berlin).
4. In a broad review of the strong interaction physics of the period, see Martin L. Perl, 1974, *High Energy Hadron Physics* (Wiley, New York). See also Frautschi, S. C., 1963, *Regge Poles and S-Matrix Theory* (Benjamin, New York).
5. G. F. Chew, S. C. Frautschi, and S. Mandelstam, 1962, Phys. Rev. 126, 1201.
6. G. Veneziano, 1968, Nuovo Cimento A 57, 190. See also J. H. Schwarz, 1973, Phys. Rep. 8, 269.
7. J. J. Sakurai, 1969, Phys. Rev. Lett. 22, 981.
8. M. Gell-Mann, 1961, C. I. T. Synchrotron Laboratory Report CTSL-20, unpublished. M. Gell-Mann, 1964a, Phys. Lett. 8, 214. The word *quork* was invented by Murray Gell-Mann, who later found quark in the novel *Finnegan's Wake*, by James Joyce, and adopted what has become the accepted spelling. Joyce apparently employed the word as a corruption of the word *quart*. The author is grateful to Murray Gell-Mann for a discussion clarifying the matter.
9. G. Zweig, 1964a, CERN-8182/Th.401 (Jan. 1964), unpublished. G. Zweig, 1964b, CERN-8419/Th.412 (Feb. 1964), unpublished.

10. M. Gell-Mann, 1961, C. I. T. Synchrotron Laboratory Report CTSL-2, unpublished. Y. Ne'eman, 1961, Nucl. Phys. 26, 222 (1961). See also M. Gell-Mann and Y. Ne'eman, 1964, *The Eightfold Way* (Benjamin, New York).
11. The quark model explained why triplet, sextet, and 27-plets of then-current SU(3) were absent of hadrons. With rough quark mass assignments, it could account for observed mass splittings within multiplets, and it provided an understanding of the anomalously long lifetime of the phi meson (discussed later in this paper).
12. Lawrence W. Jones, 1977, Rev. Mod. Phys. 49, 717.
13. ". . . We know that . . . [mesons and baryons] are mostly, if not entirely, made up out of one another. . . . The probability that a meson consists of a real quark pair rather than two mesons or a baryon and antibaryon must be quite small." M. Gell-Mann, *Proc. XIIIth International Conference on High Energy Physics*, Berkeley, California, 1967.
14. "Additional data is necessary and very welcome in order to destroy the picture of elementary constituents." J. D. Bjorken. "I think Prof. Bjorken and I constructed the sum rules in the hope of destroying the quark model." Kurt Gottfried. Both quotations from *Proc. 1967 International Symposium on Electron and Photon Interactions at High Energy*, Stanford, California, September 5–9, 1967.
15. R. H. Dalitz, 1967, Rapporteur, Session 10 of *Proceedings of the XIIIth International Conference on High Energy Physics*, Berkeley, 1966 (University of California, Berkeley), p. 215.
16. C. Becchi, and G. Morpurgo, 1965, Phys. Lett. 17, 352.
17. K. Gottfried, 1967, Phys. Rev. Lett. 18, 1174.
18. Zweig believed from the outset that the nucleon was composed of "physical" quark constituents. This was based primarily on his study of the properties of the ϕ meson. It did not decay rapidly to p-π as expected but rather decayed roughly two orders of magnitude slower to kaon-antikaon, whose combined mass was near the threshold for the decay. He saw this as a dynamical effect, one not explainable by selection rules based on the symmetry groups and explainable only by a constituent picture in which the initial quarks would "flow smoothly" into the final state. He was "severely criticized" for his views, in the period before the MIT-SLAC results were obtained. Private communication, February 1991.
19. According to a popular book on the quark search *The Hunting of the Quark* by M. Riordan (Simon and Schuster, New York, 1987) Zweig, a junior theoriest visiting at CERN when he proposed his quark theory, could not get a paper published describing his ideas until the middle 1970s, well after the constituent model was on relatively strong ground. His 1964 preprints CERN-8182/Th.401 and CERN-8419/Th.412 did, however, reach many in the physics community and helped stimulate the early quark searches.
20. A. Pais, 1986, *Inward Bound* (Oxford University Press, Oxford/New York).
21. "Throughout the 1960s, into the 1970s, papers, reviews, and books were replete with caveats about the existence of quarks" (Andrew Pickering, *Constructing Quarks*, University of Chicago Press, Chicago, 1984).
22. Further discussion of sum rules and their comparisons with data is to be found in Friedman (1991). J. I. Friedman, 1991, Nobel lecture, in *Les Prix Nobel 1990: Nobel Prizes, Presentations, Biographies and Lectures* (Almqvist & Wiksell, Stockholm/Uppsala), in press; reprinted in Rev. Mod. Phys. 63, 615.
23. "Such particles [quarks] presumably are not real but we may use them in our field theory anyway" M. Gell-Mann, 1964b, Physics 1, 63.
24. "Now what is going on? What are these quarks? It is possible that real quarks exist, but

if so they have a high threshold for copious production, many BeV; . . ." (*Proc. XIIIth International Conference on High Energy Physics*, Berkeley, California, 1967).

25. "We shall find these results [sum rules requiring cross sections of order Rutherford scattering from a point particle] so perspicuous that, by an appeal to history, an interpretation in terms of 'elementary constituents' of the nucleon is suggested." He pointed out that high-energy lepton-nucleon scattering could resolve the question of their existence and noted that "it will be of interest to look at very large inelasticity and dispose, without ambiguity, of the model completely." Lecture, "Current Algebra at Small Distances," given at International School of Physics "Enrico Fermi," Varenna, Italy, July 1967 (J. D. Bjorken, 1967, SLAC Publication 338, August 1967, unpublished; J. D. Bjorken, 1968, in *Selected Topics in Particle Physics: Proceedings of the International School of Physics "Enrico Fermi,"* Course XLI, edited by J. Steinberger (Academic, New York).

26. T. D. Lee: "I'm certainly not the person to defend the quark models, but it seems to me that at the moment the assumption of an unsubtracted dispersion relation [the subject of the discussion] is as *ad hoc* as the quark model. Therefore, instead of subtracting the quark model one may also subtract the unsubtracted dispersion relation." J. Bjorken: "I certainly agree. I would like to dissociate myself a little bit from this as a test of the quark model. I brought it in mainly as a desperate attempt to interpret the rather striking phenomena of a point-like behavior. One has this very strong inequality on the integral over the scattering cross section. It's only in trying to interpret how that inequality could be satisfied that the quarks were brought in. There may be many other ways of interpreting it." Discussion in *Proc. 1967 International Symposium on Electron and Photon Interaction at High Energy*, Stanford, September 5–9, 1967.

27. SLAC Groups A and C (and physicists from MIT and CIT), "Proposal for Spectrometer Facilities at SLAC," undated, unpublished; SLAC-MIT-CIT Collaboration, 1966, Proposal for Initial Electron Scattering Experiments Using the SLAC Spectrometer Facilities, Proposal 4b, "The Electron-Proton Inelastic Scattering Experiment," submitted 1 January 1966, unpublished.

28. W. K. H. Panofsky, 1968, in *Proceedings of the 14th International Conference on High Energy Physics*, Vienna, 1968, edited by J. Prentki and J. Steinberger (CERN, Geneva), p. 23. E. D. Bloom, D. H. Coward, H. DeStaebler, J. Drees, G. Miller, L. W. Mo, R. E. Taylor, M. Breidenbach, J. I. Friedman, G. C. Hartmann, and H. W. Kendall, 1969, Phys. Rev. Lett. 23, 930; M. Breidenbach, J. I. Friedman, H. W. Kendall, E. D. Bloom, D. H. Coward, H. DeStaebler, J. Drees, L. W. Mo, and R. E. Taylor, 1969, Phys. Rev. Lett. 23, 935.

29. W. R. Ditzler et al., 1975, Phys. Lett. 57, 201B.

30. L. W. Whitlow, SLAC Report 357, March 1990, unpublished; L.W. Whitlow, S. Rock, A. Bodek, S. Dasu, and E. M. Riordan, 1990, Phys. Lett. B 250, 193.

31. M. Rosenbluth, 1950, Phys. Rev. 79, 615.

32. P. N. Kirk, et al., 1973, Phys. Rev. D 8, 63.

33. S. D. Drell and J. D. Walecka, 1964, Ann. Phys. (NY) 28, 18.

34. L. Hand, 1963, Phys. Rev. 129, 1834.

35. Although the conjecture was published after the experimental results established the existence of scaling, the proposal that this might be true was made prior to the measurements, as discussed later in the text.

36. J. D. Bjorken, 1963, Ann. Phys. (N.Y.) 24, 201.

37. J. I. Friedman, 1959, Phys. Rev. 116, 1257.

38. The final report on the experiment is S. Klawansky, H. W. Kendall, A. K. Kerman, and D. Isabelle, 1973, Phys. Rev. C7, 795.

39. G. Miller, et al., 1972, Phys. Rev. D 5, 528; L. W. Mo, and Y. S. Tsai, 1969, Rev. Mod. Phys. 41, 205.
40. Y. S. Tsai, 1964, in *Nucleon Structure: Proceedings of the International Conference*, Stanford, 1963, edited by R. Hofstadter and L. I. Schiff (Stanford University, Stanford, CA), p. 221.
41. G. Miller, et al., 1972, Phys. Rev., D5, 528.
42. J. S. Poucher, 1971, Ph.D. thesis, Massachusetts Institute of Technology.
43. A. Bodek, M. Breidenbach, D. L. Dubin, J. E. Elias, J. I. Friedman, H. W. Kendall, J. S. Poucher, E. M. Riordan, M. R. Sogard, D. H. Coward, and D. J. Sherden, 1979, Phys. Rev. D 20, 1471.
44. W. Bartel, B. Dudelzak, H. Krehbiel, J. McElroy, U. Meyer-Berkhout, W. Schmidt, V. Walther, and G. Weber, Phys. Lett. B 28, 148 (1964).
45. E. D. Bloom, D. H. Coward, H. DeStaebler, J. Drees, G. Miller, L. W. Mo, R. E. Taylor, M. Breidenbach, J. I. Friedman, G. C. Harmann, and H. W. Kendall, 1969, Phys. Rev. Lett. 23, 930; M. Breidenbach, J. I. Friedman, H. W. Kendall, E. D. Bloom, D. H. Coward, H. DeStaebler, J. Drees, L. W. Mo, and R. E. Taylor, 1969, Phys. Rev. Lett. 23, 935.
46. J. I. Friedman and H. W. Kendall, 1972, Annu. Rev. Nucl. Sci. 22, 203.
47. W.K.H. Panofsky, 1968, *Proceedings of the 14th International Conference on High Energy Physics, Vienna, 1968*, edited by J. Prentki and J. Steinburger (CERN, Geneva), p. 23.
48. R. P. Feynman, 1969, Phys. Rev. Lett. 23, 1415.
49. C. Callen and D. J. Gross, 1968, Phys. Rev. Lett. 21, 311.
50. R. P. Feynman, 1972, *Photon-Hadron Interactions* (Benjamin, Reading, MA).
51. M. Y. and Y. Nambu, 1965, Phys. Rev. 139, 1006B.
52. D. Gross and F. Wilczek, 1973, Phys. Rev. Lett. 30, 1343.

PART THREE
NATIONAL SECURITY AND NUCLEAR WEAPONS

National defense has generated some of the most dangerous problems of the twentieth century. The security of the nation, indeed the existence of the nation, can at times directly depend on our military establishment. The public shares a powerful need for protection; threats from abroad can stimulate deep public concern and a willingness to spend hugely for the provision of adequate security. However, the responses to such concerns frequently prove excessively costly and, sometimes, excessively dangerous.

Some problems are generated by technological advances: nuclear weapons, ballistic missile defense, launch-on-warning are all examples. Others are generated by government and the defense industry's scare tactics, deception, and threat exaggeration. The controversies that result have some special features that arise from secrecy and the resulting classification of information. Paranoia is often an element. The Union of Concerned Scientists was born in a period of controversy over national security issues, and over the years it has been an active participant in many debates over major weapons systems and arms control policies.

During the early 1970s, with the heavy commitments that the ECCS hearings entailed and with the growing size and scale of UCS itself, I parted company with the Jason Panel of the Institute for Defense Analyses, which had been created to provide technical advice to the U.S. Department of Defense. The 1960s debate over antiballistic missile systems continued into the decade of the 1970s. Growth of the U.S. and Soviet nuclear arsenals continued as the Cold War and its progeny, the nuclear arms race, proceeded relentlessly.

In 1977, I and others in UCS agreed that the organization should again put a focus on the arms race. The nuclear buildup was not at all on the public's mind; this was the period of the end of the war in Southeast Asia, Richard Nixon's disgrace and departure from the presidency, and two successive Arab oil embargoes, all profoundly unsettling to the country. In this environment, it would be necessary to raise the issue of nuclear arms and attempt to place it, foursquare, in the public consciousness.

We took what appeared to be a natural and potentially successful road: set out the facts, issues, and threats in a declaration supported by a heavyweight representation of the country's scientific community. We were to learn some surprising and powerful lessons.

The preparation of the document was not difficult nor was the task of enlisting distinguished and knowledgeable sponsors. After a mass mailing of the appeal, over 12,000 supporters from the scientific community signed on. An edited version of the declaration and the accompanying press release are reprinted here. It was one of the strongest statements from this community that had been made public up to that time, and we expected it to generate the beginnings of a movement against the arms race, following its release in November 1977.

The delcaration was mentioned briefly in the next day's press and then disappeared promptly and totally from public view. It was not picked up by the country's peace groups. There was nothing we could conceive of to resurrect it; it proved dead on arrival. It was an unexpected and upsetting experience. Evidently, at that time, the public did not perceive a threat from the nuclear buildup. It was apparent from this experience that major difficulties can arise in trying to gain active public support even for an issue of grave importance if the public cannot sense the threat or has other pressing matters on its mind.

It followed that to stimulate immediate change, an organization such as UCS would have to capitalize on existing concerns, give them flesh and blood and immediacy in the public's mind, and help guide, in careful ways, the search for and implementation of solutions. Sustained activity is essential for success. Subsequent observations of other environmental controversies have confirmed these conclusions.

There are clearly circumstances where an attempt to generate concerns and educate the public on a little-known, emerging threat can be very important. There should be no illusion about the difficulties to be faced or the time that may be required. It is clearly an uphill task to generate broad concerns if the times and the issues are not, as the lawyers say, "ripe." On the other hand, should an issue become ripe, it is possible to draw in large numbers of people and to generate major political force.

A public opinion expert I later talked to, in a discussion of a just-published list of the public's priorities on some 40 issues of general concern, said that the public could keep no more than 2.5 issues in mind at one time. If a new matter became unusually vexing, such as crime, gasoline shortages, political scandal, or the like, then it would displace one of the issues earlier at center stage. An issue well down in such a list has little chance, if any, of stimulating public interest for change as long as circumstances remain unaltered.

In the later 1970s, extending through the 1980s and into the early 1990s, the nuclear arms threat did indeed become ripe and through a quite unexpected means: Ronald Reagan and the country's right wing. In the campaign for the presidency in 1979 and well into his two terms as president, Reagan raised the fear that Soviet domination of the United States and the free world would be gained by Soviet attainment of nuclear superiority. The military buildup that he and his fellow conservatives supported and to a considerable extent implemented in the 1980s included asserting that in a nuclear war, including all-out nuclear war, one side could "prevail."[1] This program and the fear that it created was all that was required for the issues to galvanize the public.

At the outset of the campaign, my colleagues and I speculated that the nuclear and conventional buildup, by then generating considerable public dismay, might dwindle during the Reagan administration, that an astute administration would drop the issue, replacing it with less worrisome matters. It was not to be; the saber rattling went on and on, spawning the nuclear freeze,[2] revitalizing the peace movement, and giving life to considerable counterforces to the arms buildup. These activities helped generate public support for the arms control treaties with the Soviet Union that have, in the last ten years, come into being.

In the new climate of fear of the 1980s, UCS and other groups' arms control programs flourished as public interest and support climbed. We prepared and released a declaration urging the ratification of SALT (Strategic Arms Limitation Treaty) in October 1979 with nearly 3500 supporters and, in November 1981, a "Scientists' Declaration on the Threat of Nuclear War." The last document rode on the authority of only nine sponsors; we were learning that large numbers of names are far from necessary on a declaration, for the impact depends generally on other circumstances and conditions, including especially the prestige of the signers. It was released as part of the National Convocation on the Threat of Nuclear War, an activity that I had initiated and that UCS organized and sponsored. It was held on November 11, 1981, at 150 colleges and universities in 41 states, and it was a great success. The following year it had expanded to 500 campuses and in 1983 to 750 campuses, high school and college, in the United States, Canada, and Europe. The activity continued for many years.

One consequence of these teach-ins was of lasting impact: the establishment of peace study and international security programs in a number of colleges and universities. Such programs started to flourish in the late 1980s, stimulated in part by the teach-ins and the presence of scientist activists on many campuses.

In the late 1980s, with arms issues waning in public attention, we shifted the focus of our teach-ins to global environmental problems. We altered the schedules to emphasize programs lasting longer than a day or a week. The 1991 activities involved programs centered on global warming at over 1000 colleges, high schools, and communities, some of which extended over an academic year. The following two years' activities, now centered on energy issues, were about the same size. We discontinued the programs by 1994.

In 1982 Kurt Gottfried initiated and organized a UCS-sponsored study group to study and support a move by the United States and its NATO allies toward a policy of no first use of nuclear weapons. Fourteen of us, scientists and military and defense experts, signed the resulting declaration, and Kurt, a retired senior military officer, Vice Admiral John M. Lee, and I prepared a UCS report and later an article setting out the case for the new policy.[3]

Kurt was the initiator of another large and effective UCS program that developed in two phases, the first directed against the proposed U.S. deployment of antisatellite weaponry, generally referred to as ASAT, the second, following Reagan's 1983 proposal that the United States develop and deploy a space-based ballistic missile defense system, directed against that proposal, widely known as "Star Wars." In this program of analysis, writing, and public outreach, we had substantial help from two senior scientists, Hans Bethe and Richard L. Garwin. Bethe, a Nobel laureate who first discerned the nuclear reactions responsible for the sun's energy production, had been head of the Theoretical Division at Los Alamos during the Manhattan Project and knew a great deal about defense issues. Garwin, one of the world's top experimental physicists, had long experience with military systems and was regarded as possessing the keenest analytical sense on weapons matters of anyone in the country. Their contributions were invaluable.

The UCS ASAT program began with a study[4] that in turn formed the basis for a proposal for a space weapons treaty set before the Senate Foreign Relations Committee.[5] A *Scientific American* article[6] ended the ASAT involvement as UCS shifted gears to criticize Reagan's Star Wars plans.

Owing to the head start provided by the work on ASAT space weapons, this UCS program generated the earliest and by far the most powerful criticism of the mistaken Star Wars enterprise.[7] The first report had substantial public impact. In the continuation of the work, two UCS books were prepared,[8] an article, included here, and a great deal of public education and outreach. The UCS effort was, I believe, the main public program that led to the eventual defeat of Star Wars.

The Star Wars debate was characterized by an unusual level of deception and bad science employed by its proponents, both from within the National Laboratories and from outside.[9] This always presents acute problems to critics such as UCS who do not have access to classified material and whose resources are minute compared with those of the groups arrayed against them. One resource we exploited, as in the nuclear power controversy, was experts with classified access working in government, corporations, or universities who shared our critical views. Such experts, who must keep careful track of what information is and is not classified so that they do not breach security constraints, can provide enormous help to outsiders in public debate. Sometimes referred to as "insiders," they fill a niche that otherwise can leave outsiders seriously vulnerable. Still, there are no easy solutions to difficulties posed by deception. About all that can be done is to keep a steady course and respond with thoughtful evidence to as many of the wayward arguments as is possible. It can take a thick skin, a lot of time and energy, and considerable persistence.

Two reprints included here, stemmed from the Star Wars controversy. The first, "Space-Based Ballistic Missile Defense," is largely devoted to the technical problems and weaknesses of such defense systems. The second, excerpts from the UCS book *The Fallacy of Star Wars*, focuses on other issues related to space-based defense.

In 1990, Bruce G. Blair, a senior arms analyst at Brookings Institution, and I teamed up to prepare "Accidental Missile Launch," an article reprinted here, setting out the risks and consequences of accidental nuclear war and suggesting changes that would lower the risk. He and I had met during the course of a study organized by Kurt on nuclear crisis stability[10] that was held in the 1985–1986 period. We did not leave the subject, for two years later we prepared a *New York Times* op-ed piece, also reprinted in this volume, pointing out the lingering risks of the United States and the Soviet Union retaining missiles kept on hair-trigger alert and aimed at one another and suggesting some useful dealerting procedures. Bruce, with other collaborators, continued to press for dealerting the strategic nuclear forces.[11] He has had increasing success, attracting some powerful supporters, including former Senator Sam Nunn, and has gotten dealerting on the nation's national security agenda.

It is, of course, impossible to determine at all well the consequences of my and UCS's efforts in national security matters. Clearly they had a broad reach. They have, along with the remainder of the domestic and foreign peace movements and pressure from the U.S. arms buildup, aided by the breakup of the Soviet Union, led to a great reduction of nuclear risks.

References

1. Robert Scheer, *With Enough Shovels: Reagan, Bush and Nuclear War* (Random House, New York, 1982).
2. D. Ford, H. Kendall, and S. Nadis (Union of Concerned Scientists), *Beyond the Freeze*, (Beacon Press, Boston, 1982).
3. Kurt Gottfried, Henry W. Kendall, and John M. Lee, "'No First Use' of Nuclear Weapons," *Scientific American* **250**(3), (March 1984).
4. *Antisatellite Weapons: Arms Control or Arms Race?* Report of a panel cochaired by Kurt Gottfried and Richard L. Garwin, Union of Concerned Scientists, 1983.
5. *Controlling Space Weapons*, U. S. Senate Committee on Foreign Relations, 98th Congress, 1st Session, May 18, 1983, pp. 112–129, 144–149.
6. Richard L. Garwin, Kurt Gottfried, and D. L. Hafner, "Antisatellite Weapons," *Scientific American* **250**(5) (June 1984) 45–56.
7. *Space-based Missile Defense*, report of a panel cochaired by Kurt Gottfried and Henry W. Kendall, Union of Concerned Scientists, 1984. Extensive excerpts from this report were reprinted in *New York Review of Books*, April 26, 1984, pp. 47–52.
8. *The Fallacy of Star Wars*, based on studies conducted by the Union of Concerned Scientists and cochaired by Richard L. Garwin, Kurt Gottfried, and Henry Kendall, edited by John Tirman (Vintage Books, New York, 1984). *Empty Promise—The Growing Case against Star Wars*, Union of Concerned Scientists, edited by John Tirman (Beacon Press, Boston, 1986).
9. P. M. Boffey, W. J. Broad, L. H. Gelb, C. Mohr, and H. B. Noble, *Claiming the Heavens* (Times Books, New York, 1988). William J. Broad, *Teller's War—The Top-Secret Story Behind the Star Wars Deception* (Simon & Schuster, New York, 1992).
10. Kurt Gottfried and Bruce G. Blair, eds., *Crisis Stability and Nuclear War* (Oxford University Press, New York and Oxford, 1988).
11. Bruce G. Blair, Harold A. Feiveson, and Frank N. von Hippel, "Taking nuclear weapons off hair-trigger alert," *Scientific American* **277**(5), (November 1997).

NEWS RELEASE

Union of Concerned Scientists, Cambridge, Mass.,
December 20, 1977

Announcing Declaration on the Nuclear Arms Race

More than 12,000 scientists, engineers, and other professionals, many of them leaders in their fields, as well as more than 1000 other signatories, called on the United States to take the initiative in halting the nuclear arms race, "a grim feature of modern life." In a Declaration on the Nuclear Arms Race, the many signers call the continuing competition "increasingly a mortal threat to all mankind." They urged the administration to halt the testing and deployment of new nuclear weapons and nuclear weapons systems for a period of 2 to 3 years and challenged the Soviet Union to reciprocate. They called also for a rapid move toward a comprehensive nuclear test ban. The declaration emphasized that these bold steps would *not* jeopardize U.S. security owing to the immense destructive capacity of the U.S. nuclear arsenal. There is no defense against nuclear missiles.

The signatories criticized continued competition for nuclear superiority and strategic advantage by building new and more powerful weapons and delivery systems. Technical developments in armaments have consistently undercut past negotiations. The competition undermines the present state of mutual deterrence or balance of terror and increases the risk of a nuclear war that "would unhinge and devastate the target nations so effectively that they would no longer function as modern industrial states." The nuclear arms race is "in full swing," even though it is no longer possible for either side to gain decisive military superiority.

The recommendations would set the stage for gradual reductions in strategic arms. They "would represent a policy of restraint of the greatest political significance . . . . Should the Soviet Union reciprocate—and they, like the United States, have so much to gain in so doing—a great step forward would be taken to diminish the threat of nuclear war."

* * * * * * * *

The declaration was sponsored by 21 knowledgeable and experienced people including four former Presidential Science Advisors. Among the sponsors and signers are a former Commissioner of the Atomic Energy Commission, the Director of the Fermi National Accelerator Laboratory, recipients of presidential medals, several present or former presidents of major universities, numerous winners of scientific honors and awards—including 27 Nobel prizes—and members of honorary scientific societies—including 400 members of the National Academy of Sciences—both here and abroad. There are many department and division heads from universities and industry throughout the United States.

Declaration on the Nuclear Arms Race

I. The Arms Race

The nuclear arms race, a grim feature of modern life, has been with us since the devastation of Hiroshima and Nagasaki. Driven by a futile search for security through nuclear "superiority," this vastly expensive competition is increasingly a mortal threat to all humanity. Control of the arms race is one of mankind's most urgent needs.

The inventories of nuclear warheads are coupled with accurate, long range and relatively invulnerable delivery systems. Together their destructive capacity is immense. If used in war they would kill hundreds of millions of persons, carry radioactive injury and death to many of the world's nations, profoundly damage the environment of the Earth we live and depend on, and unhinge and devastate the target nations so effectively that they would no longer function as modern industrial states. Indeed, the superpowers' inventories are so great that even small fractions of them could produce damage far beyond any past experience.

Neither superpower could now launch a counterforce surprise nuclear attack to disarm the other. Enough of each side's forces will survive or are invulnerable that the return blow would still produce prodigious damage, so shattering that no first strike is remotely rewarding. Thus, a relatively stable but uneasy balance has resulted in the recent past, the state of Mutually Assured Destruction. This balance of terror, while morally repugnant to many, is a fact of modern life. The superpowers have recognized that the populations of the United States and the Soviet Union have become unavoidably hostage because of the ineffectiveness of defenses against nuclear-armed strategic weapons systems—and so their 1972 agreement and treaty in effect terminated efforts at active antimissile defenses.

Union of Concerned Scientists, 1977.

The security of the United States and the Soviet Union, and of the other nations across the globe, is only as high as the expectation that the nuclear arsenals will never be used.

Strategic nuclear arsenals could be drastically reduced while still retaining ample deterrent forces in a world devoid of real peace. However, the superpowers—while professing commitment to the principle of nuclear parity—continue to reach for nuclear superiority or at least for unilateral advantage through the development and deployment of new weapons and weapons systems. By and large the U.S. has been ahead in the superpowers' arms technology race: the first in nuclear and thermonuclear weapons, nuclear submarines, solid fuel missiles, and many other developments. The U.S. continues to forge ahead, developing MX—an advanced, perhaps mobile, land-based intercontinental missile—and multiple independently targetable reentry vehicles (MIRV) of extreme accuracy; the neutron bomb; the air- and sea-launched strategic range cruise missiles.

Many of these innovations have been stimulated by uncertainty about what the Soviet Union was deploying. Soviet responses are clouded in secrecy and are possibly highly threatening. They have forced the U.S. to proceed. In general, the Soviet Union has responded to U.S. technological innovations but with a lag—averaging over 4 years—in reaching U.S. levels. Their deployments then continue, exceeding the United States and so raising fears of their gaining "superiority." The Soviet Union is developing and deploying MIRV'ed missiles of ever greater range, accuracy, and explosive power, perhaps greatly intensifying the civil defense of its population, and continuing other developments. The Soviet Union now has more strategic missiles and a greater nuclear throw-weight, while the United States exceeds in the accuracy of delivery systems as well as in numbers of nuclear warheads. The Soviets continue also to increase their conventional forces raising fears of aggression aimed at Western Europe. This has stimulated responses in conventional arms, and, especially grave, in dependence on nuclear weapons among the NATO nations. The United States and the Soviet Union both are engaged in vigorous underground nuclear warhead test programs. The responsibility for the race is unmistakably shared.

The arms race is in full swing! The roughly twelve thousand strategic warheads of today are likely to become thirty thousand long before the end of the century and the tens of thousands of tactical weapons augmented also. These increases and the improvements in missile accuracy, retargeting capability and invulnerability lead to greater "flexibility"—and so to the greater likelihood of starting nuclear weapons use. What results is the undermining of the balance of terror. New weapons now in sight will further decrease the stability of this delicate balance and will make the monitoring of future arms agreements more difficult, if not impossible, without gaining decisive military superiority for either side.

The superpowers' belief that security rests with potent nuclear armaments is more and more shared by other nations. The strategic arms race stimulates the proliferation of nuclear weapons among nations some of which may be weak or irresponsible, and thus more likely to resort to the use of nuclear weapons in a local war. Such wars could easily widen, thus adding to the likelihood of a final immense nuclear holocaust between the superpowers.

More than ever it is urgent now to slow down and ultimately to stop the nuclear arms race, thus improving the stability of the nuclear standoff and setting the stage for reduction of the great inventories of weapons.
We hereby recommend that:

1. The United States announce that it will halt underground testing of nuclear explosives provided that the Soviet Union follows suit within a reasonable time.

2. The United States announce that it will not field test or deploy *new* strategic nuclear weapons, nuclear weapons systems, or missile defense systems for a period of 2 to 3 years provided the Soviet Union demonstrably does likewise.

These measures, carried out with due care, *do not jeopardize our security*. The recommendations do not stem from blind faith in the good intentions of the Soviet Union. We already can detect Soviet tests of nuclear weapons smaller than the Hiroshima bomb with existing seismic sensors and clearly distinguish them from earthquakes. Hence underground tests of strategic warheads cannot escape detection. Our satellites already inspect Soviet missile launches, missile site construction, and submarine arrivals and departures; thus we would know if the Soviet Union were not following our lead. Should the recommended initiatives not bear fruit, the interruption in testing would hardly degrade our security. It takes many years to develop and deploy strategic weapons systems and our strength is such that a short delay of the sort we recommend cannot put the U.S. at risk.

These measures, carried out with due care, *can restrain the technological arms race*. Without underground tests there is not enough confidence in the new warhead designs to allow deployment. New missiles also depend on more accurate guidance systems, and these can be tried and perfected only in repeated test firings. By reducing the number of missile test firings to those needed for maintenance a major hurdle to new deployments would be created.

This is the moment for such moves. We are, once again, at a turning point in the nuclear arms race. Because SALT I succeeded in placing ceilings on the number of missile launchers, it stimulated an intense race towards more accurate and powerful missiles, more warheads per launcher, and the development of new and more potent bombers and submarines to replace existing fleets. Most importantly President Carter has displayed a more penetrating understanding of the dangers of the arms race than the previous leaders of the U.S. and U.S.S.R. and has indicated a readiness to consider imaginative policies. Our recommendations do not only meet a current need—they come at a propitious moment.

The United States should take the initiative. The U.S. lead in new weapons technology in the nuclear era is a reflection of our overall superiority in creating new technologies and sophisticated industries. Under these circumstances, we cannot expect the U.S.S.R. to take the initiative.

Our proposals would be *an important step* toward the controls of strategic weapons technology which are so essential to our short term security. They would thereby create that base of confidence and stability which is a prerequisite to overall reductions of the nuclear arsenals.

We *urge* the government of the United States to demonstrate its dedication to arms control by initiating the unilateral reciprocal steps we have recommended, that represent the first steps leading to gradual disarmament. These actions, if carried out by the United States, would represent a policy of restraint of the greatest political significance and yet, for an interim period, be devoid of military risk. Should the Soviet Union reciprocate—and they, like the United States, have much to gain in so doing—a great step forward would be taken to diminish the threat of nuclear war.

Space-based Ballistic Missile Defense

For two decades both the U.S. and the U.S.S.R. have been vulnerable to a devastating nuclear attack, inflicted by one side on the other in the form of either a first strike or a retaliatory second strike. This situation did not come about as the result of careful military planning. "Mutual assured destruction" is not a policy or a doctrine but rather a fact of life. It simply descended like a medieval plague—a seemingly inevitable consequence of the enormous destructive power of nuclear weapons, of rockets that could hurl them across almost half of the globe in 30 minutes and of the impotence of political institutions in the face of such momentous technological innovations.

This grim development holds different lessons for different people. Virtually everyone agrees that the world must eventually escape from the shadow of mutual assured destruction, since few are confident that deterrence by threat of retaliation can avert a holocaust indefinitely. Beyond this point, however, the consensus dissolves. Powerful groups in the governments of both superpowers apparently believe that unremitting competition, albeit short of war, is the only realistic future one can plan for. In the face of much evidence to the contrary they act as if the aggressive exploitation for military purposes of anything technology has to offer is critical to the security of the nation they serve. Others seek partial measures that could at least curb the arms race, arguing that this approach has usually been sidetracked by short-term (and shortsighted) military and political goals. Still others have placed varying degress of faith in radical solutions: novel political moves, revolutionary technological advances or some combination of the two.

President Reagan's Strategic Defense Initiative belongs in this last category. In his televised speech last year calling on the nation's scientific community "to give us the means of rendering these nuclear weapons impotent and obsolete" the president expressed the hope that a technological revolution would enable the U.S. to

---

By Hans A. Bethe, Richard L. Garwin, Kurt Gottfried, and Henry W. Kendall, 1984. Reprinted with permission from *Scientific American* **251**(4), 39–49 (October 1984).

"intercept and destroy strategic ballistic missiles before they reached our own soil or that of our allies." If such a breakthrough could be achieved, he said, "free people could live secure in the knowledge that their security did not rest upon the threat of instant U.S. retaliation."

Can this vision of the future ever become reality? Can any system for ballistic-missile defense eliminate the threat of nuclear annihilation? Would the quest for such a defense put an end to the strategic-arms race, as the president and his supporters have suggested, or is it more likely to accelerate that race? Does the president's program hold the promise of a secure and peaceful world or is it perhaps the most grandiose manifestation of the illusion that science can re-create the world that disappeared when the first nuclear bomb was exploded in 1945?

These are complex questions, with intertwined technical and political strands. They must be examined carefully before the U.S. commits itself to the quest for such a defense, because if the president's dream is to be pursued, space will become a potential field of confrontation and battle. It is partly for this reason the Strategic Defense Initiative is commonly known as the "Star Wars" program.

This article, which is based on a forth-coming book by a group of us associated with the Union for Concerned Scientists, focuses on the technical aspects of the issue of space-based ballistic-missile defense. Our discussion of the political implications of the president's Strategic Defense Initiative will draw on the work of two of our colleagues, Peter A. Clausen of the Union for Concerned Scientists and Richard Ned Lebow of Cornell University.

The search for a defense against nuclear-armed ballistic missiles began three decades ago. In the 1960's both superpowers developed antiballistic missile (ABM) systems based on the use of interceptor missiles armed with nuclear warheads. In 1968 the U.S.S.R. began to operate an ABM system around Moscow based on the Galosh interceptor, and in 1974 the U.S. completed a similar system to protect Minuteman missiles near Grand Forks Air Force Base in North Dakota. (The U.S. system was dismantled in 1975.)

Although these early efforts did not provide an effective defense against a major nuclear attack, they did stimulate two developments that have been dominant features of the strategic landscape ever since: the ABM Treaty of 1972 and the subsequent deployment of multiple independently targetable reentry vehicles (MIRV's), first by the U.S. and later by the U.S.S.R.

In the late 1960's a number of scientists who had been involved in investigating the possibility of ballistic-missile defense in their capacity as high-level advisers to the U.S. Government took the unusual step of airing their criticism of the proposed ABM systems both in congressional testimony and in the press [see "Anti-Ballistic-Missile Systems," by Richard L. Garwin and Hans A. Bethe *Scientific American*, March, 1968]. Many scientists participated in the ensuing debate, and eventually a consensus emerged in the scientific community regarding the flaws in the proposed systems.

The scientists' case rested on a technical assessment and a strategic prognosis. On the technical side they pointed out that the systems then under consideration were inherently vulnerable to deception by various countermeasures and to preemptive attack on their exposed components, particularly their radars. On the strategic side the scientists argued that the U.S.S.R. could add enough missiles to its attacking force to ensure penetration of any such defense. These arguments eventually carried the day, and they are still germane. They were the basis for the ABM Treaty, which was signed by President Nixon and General Secretary Brezhnev in Moscow in May, 1972. The ABM Treaty formally recognized that not only the deployment but also the development of such defensive systems would have to be strictly controlled if the race in offensive missiles was to be contained.

MIRV's were originally conceived as the ideal countermeasure to ballistic-missile defense, and in a logical world they would have been abandoned with the signing of the ABM Treaty. Nevertheless, the U.S. did not try to negotiate a ban on MIRV's. Instead it led the way to their deployment in spite of repeated warnings by scientific advisers and the Arms Control and Disarmament Agency to senior government officials that MIRV's would undermine the strategic balance and ultimately be to the advantage of the U.S.S.R. because of its larger ICBM's. The massive increase in the number of nuclear warheads in both strategic arsenals during the 1970's is largely attributable to the introduction of MIRV's. The result, almost everyone now agrees, is a more precarious strategic balance.

The president's Strategic Defense Initiative is much more ambitious than the ABM proposals of the 1960's. To protect an entire society a nationwide defense of "soft" targets such as cities would be necessary; in contrast, the last previous U.S. ABM plan—the Safeguard system proposed by the Nixon Administration in 1969—was intended to provide only a "point" defense of "hard" targets such as missile silos and command bunkers. The latter mission could be accomplished by a quite permeable terminal-defense system that intercepted warheads very close to their targets, since a formidable retaliatory capability would remain even if most of the missile silos were destroyed. A large metropolitan area, on the other hand, could be devastated by a handful of weapons detonated at high altitude; if necessary, the warheads could be designed to explode on interception.

To be useful a nationwide defense would have to intercept and eliminate virtually all the 10,000 or so nuclear warheads that each side is currently capable of committing to a major strategic attack. For a city attack it could not wait until the atmosphere allowed the defense to discriminate between warheads and decoys. Such a high rate of attrition would be conceivable only if there were several layers of defense, each of which could reliably intercept a large percentage of the attacking force. In particular, the first defensive layer would have to destroy most of the attacking warheads soon after they left their silos or submerged submarines, while the booster rockets were still firing. Accordingly boost-phase interception would be an indispensable part of any defense of the nation as a whole.

Booster rockets rising through the atmosphere thousands of miles from U.S. territory could be attacked only from space. That is why the Strategic Defense Initiative is regarded primarily as a space-weapons program. If the president's plan is actually pursued, it will mark a turning point in the arms race perhaps as significant as the introduction of ICBM's.

Several quite different outcomes of the introduction of space weapons have been envisioned. One view (apparently widely held in the Reagan Administration) has been expressed most succinctly by Robert S. Cooper, director of the Defense Advanced Research Projects Agency. Testifying last year before the Armed Services Committee of the House of Representatives, Cooper declared: "The policy for the first time recognizes the need to control space as a military environment." Indeed, given the intrinsic vulnerability of space-based systems, the domination of space by the U.S. would be a prerequisite to a reliable ballistic-missile defense of the entire nation. For that reason, among others, the current policy also calls for the acquisition by the U.S. of antisatellite weapons [see "Antisatellite Weapons," by Richard L. Garwin, Kurt Gottfried and Donald L. Hafner *Scientific American*, June 1987].

The notion that the U.S. could establish and maintain supremacy in space ignores a key lesson of the post-Hiroshima era: a technological breakthrough of even the most dramatic and unexpected nature can provide only a temporary advantage. Indeed, the only outcome one can reasonably expect is that both superpowers would eventually develop space-based ballistic-missile-defense systems. The effectiveness of these systems would be uncertain and would make the strategic balance more precarious than it is today. Both sides will have expanded their offensive forces to guarantee full confidence in their ability to penetrate defenses of unknown reliability, and the incentive to cut one's own losses by striking first in a crisis will be even greater than it is now. Whether or not weapons deployed in space could ever provide a reliable defense against ballistic missiles, they would be potent antisatellite weapons. As such they could be used to promptly destroy an opponent's early-warning and communications satellites, thereby creating a need for critical decisions at a tempo ill suited to the speed of human judgment.

Our analysis of the prospects for a space-based defensive system against ballistic-missile attack will focus on the problem of boost-phase interception. It is not only an indispensable part of the currently proposed systems but also what distinguishes the current concept from all previous ABM plans. On the basis of our technical analysis and our assessment of the most likely response of the U.S.S.R. we conclude that the pursuit of the president's program would inevitably stimulate a large increase in the Russian strategic offensive forces, further reduce the chances of controlling events in a crisis and possibly provoke the nuclear attack it was designed to prevent. In addition the reliability of the proposed defense would remain a mystery until the fateful moment at which it was attacked.

Before assessing the task of any defense one must first examine the likely nature of the attack. In this case we shall concentrate on the technical and military attributes of the land-based ICBM and on how a large number of such missiles could be used in combination to mount a major strategic attack.

The flight of an ICBM [Figure 1 and Table 1] begins when the silo door opens and hot gases eject the missile. The first-stage booster then ignites. After exhausting its fuel the first stage falls away as the second stage takes over; this sequence

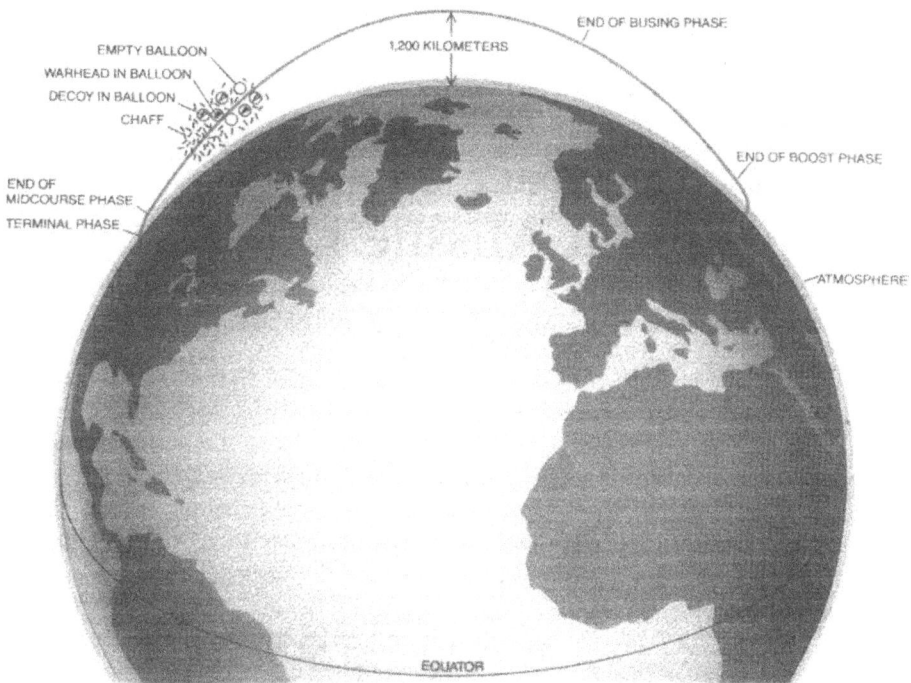

Illustration © George V. Kelvin/Scientific American

Figure 1. Four distinct phases are evident in the flight of an intercontinental ballistic missile (ICBM). In boost phase the missile is carried above the atmosphere by a multistage booster rocket. Most modern strategic missiles carry multiple independently targetable reentry vehicles (MIRV's), which are released sequentially by a maneuverable "bus" during the busing, or postboost, phase. If the country under attack had a ballistic-missile-defense system, the bus would also dispense a variety of "penetration aids," such as decoys, balloons enclosing MIRV's and decoys, empty balloons, radar-reflecting wires called chaff and infrared-emitting aerosols. During the midcourse phase the heavy MIRV's and the light penetration aids would follow essentially identical trajectories. In the terminal phase this "threat cloud" would reenter the atmosphere, and friction with the air would retard the penetration aids much more than the MIRV's. For ICBM's the flight would last between 25 and 30 minutes; for submarine-launched ballistic missiles (SLBM's) it could be as short as eight to 10 minutes.

Table 1 Characteristics of first two phases

| Missile | Gross weight (kilograms) | End of boost phase | | End of busing | | Usual payload |
		Time (seconds)	Altitude (kilometers)	Time (seconds)	Altitude (kilometers)	
SS-18	220,000	300	400	?	?	10 MIRV's on one bus
MX	89,000	180	200	650	1,100	10 MIRV's on one bus
MX with fast-burning booster	87,000	50	90	60	110	Several microbuses with MIRV's and penetration aids
Midgetman	19,000	220	340	—	—	Single warhead
Midgetman with fast-burning booster	22,000	50	80	—	—	Single warhead with penetration aids

is usually repeated at least one more time. The journey from the launch point to where the main rockets stop burning is the boost phase. For the present generation of ICBM's the boost phase lasts for 3 to 5 minutes and ends at an altitude of 300 to 400 kilometers, above the atmosphere.

A typical ICBM in the strategic arsenal of the U.S. or the U.S.S.R. is equipped with MIRV's, which are dispensed by a maneuverable carrier vehicle called a bus after the boost phase ends. The bus releases the MIRVs one at a time along slightly different trajectories toward their separate targets. If there were defenses, the bus could also release a variety of penetration aids, such as light-weight decoys, reentry vehicles camouflaged to resemble decoys, radar-reflecting wires called chaff and infrared-emitting aerosols. Once the bus had completed its task the missile would be in midcourse. At that point the ICBM would have proliferated into a swarm of objects, each of which, no matter how light, would move along a ballistic trajectory indistinguishable from those of its accompanying objects. Only after the swarm reentered the atmospheres would the heavy, specially shaped reentry vehicles be exposed as friction with the air tore away the screen of lightweight decoys and chaff.

This brief account reveals why boost-phase interception would be crucial: every missile that survived boost phase would become a complex "threat cloud" by the time it reached midcourse. Other factors also amplify the importance of boost-phase interception. For one thing, the booster rocket is a much larger and more fragile target than the individual reentry vehicles are. For another, its flame is an abundant source of infrared radiation, enabling the defense to get an accurate fix on the missile. It is only during boost phase that a missile reveals itself by emitting an intense signal that can be detected at a large distance. In midcourse it must first be found by illuminating it with microwaves (or possibly laser light) and then sensing the reflected radiation or by observing its weak infrared signal, which is due mostly to reflection of the earth's infrared radiation.

Because a nationwide defense must be capable of withstanding any kind of strategic attack, the exact nature of the existing offensive forces is immaterial to the evaluation of the defense. At present a full-scale attack by the U.S.S.R. on the U.S. could involve as many as 1,400 land-based ICBM's. The attack might well begin with submarine-launched ballistic missiles (SLBM's), since their unpredictable launch points and short flight times (10 minutes or less) would lend the attack an element of surprise that would be critical if the national leadership and the ground-based bomber force were high-priority targets.

SLBM's would be harder to intercept than ICMB's, which spend 30 minutes or so on trajectories whose launch points are precisely known. Moreover, a space-based defense system would be unable to intercept ground-hugging cruise missiles, which can deliver nuclear warheads to distant targets with an accuracy that is independent of range. Both superpowers are developing sea-launched cruise missiles, and these weapons are certain to become a major part of their strategic forces once space-based ballistic-missile-defense systems appear on the horizon.

The boost-phase layer of the defense would require many components that are not weapons in themselves. They would provide early warning of an attack by sensing the boosters' exhaust plumes; ascertain the precise number of the attacking missiles and, if possible, their identities; determine the trajectories of the missiles and get a fix on them; assign, aim and fire the defensive weapons; assess whether or not interception was successful, and, if time allowed, fire additional rounds. This intricate sequence of operations would have to be automated, because the total duration of the boost phase, now a few minutes, is likely to be less than 100 seconds by the time the proposed defensive systems are ready for deployment.

If a sizable fraction of the missiles were to survive boost-phase interception, the midcourse defensive layer would have to deal with a threat cloud consisting of hundreds of thousands of objects. For example, each bus could dispense as many as 100 empty aluminized Mylar balloons weighing only 100 grams each. The bus would dispense reentry vehicles (and possibly some decoy reentry vehicles of moderate weight) enclosed in identical balloons. The balloons and the decoys would have the same optical and microwave "signature" as the camouflaged warheads, and therefore the defensive system's sensors would not be able to distinguish between them. The defense would have to disturb the threat cloud in some way in order to find the heavy reentry vehicles, perhaps by detonating a nuclear explosive in the path of the cloud. To counteract such a measure, however, the reentry vehicles could be designed to release more balloons. Alternatively, the midcourse defense could be designed to target everything in the threat cloud, a prodigious task that might be beyond the supercomputers expected a decade from now. In short, the midcourse defense would be overwhelmed unless the attacking force was drastically thinned out in the boost phase.

Because the boosters would have to be attacked while they could not yet be seen from any point on the earth's surface accessible to the defense, the defensive system would have to initiate boost-phase interception from a point in space, at a range measured in thousands of kilometers. Two types of "directed energy" weapon are currently under investigation for this purpose: one type based on the use of laser beams, which travel at the speed of light (300,000 kilometers per second), and the other based on the use of particle beams, which are almost as fast. Nonexplosive projectiles that home on the booster's infrared signal have also been proposed.

There are two alternatives for basing such weapons in space. They could be in orbit all the time or they could be "popped up" at the time of the attack. There are complementary advantages and disadvantages to each approach. With enough weapons in orbit some would be "on station" whenever they were needed, and they could provide global coverage; on the other hand, they would be inefficient because of the number of weapons that would have to be actively deployed, and they would be extremely vulnerable. Pop-up weapons would be more efficient and less vulnerable, but they would suffer from formidable time constraints and would offer poor protection against a widely dispersed fleet of strategic submarines.

Pop-up interceptors of ICMB's would have to be launched from submarines, since the only accessible points close enough to the Russian ICBM silos are in the Arabian Sea and the Norwegian Sea, at a distance of more than 4,000 kilometers [Figure 2]. An interceptor of this type would have to travel at least 940 kilometers before he could "see" an ICBM just burning out at an altitude of 200

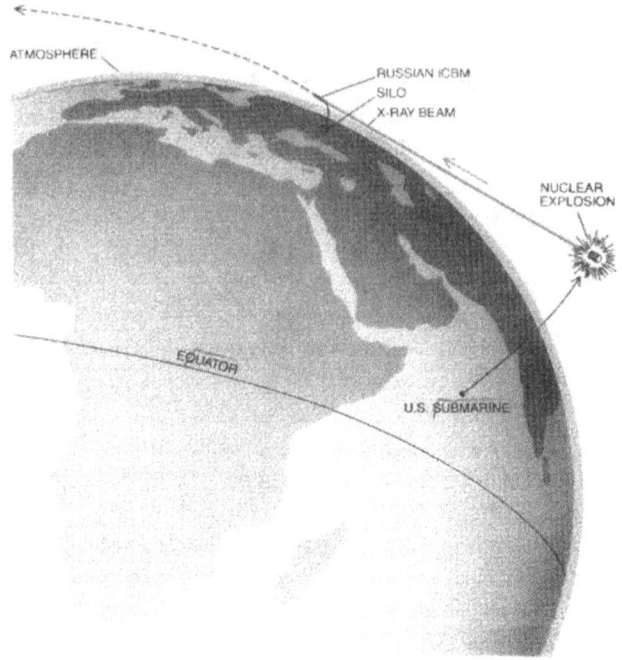

Illustration © George V. Kelvin/Scientific American

Figure 2. "Pop-up" defensive system would rely on a comparatively light interceptor launched from a submarine stationed in waters as close to the Russian ICBM fields as possible (in this case in the northern Indian Ocean). At present the leading candidate for this mission is the X-ray laser, a device consisting of a nuclear explosive surrounded by a cylindrical array of thin metallic fibers. Thermal X rays from the nuclear explosion would stimulate the emission of a highly directed beam to X-radiation from the fibers in the microsecond before the device was destroyed. In order to engage ICBM's similar to the MX rising out of the closest missile silos in the U.S.S.R. while they were still in their boost phase, the interceptor would have to travel at least 940 kilometers from the submarine to the point where the device would be detonated.

kilometers. If the interceptor were lofted by an ideal instant-burn booster with a total weight-to-payload ratio of 14 to one, it could reach the target-sighting point in about 120 seconds. For comparison, the boost phase of the new U.S. MX missile (which has a weight-to-payload ratio of 25 to one) is between 150 and 180 seconds. In principle, therefore, it should just barely be possible by this method to intercept a Russian missile comparable to the MX, provided the interception technique employed a beam that moves at the speed of light. On the other hand, it would be impossible to intercept a large number of missiles, since many silos would be more than 4,000 kilometers away, submarines cannot launch all their missiles simultaneously and 30 seconds would leave virtually no time for the complex sequence of operations the battle-management system would have to perform.

A report prepared for the Fletcher panel, the study team set up [in 1983] by the Department of Defense under the chairmanship of James C. Fletcher of the University of Pittsburgh to evaluate the Strategic Defense Initiative for the president, bears on this question. According to the report, it is possible to build ICBMs that could complete the boost phase and disperse their MIRVs in only 60 seconds, at a sacrifice of no more than 20 percent of payload. Even with zero decision time a hypothetical instant-burn rocket that could pop up an interceptor system in time for a speed-of-light attack on such an ICBM would need an impossible weight-to-payload ratio in excess of 800 to one! Accordingly all pop-up interception schemes, no matter what kind of antimissile weapon they employ, depend on the assumption that the U.S.S.R. will not build ICBM's with a boost phase so short that no pop-up system could view the burning booster.

The time constraint faced by pop-up schemes could be avoided by putting at least some parts of the system into orbit [Figure 3]. An antimissile satellite in a low orbit would have the advantage of having the weapon close to its targets, but it would suffer from the "absentee" handicap: because of its own orbital motion, combined with the earth's rotation, the ground track of such a satellite would pass close to a fixed point on the earth's surface only twice a day. Hence for every low-orbit weapon

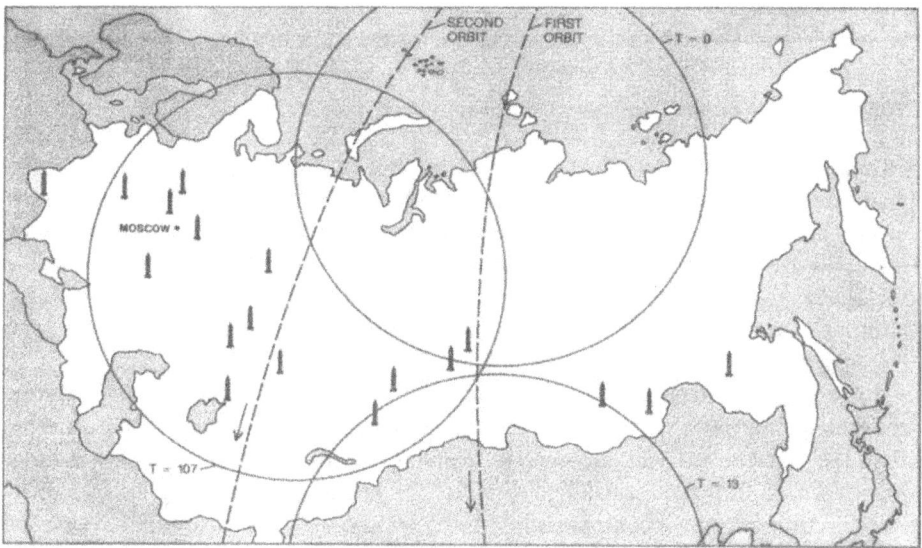

Figure 3. Coverage of the U.S.S.R. by an antimissile weapon with a range of 3,000 kilometers deployed in a polar orbit at an altitude of 1,000 kilometers is indicated by the three circles on this map. The circles show the extent of the weapon's effect at two times separated by 13 minutes on one circuit of the earth and at another time 94 minutes later, on the next circuit. The orbiting weapon could be either a laser or a "fighting mirror" designed to reflect the light sent to it by a mirror stationed at an altitude of 36,000 kilometers above the equator.

that was within range of the ICBM silos many others would be "absentees": they would be below the horizon and unable to take part in the defense. This unavoidable replication would depend on the range of the defensive weapon, the altitude and inclination of its orbit and the distribution of the enemy silos.

The absentee problem could be solved by mounting at least some components of the defensive system on a geosynchronous satellite, which remains at an altitude of some 36,000 kilometers above a fixed point on the Equator, or approximately 39,000 kilometers from the Russian ICBM fields. Whichever weapon were used, however, this enormous range would make it virtually impossible to exploit the radiation from the booster's flame to accurately fix an aim point on the target. The resolution of any optical instrument, whether it is an observing telescope or a beam-focusing mirror, is limited by the phenomenon of diffraction. The smallest spot on which a mirror can focus a beam has a diameter that depends on the wavelength of the radiation, the aperture of the instrument and the distance to the spot. For infrared radiation from the booster's flame the wavelength would typically be one micrometer, so that targeting on a spot 50 centimeters across at a range of 39,000 kilometers would require a precisely shaped mirror 100 meters across—roughly the length of a football field. (For comparison, the largest telescope mirrors in the world today are on the order of five meters in diameter.)

The feasibility of orbiting a high-quality optical instrument of this stupendous size seems remote. The wavelengths used must be shortened, or the viewing must be reduced, or both. Accordingly it has been suggested that a geosynchronous defensive system might be augmented by other optical elements deployed in low orbits.

One such scheme that has been proposed calls for an array of groundbased excimer lasers designed to work in conjunction with orbiting optical elements. The excimer laser incorporates a pulsed electron beam to excite a mixture of gases such as xenon and chlorine into a metastable molecular state, which spontaneously reverts to the molecular ground state; the latter in turn immediately dissociates into two atoms, emitting the excess energy in the form of ultraviolet radiation at a wavelength of .3 micrometer.

Each ground-based excimer laser would send its beam to a geosynchronous mirror with a diameter of five meters, and the geosynchronous mirror would in turn reflect the beam toward an appropriate "fighting mirror" in low orbit. The fighting mirror would then redirect and concentrate the beam onto the rising booster rockets, depending on an accompanying infrared telescope to get an accurate fix on the boosters.

The main advantage of this scheme is that the intricate and heavy lasers, together with their substantial power supplies, would be on the ground rather than in orbit. The beam of any groundbased laser, however, would be greatly disturbed in an unpredictable way by ever present fluctuations in the density of the atmosphere, causing the beam to diverge and lose its effectiveness as a weapon. One of us (Garwin) has described a technique to compensate for these disturbances, making it possible, at least in principle, to intercept boosters by this scheme [Figure 4].

Assuming that such a system could be made to work perfectly, its power requirement can be estimated. Such an exercise is illuminating because it gives an impression of the staggering total cost of the system. Again information from the Fletcher panel provides the basis for our estimate. Apparently the "skin" of a booster can be "hardened" to withstand an energy deposition of 200 megajoules per square meter, which is roughly what is required to evaporate a layer of carbon three millimeters thick. With the aid of a geosynchronous mirror five meters in diameter and a fighting and viewing mirror of the same size, the beam of the excimer laser de-

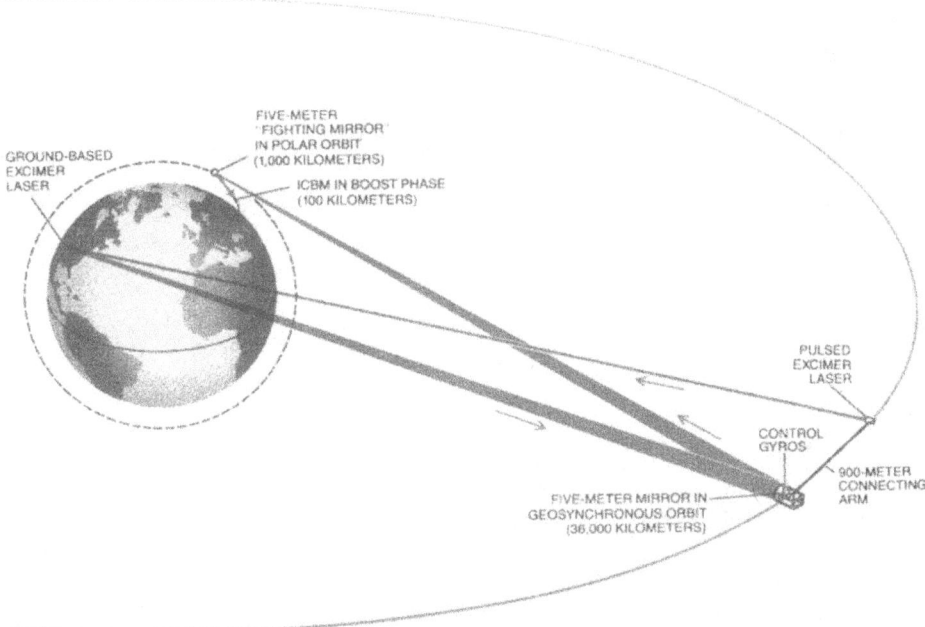

Figure 4. Ground-based laser weapon with orbiting optical elements is designed to intercept ICBM's in boost phase. The excimer laser produces an intense beam of ultraviolet radiation at a wavelength of .3 micrometer. The ground-based mirror would send its beam to a five-meter geosynchronous mirror, which would in turn reflect the beam toward a similar fighting and viewing mirror in a comparatively low orbit; this mirror would then reflect the beam toward the rising booster, depending on its ability to form an image of the infrared radiation from the booster's exhaust plume to get a fix on the target (*diagram at left*). In order to compensate for fluctuations in the density of the atmosphere the geosynchronous satellite would be equipped with smaller excimer laser mounted on a 900-meter connecting arm ahead of the main mirror. A pulse of ultraviolet radiation from this laser would be directed at the ground-based laser, which would reverse the phase of the incoming beam and would emit a much more intense outgoing beam that would exactly precompensate for the atmospheric disturbance encountered by the incoming beam (*diagram at right*). The gain cells would be powered by pulsed electron beams synchronized with the outgoing beam. Such difficulties as mirror vulnerability must be resolved if such a device is ever to be effective.

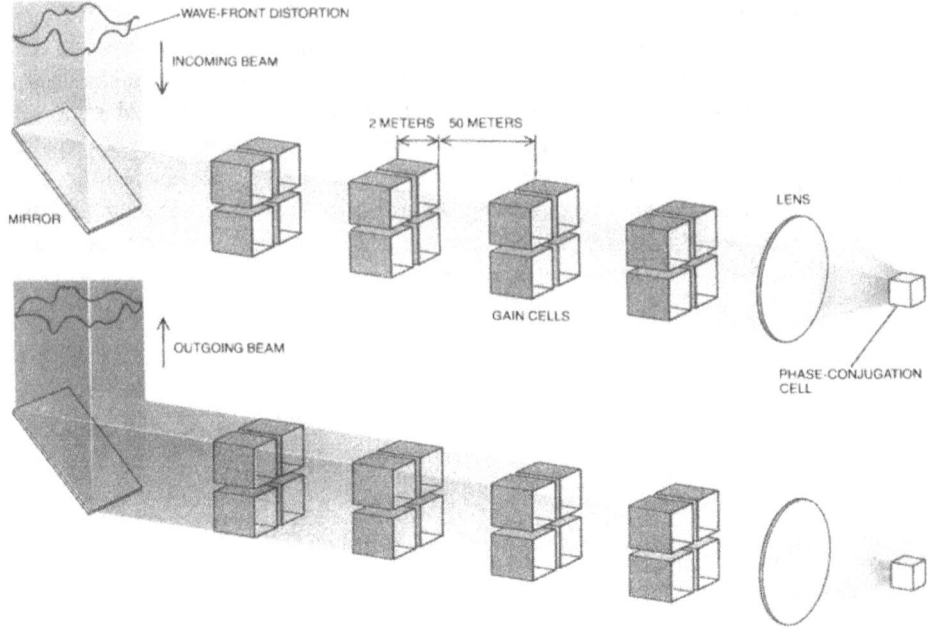

Figure 4. (continued)

scribed above would easily be able to make a spot one meter across on the skin of a booster at a range of 3,000 kilometers from the fighting mirror; the resulting lethal dose would be about 160 megajoules.

A successful defense against an attack by the 1,400 ICBMs in the current Russian force would require a total energy deposition of 225,000 megajoules. (A factor of about 10 is necessary to compensate for atmospheric absorption, reflection losses at the mirrors and overcast skies.) If the time available for interception were 100 seconds and the lasers had an electrical efficiency of 6 percent the power requirement would be more than the output of 300 1,000-megawatt power plants, or more than 60 percent of the current electrical generating capacity of the entire U.S. Moreover, this energy could not be extracted instantaneously from the national power grid, and it could not be stored by any known technology for instantaneous discharge. Special power plants would have to be built; even though they would need to operate only for minutes, an investment of $300 per kilowatt is a reasonable estimate, and so the outlay for the power supply alone would exceed $100 billion.

This partial cost estimate is highly optimistic. It assumes that all the boosters could be destroyed on the first shot that the Russians would not have shortened the boost phase of their ICBM's, enlarged their total strategic-missile force or installed enough countermeasures to degrade the defense significantly by the time this particular defensive system was ready for deployment at the end of the century. Of course the cost of the entire system of lasers, mirrors, sensors and computers would far exceed the cost of the power plant, but at this stage virtually all the required technologies are too immature to allow a fair estimate of their cost.

The exact number of mirrors in the excimer scheme depends on the intensity of the laser beams. For example, if the lasers could deliver a lethal dose of heat in just five seconds, one low-orbit fighting mirror could destroy 20 boosters in the assumed time of 100 seconds. It follows that 70 mirrors would have to be within range of the Russian silos to handle the entire attack, and each mirror would need to have a corresponding mirror in a geosynchronous orbit. If the distance at which a fighting mirror could focus a small enough spot of light was on the order of 3,000 kilometers, there would have to be about six mirrors in orbit elsewhere for every one "on station" at the time of the attack, for a total of about 400 fighting mirrors. This allowance for absenteeism is also optimistic, in that it assumes the time needed for targeting would be negligible, there would be no misses, the Russian countermeasures would be ineffective and excimer lasers far beyond the present state of the art could be built.

The second boost-phase interception scheme we shall consider is a pop-up system based on the X-ray laser, the only known device light enough to be a candidate for this role. As explained above, shortening the boost phase of the attacking missiles would negate any pop-up scheme. In this case a shortened boost phase would be doubly crippling, since the booster would stop burning within the atmosphere, where X rays cannot penetrate. Nevertheless, the X-ray laser has generated a good deal of interest, and we shall consider it here even though it would be feasible only if the Russians were to refrain from adapting their ICBM's to thwart this threat.

The X-ray laser consists of a cylindrical array of thin fibers surrounding a nuclear explosive. The thermal X rays generated by the nuclear explosion stimulate the emission of X-radiation from the atoms in the fibers. The light produced by an ordinary optical laser can be highly collimated, or directed, because it is reflected back and forth many times between the mirrors at the ends of the laser. An intense X-ray beam, however, cannot be reflected in this way, and so the proposed X-ray laser would emit a rather divergent beam; for example, at a distance of 4,000 kilometers it would make a spot about 200 meters across.

The U.S. research program on X-ray lasers is highly classified. According to a Russian technical publication, however, such a device can be expected to operate at an energy of about 1,000 electron volts. Such a "soft" X-ray pulse would be absorbed in the outermost fraction of a micrometer of a booster's skin, "blowing off" a thin surface layer. This would have two effects. First, the booster as a whole would recoil. The inertial-guidance system would presumably sense the blow, however, and it could still direct the warheads to their targets. Second, the skin would be subjected to an abrupt pressure wave that, in a careless design, could cause the skin to shear at its supports and damage the booster's interior. A crushable layer installed under the skin could prolong and weaken the pressure wave, however, thereby protecting both the skin and its contents.

Other interception schemes proposed for ballistic-missile defense include chemical-laser weapons, neutral-particle-beam weapons and nonexplosive homing vehicles, all of which would have to be stationed in low orbits.

The brightest laser beam attained so far is an infrared beam produced by a chemical laser that utilizes hydrogen fluoride. The U.S. Department of Defense plans to demonstrate a two-megawatt version of this laser by 1987 [Figure 5]. Assuming that 25-megawatt hydrogen fluoride lasers and optically perfect 10-meter mirrors eventually become available, a weapon with a "kill radius" of 3,000 kilometers would be at hand. A total of 300 such lasers in low orbits could destroy 1,400 ICBM boosters in the absence of countermeasures if every component worked to its theoretical limit.

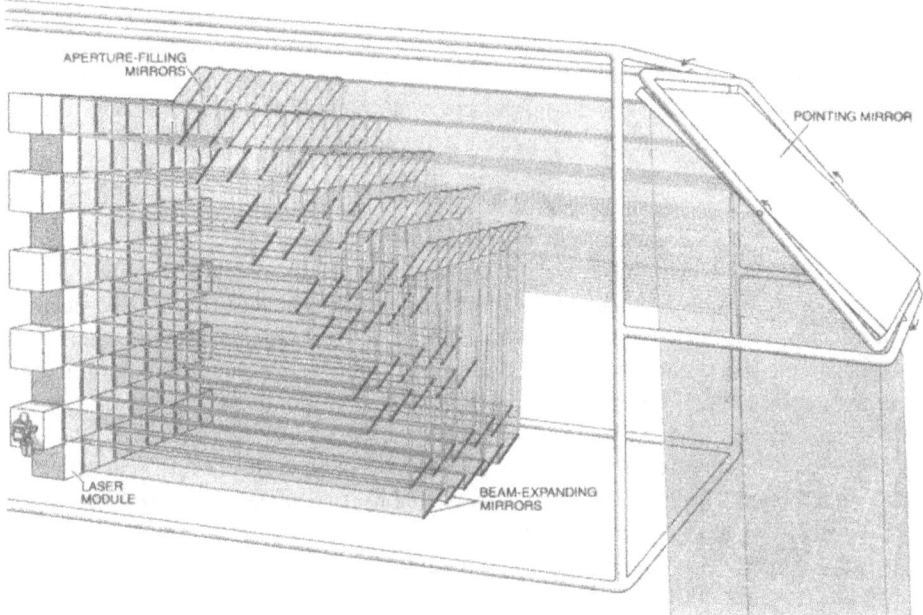

Figure 5. Orbiting laser weapon is shown in this highly schematic diagram based on several assumptions about the physical requirments of such a system. The weapon, which is designed to intercept ICBM's in boost phase from a comparatively low earth orbit, is scaled to generate a total of 25 megawatts of laser light at a wavelength of 2.7 micrometers from a bank of 50 chemical lasers, utilizing hydrogen fluoride as the lasing medium. The lasers, each of which occupies a cubic volume approximately two meters on a side, are arranged to produce an output beam with a square cross section 10 meters on a side. Assuming that the light from the entire bank or laser modules is in phase and that all the mirrors are optically perfect, it can be calculated that a weapon of this type could deliver a lethal dose of heat in seven seconds to a booster at a "kill radius" of some 3,000 kilometers. Some 300 such lasers would be needed in orbit to destroy the 1,400 ICBM's in the current Russian arsenal, assuming no counter-measures were taken other than "hardening" the missiles. Only the front of the weapon is shown; the fuel supply and other components would presumably be mounted behind the laser modules, to the left.

A particle-beam weapon could fire a stream of energetic charged particles, such as protons, that could penetrate deep into a missile and disrupt the semiconductors in its guidance system. A charged-particle beam, however, would be bent by the earth's magnetic field and therefore could not be aimed accurately at distant targets. Hence any plausible particle-beam weapon would have to produce a neutral beam, perhaps one consisting of hydrogen atoms (protons paired with oppositely charged electrons). This could be done, although aiming the beam would still present formidable problems. Interception would be possible only above the atmosphere at an altitude of 150 kilometers or more, since collisions with air molecules would disintegrate the atoms and the geomagnetic field would then fan out the beam. Furthermore, by using gallium arsenide semiconductors, which are about 1,000 times more resistant to radiation damage than silicon semiconductors, it would be possible to protect the missile's guidance computer from such a weapon.

Projectiles that home on the booster's flame are also under discussion. They have the advantage that impact would virtually guarantee destruction, whereas a beam weapon would have to dwell on the fast-moving booster for some time. Homing weapons, however, have two drawbacks that preclude their use as boost-phase interceptors. First, they move at less than .01 percent of the speed of light, and therefore they would have to be deployed in uneconomically large numbers. Second, a booster that burned out within the atmosphere would be immune to them, since friction with the air would blind their homing sensors.

That such a homing vehicle can indeed destroy an object in space was demonstrated by the U.S. Army in its current Homing Overlay test series. On June 10 a projectile launched from Kwajalein Atoll in the Pacific intercepted a dummy Minuteman warhead at an altitude of more than 100 miles. The interceptor relied on a homing technique similar to that of the Air Force's aircraft-launched antisatellite weapon. The debris from the collision was scattered over many tens of kilometers and was photographed by tracking telescopes [Figure 6]. The photographs show, among other things, the difficulty of evading a treaty that banned tests of weapons in space.

In an actual ballistic-missile-defense system such an interceptor might have a role in midcourse defense. It would have to be guided to a disguised reentry vehicle hidden in a swarm of decoys and other objects designed to confuse its infrared sensors. The potential of this technique for midcourse interception remains to be demonstrated, whereas its potential for boost-phase interception is questionable in view of the considerations mentioned above. On the other hand, a satellite is a larger and more fragile target than a reentry vehicle, and so the recent test shows the U.S. has a low-altitude antisatellite capability at least equivalent to the U.S.S.R.

The importance of countermeasures in any consideration of ballistic-missile defense was emphasized recently by Richard D. DeLauer, Under Secretary of Defense for Research and Engineering. Testifying on this subject before the House Armed Services Committee, DeLauer stated that "any defensive system can be overcome with proliferation and decoys, decoys, decoys, decoys."

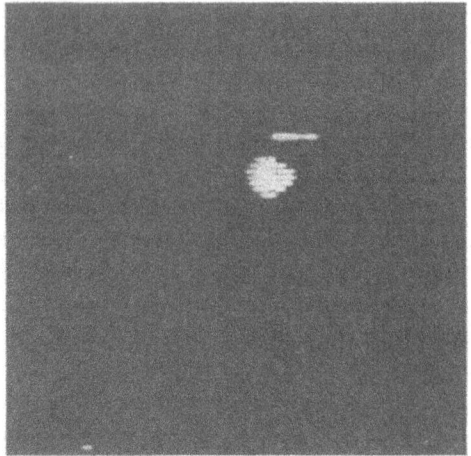
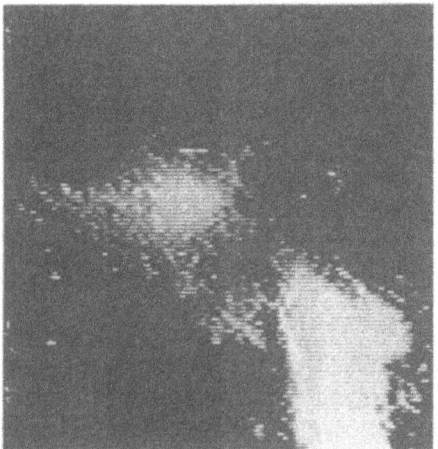
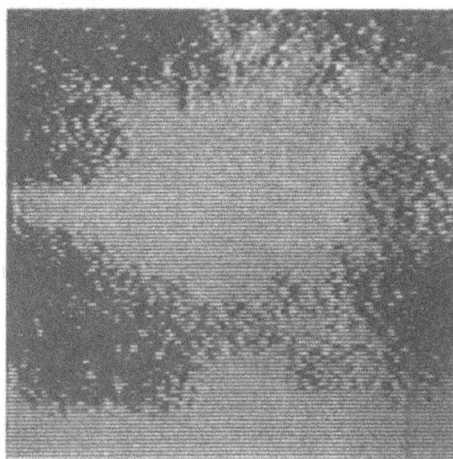

Figure 6. Successful interception of a ballistic-missile warhead was achieved on June 10 in the course of the U.S. Army's Homing Overlay Experiment: The target was a dummy warhead mounted on a Minuteman ICBM that was launched from Vandenberg Air Force Base in California. The interceptor was a nonexplosive infrared-homing vehicle that was launched 20 minutes later from Kwajalein Atoll in the western Pacific. This sequence of photographs was made from a video display of the interception recorded through a 24-inch tracking telescope on Kwajalein. The first frame shows the rocket plume from the homing vehicle a fraction of a second before its collision with the target above the atmosphere. The short horizontal bar above the image of the plume is a tracking marker. The smaller point of light at the lower left is a star. The second and third frames show the spreading clouds of debris from the two vehicles moments after the collision. Within seconds more than a million fragments were strewn over an area of 40 square kilometers. Just before the collision the homing vehicle had deployed a weighted steel "net" 15 feet across to increase the chances of interceptions; as it happened, the vehicle's infrared sensor guided it to a direct, body-to-body collision with the target. According to the authors, the test demonstrates that the U.S. now has a low-altitude antisatellite capability at least equivalent to that of the U.S.S.R.

One extremely potent countermeasure has already been mentioned, namely that shortening the boost phase of the offensive missiles would nullify any boost-phase interception scheme based on X-ray lasers, neutral-particle beams or homing vehicles. Many other potent countermeasures that exploit existing technologies can also be envisioned. All of them rely on generic weaknesses of the defense. Among these weaknesses four stand out: (1) Unless the defensive weapons were cheaper than the offensive ones, any defense could simply be overwhelmed by a missile buildup, (2) the defense would have to attack every object that behaves like a booster, (3) any space-based defensive component would be far more vulnerable than the ICBM's it was designed to destroy, (4) since the booster, not the flame, would be the target, schemes based on infrared detection could be easily deceived.

Countermeasures can be divided into three categories: those that are threatening, in the sense of manifestly increasing the risk to the nation deploying the defensive system; those that are active, in the sense of attacking the defensive system itself and those that are passive, in the sense of frustrating the system's weapons. These distinctions are politically and psychologically significant.

The most threatening response to a ballistic-missile-defense system is also the cheapest and surest: a massive buildup of real and fake ICBM's. The deployment of such a defensive system would violate the ABM Treaty, almost certainly resulting in the removal of all negotiated constraints on offensive missiles. Therefore many new missile silos could be constructed. Most of them could be comparatively inexpensive fakes arrayed in clusters about 1,000 kilometers across to exacerbate the satellites' absentee problem. The fake silos could house decoy ICBM's—boosters without expensive warheads or guidance packages—that would be indistinguishable from real ICBM's during boost phase. An attack could begin with a large proportion of decoys and shift to real ICBM's as the defense exhausted its weapons.

All space systems would be highly vulnerable to active countermeasures. Few targets could be more fragile than a large, exquisitely made mirror whose performance would be ruined by the slightest disturbance. If an adversary were to put a satellite into the same orbit as that of the antimissile weapon but moving in the opposite direction, the relative velocity of the two objects would be about 16 kilometers per second, which is eight times faster than that of a modern armor-piercing antitank projection. If the satellite were to release a swarm of one-ounce pellets, each pellet could penetrate 15 centimeters of steel (and much farther if it were suitably shaped). Neither side could afford to launch antimissile satellites strong enough to withstand such projectiles. Furthermore, a large number of defensive satellites in low or geosynchronous orbits could be attacked simultaneously by "space mines": satellites parked in orbit near their potential victims and set to explode by remote control or when tampered with.

Passive countermeasures could be used to hinder targeting or to protect the booster. The actual target would be several meters above the flame, and the defensive weapon would have to determine the correct aim point by means of an algorithm stored in its computer. The aim point could not be allowed to drift by more than a fraction of a meter, because the beam weapon would have to dwell on one spot for at least several seconds as the booster moved several tens of kilometers.

Aiming could therefore be impeded if the booster flame were made to fluctuate in an unpredictable way. This effect could be achieved by causing additives in the propellant to be emitted at random from different nozzles or by surrounding the booster with a hollow cylindrical "skirt" that could hide various fractions of the flame or even move up and down during boost phase.

Booster protection could take different forms. A highly reflective coating kept clean during boost phase by a stripable foil wrapping would greatly reduce the damaging effect of an incident laser beam. A hydraulic cooling system or a movable heat-absorbing ring could protect the attacked region at the command of heat sensors. Aside from shortening the boost phase the attacking nation could also equip each booster with a thin metallic sheet that could be unfurled at a high altitude to absorb and deflect an X-ray pulse.

Finally, as DeLauer has emphasized, all the proposed space weapons face formidable systemic problems. Realistic testing of the system as a whole is obviously impossible and would have to depend largely on computer simulation. According to DeLauer, the battle-management system would face a task of prodigious complexity that is "expected to stress software-development technology"; in addition it would have to "operate reliably even in the presence of disturbances caused by nuclear-weapons effects or direct-energy attack." The Fletcher panel's report states that the "survivability of the system components is a critical issue whose resolution requires a combination of technologies and tactics that remain to be worked out." Moreover, nuclear attacks need not be confined to the battle-management system. For example, airbursts from a precursor salvo of SLBM's could produce atmospheric disturbances that would cripple an entire defensive system that relied on the ground-based laser scheme.

Spokesmen for the Reagan Administration have stated that the Strategic Defense Initiative will produce a shift to a "defense-dominated" world. Unless the move toward ballistic-missile defense is coupled with deep cuts in both sides' offensive forces, however, there will be no such shift. Such a coupling would require one or both of the following conditions: a defensive technology that was so robust and cheap that countermeasures or an offensive buildup would be futile, or a political climate that would engender arms-control agreements of unprecedented scope. Unfortunately neither of these conditions is in sight.

What shape, then, is the future likely to take if attempts are made by the U.S. and the U.S.S.R. to implement a space-based system aimed at thwarting a nuclear attack? Several factors will have a significant impact. First, the new technologies will at best take many years to develop, and, as we have argued, they will remain vulnerable to known countermeasures. Second, both sides are currently engaged in "strategic modernization" programs that will further enhance their already awesome offensive forces. Third, in pursuing ballistic-missile defense both sides will greatly increase their currently modest antisatellite capabilities. Fourth, the ABM Treaty, which is already under attack, will fall by the wayside.

These factors, acting in concert, will accelerate the strategic-arms race and simultaneously diminish the stability of the deterrent balance in a crisis. Both superpowers have always been inordinately sensitive to real and perceived shifts in the strategic balance. A defense that could not fend off a full-scale strategic attack but might be quite effective against a weak retaliatory blow following an all-out preemptive strike would be particularly provocative. Indeed, the leaders of the U.S.S.R. have often stated that any U.S. move toward a comprehensive ballistic-missile-defense system would be viewed as an attempt to gain strategic superiority, and that no effort would be spared to prevent such an outcome. It would be foolhardy to ignore these statements.

The most likely Russian response to a U.S. decision to pursue the president's Strategic Defense Initiative should be expected to rely on traditional military "worse case" analysis; in this mode of reasoning one assigns a higher value to the other side's capabilities than an unbiased examination of the evidence would indicate, while correspondingly undervaluing one's own capabilities. In this instance the Russians will surely overestimate the effectiveness of the U.S. ballistic-missile defense and arm accordingly. Many near-term options would then be open to them. They could equip their large SS-18 ICBM's with decoys and many more warheads; they could retrofit their deployed ICBM's with protective countermeasures; they could introduce fast-burn boosters; they could deploy more of their current-model ICBM's and sea-launched cruise missiles. The latter developments would be perceived as unwarranted threats by U.S. military planners, who would be quite aware of the fragility of the nascent U.S. defensive system. A compensating U.S. buildup in offensive missiles would then be inevitable. Indeed, even if both sides bought identical defensive systems from a third party, conservative military analysis would guarantee an accelerated offensive-arms race.

Once one side began to deploy space-based antimissile beam weapons the level of risk would rise sharply. Even if the other side did not overrate the system's antimissile capability, it could properly view such a system as an immediate threat to its strategic satellites. A strategy of "launch on warning" might then seem unavoidable, and attempts might also be made to position space mines alongside the antimissile weapons. The last measure might in itself trigger a conflict since the antimissile system should be able to destroy a space mine at a considerable distance if it has any capability for its primary mission. In short, in a hostile political climate even a well-intentioned attempt to create a strategic defense could provoke war, just as the mobilizations of 1914 precipitated World War I.

Even if the space-based ballistic-missile defense did not have a cataclysmic birth, the successful deployment of such a defense would create a highly unstable strategic balance. It is difficult to imagine a system more likely to induce catastrophe than one that requires critical decisions by the second, is itself untested and fragile and yet is threatening to the other side's retaliatory capability.

In the face of mounting criticism Administration spokesmen have in recent months offered less ambitious rationales for the Strategic Defense Initiative than the president's original formulation. One theme is that the program is just a research effort and that no decision to deploy will be made for many years. Military research programs are not normally announced from the Oval Office, however, and there is no precedent for a $26-billion, five-year military-research program without any commitment to deployment. A program of this magnitude, launched under such auspices, is likely to be treated as an essential military policy by the U.S.S.R. no matter how it is described in public.

Another more modest rationale of the Strategic Defense Initiative is that it is intended to enhance nuclear deterrence. That role, however, would require only a terminal defense of hard targets, not weapons in space. Finally, it is contended that even an imperfect antimissile system would limit damage to the U.S.; the more likely consequence is exactly the opposite, since it would tend to focus the attack on cities, which could be destroyed even in the face of a highly proficient defense.

In a background report titled *Directed Energy Missile Defense in Space*, released earlier this year by the Congressional Office of Technology Assessment, the author, Ashton B. Carter of the Massachusetts Institute of Technology, a former Defense Department analyst with full access to classified data on such matters, concluded that "the prospect that emerging 'Star Wars' technologies, when further developed, will provide a perfect or near-perfect defense system . . . is so remote that it should not serve as the basis of public expectation or national policy." Based on our assessment of the technical issues, we are in complete agreement with this conclusion.

In our view the questionable performance of the proposed defense, the case with which it could be overwhelmed or circumvented and its potential as an antisatellite system would cause grievous damage to the security of the U.S. if the Strategic Defense Initiative were to be pursued. The path toward greater security lies in quite another direction. Although research on ballistic-missile defense should continue at the traditional level of expenditure and within the constraints of the ABM Treaty, every effort should be made to negotiate a bilateral ban on the testing and use of space weapons.

It is essential that such an agreement cover all altitudes, because a ban on high-altitude antisatellite weapons alone would not be viable if directed energy weapons were developed for ballistic-missile defense. Once such weapons were tested against dummy boosters or reentry vehicles at low altitude, they would already have the capability of attacking geosynchronous satellites without testing at high altitude. The maximum energy density of any such beam in a vacuum is inversely proportional to the square of the distance. Once it is demonstrated that such a weapon can deliver a certain energy dose in one second at a range of 4,000 kilometers, it is established that the beam can deliver the same dose at a range of 36,000 kilometers in approximately 100 seconds. Since the beam could dwell on a satellite indefinitely, such a device could be a potent weapon against satellites in geosynchronous orbits even if it failed in its ballistic-missile-defense mode.

As mentioned above, the U.S. interception of a Minuteman warhead over the Pacific shows that both sides now have a ground-based antisatellite weapon of roughly equal capability. Hence there is no longer an asymmetry in such antisatellite weapons. Only a lack of political foresight and determination blocks the path to agreement. Such a pact would not permanently close the door on a defense-dominated future. If unforeseen technological developments were to take place in a receptive international political climate in which they could be exploited to provide greater security than the current conditions of deterrence by threat of retaliation provides, the renegotiation of existing treaties could be readily achieved.

The Fallacy of Star Wars

Part I 1: The Evolution of Space Weapons

The invention of nuclear weapons was the most dramatic technological breakthrough in military history. The destruction of Hiroshima, just six years after the discovery of nuclear fission, inaugurated the nuclear era. In 1945 only the United States could make atomic weapons, and the military muscle built to defeat fascism appeared ready to enforce a Pax Americana: the U.S. armed forces controlled the seas and airways that any potential attacker would have to cross. America was totally secure.

The United States did not rest on its laurels. The nation's scientific knowledge, industrial skills, and unmatched resources were quickly exploited to expand the U.S. nuclear arsenal with growing numbers of hydrogen bombs, intercontinental bombers, and nuclear-tipped missiles launched from land or sea. Yet despite this stream of ingenious military innovations, American security declined precipitously because the Soviet Union, though devastated by World War II and technologically inferior to America, acquired its own formidable nuclear force in less than twenty years. The Russians showed that they had the will and the means to duplicate Western technical advances.

The United States maintained unquestioned nuclear superiority throughout the 1950s and 1960s, but the political utility of that superiority was difficult to see. The Soviet Union converted Czechoslovakia into its puppet, suppressed revolts in Hungary and Poland, and supported the invasion of South Korea and insurrections in Southeast Asia. Still, an obsessive drive toward more sophisticated weapons impelled both nations, as it does today, to accumulate nuclear arsenals whose dimensions already exceeded any rational political or military requirements long ago.

The result of that obsession has been two decades of living under the threat of "mutual assured destruction," the guarantee that a crushing nuclear attack by one superpower will spark a crushing nuclear retaliation in return. This demonic pact is

---

Excerpts based on studies conducted by the Union of Concerned Scientists and cochaired by Richard L. Garwin, Kurt Gottfried, and Henry W. Kendall. Edited by John Tirman, 1984. Reprinted by permission of Vintage Books, a division of Random House, New York.

not the product of careful military planning; it is not a policy or a doctrine. Rather, it is a fact of life. It descended like a medieval plague—momentous, inexorable, and somewhat mysterious. And even as the specter of atomic annihilation settled over the political terrain of East and West, it became common to regard the threat of mutual suicide as neither logical nor moral. But beyond that bare consensus, there were sharply divergent views of how to end the nightmare of "The Bomb."

Opinions are no less divergent today. Powerful groups within the governments of both superpowers believe that unremitting competition, though short of war, is the only realistic future; they believe that aggressive military exploitation of whatever technology offers is critical to the security of the nations they serve. Still others have placed faith in radical steps—novel political proposals, revolutionary technological breakthroughs, or both.

President Reagan's Star Wars initiative—dramatically introduced in a televised speech on March 23, 1983—belongs to this last category. In calling on the scientific community "to give us the means of rendering these [nuclear] weapons impotent and obsolete," the president was urging a technological revolution that would enable the United States to "intercept and destroy strategic ballistic missiles before they could reach our own soil." His words envisaged a people liberated from deterrence, "secure in the knowledge that their security did not rest on the threat of instant retaliation." With unassailable hope, the president said: "It is better to save lives than to avenge them."

The vision of a perfect defense is immensely attractive, but it raises complex questions. Can Star Wars eliminate the threat of nuclear annihilation and end the arms race? Or is it an illusion that science can re-create the world that disappeared when the nuclear bomb was born? Will the attempt to install space defenses instead precipitate a nuclear conflict that could not be confined to space? These questions have intertwined political and technical strands. They must be examined carefully before the United States commits itself to the quest for a missile defense, because such a commitment would carry the nation—and the world—across a great divide.

If the president's vision is pursued, outer space could become a battlefield. An effective defense against a missile attack must employ weapons operating in space. This now peaceful sanctuary, so long a symbol of cooperation, would be violated. And the arduous process of arms control, which has scored so few genuine successes in the nuclear era, would also be imperiled—perhaps terminated—by the deployment of weapons in space.

The competition for Star Wars weapons is now in its embryonic stages. The United States and the Soviet Union are developing and testing anti-satellite weapons (ASATs) with modest capabilities. Much more lethal weapons, able to destroy satellites that warn of nuclear attack and others that command U.S. nuclear forces, will be built if the space arms race is not slowed down.

Anti-satellite warfare, though less exotic and far less threatening to peace, is still part and parcel of the space-based missile defense demanded by President Reagan. Determined pursuit of the latter will surely foreclose attempts to restrain the former. Similarly, negotiated constraints on ASAT weapons would seriously handicap the development of missile defenses. In this sense, space weapons form a seamless web, offering a choice between two very different paths: a major national commitment to struggle for military dominance on the "new high ground," or a broad agreement with the Soviets to ban space weapons.

The belief that science and technology can relieve America and Russia of the burden of nuclear deterence has gripped many policymakers and the public alike for nearly forty years. Sadly, this belief has usually propelled both nations toward more weapons-building, not less. And the technological revolutions so earnestly sought by successive presidencies have undermined American security when the Soviets copied our lead.

Perhaps the most virulent form of this marriage of technological hope and ideological fear is the search for the perfect defense, the protective shield that would recreate the days before atomic weapons menaced every hour. With the advances in space technology and the growing reluctance to seek political accommodations, it has returned again.

* * * * * * * *

Part II 3: Physics and Geometry versus Secrecy and Clairvoyance

Two prevalent misconceptions must be dispelled before analyzing the technical prospects for space-based missile defense. The first is the belief that complex military systems cannot be assessed without access to highly secret information. The second is that scientists can do anything given enough money. These are related misconceptions; both see science as a somewhat occult activity.

A sound appraisal of the prospects for space-based missile defense *can* be given on the basis of firmly established scientific laws and facts. To a large extent, such an assessment can be understood by inquisitive citizens without specialized knowledge. No one should shrink from forming a judgment on this issue.

Highly classified information is sometimes indispensable to understanding the performance of a particular weapon system. Without disclosures by the U.S. government, the public would know very little about the explosive energy ("yield") of nuclear weapons or the accuracy of intercontinental missiles. Assessing ballistic missile defense, however, does not require such information. In particular, we are evaluating *total* ballistic missile defense—a defense of our society against an armada of some 10,000 nuclear weapons—not merely the defense of highly reinforced (or "hard") military targets, such as missile silos and command bunkers.

To know whether a defense can be expected to protect, say, 20 percent versus 70 percent of our missile silos would require detailed classified data about missile accuracy, weapon yield, and silo hardness. In contrast, a population defense must be virtually leakproof. Even a handful of thermonuclear explosions over our large cities would kill millions instantly, and would disrupt our society beyond recognition. A missile defense must also intercept enemy warheads far from their targets. Interception of warheads as they are dropping onto our cities or other "soft" targets will not do: nuclear charges can be set to explode as soon as the warhead is struck by a defensive weapon, and this "salvage-fused" warhead would still destroy its target if its yield is sufficiently high.[1]

To defend U.S. society totally then, attacking missiles must be intercepted soon after they leave their launch sites or while they are traversing space. Interception means the delivery of a blow to the enemy missile powerful enough to disrupt it in some way. This is precisely where the constraints of geometry and physics enter, for it is only in science fiction that unlimited amounts of energy can be transmitted over arbitrarily large distances against a tiny target moving at great speed.

The real-life problems of missile defense have been studied intensively by the U.S. defense establishment for a quarter of a century, and some of the authors of this book have contributed to many phases of this effort. These investigations have made it clear that a total missile defense must overcome a number of daunting obstacles that involve immutable laws of nature and basic scientific principles. If the attacker sets out to evade and overwhelm the defense, these same laws and principles are, almost without exception, on the attacker's side. Accordingly, to erect a total BMD one must leap over these obstacles by some combination of novel, ingenious measures, and do so with enough of a margin to compensate for the attacker's innate advantage that just one large bomb can destroy a city.

What are these immutable laws of nature and basic scientific principles? A few of the most important examples:

- The earth rotates about its axis and satellites move along prescribed orbits. In general, therefore, a satellite cannot hover above a given spot, such as a missile silo complex in Siberia.
- Even a thin layer of atmosphere absorbs X rays.
- Electrically charged particles follow curved paths in the earth's magnetic field.
- A laser can emit a perfectly parallel pencil of light, but the wave nature of light guarantees that this beam will eventually flare outwards and become progressively more diffuse.
- The earth is round; a weapon must therefore be far above the United States to see a silo in Siberia.

* * * * * * * *

In assessing any particular proposed system, we will give it the benefit of the doubt and assume that it performs as well as the constraints of science permit. What is not yet known by anyone, and will usually be veiled in secrecy, is how close the effectiveness of the proposed system comes to its maximum possible value. But this ignorance has no impact on our assessments since, for the sake of clarity, we assume that this limit will be reached.

All of these theoretical BMD systems are exceedingly optimistic extrapolations beyond current state-of-the-art technology. To a large extent, the status of current technology is reported in the open literature, and the Fletcher panel's assessments of the gaps between our current abilities and the requirements for a space-based

BMD system, among other evaluations, have been widely reported. Every total BMD system that has been proposed is comprised of a large set of distinct devices. In virtually every instance, existing prototypes for these individual devices perform far below the theoretical limit. So a prodigious gap exists between the current capabilities of the proposed systems and the theoretical limits that are granted in this book. For that reason, it is not possible to make sound estimates of the true cost of these systems, for that would require a level of engineering knowledge that does not exist at either the public or the classified level.

Nevertheless, it is important to gain a rough understanding of the costs that would be involved. Consider the person who has thought of the automobile and wishes to estimate its cost before inventing the internal combustion engine. Knowing how to construct wagons, the inventor devotes considerable effort to estimating the cost of the automobile's body to obtain a lower limit, perhaps an order-of-magnitude estimate. In a similar vein, we rely on the law of energy conservation: if a beam weapon in space must deliver a certain amount of energy to disrupt enemy missiles, that energy must come from a fuel supply. The cost of the whole system, therefore, is certainly far greater than the cost of lifting that fuel into orbit.

A common perception, and one that is often voiced in the debate over missile defense, is that science and technology can accomplish any designated task. That is certainly untrue. The laws of thermodynamics tell us that one cannot build a perpetual motion machine, and the principle of relativity implies that it would be futile to try to design spaceships that move faster than the speed of light. The laws of nature set limits on what human beings can do. Nevertheless, it is true that the advances scored by science and technology have been remarkable and often unpredictable. But none of these advances violated firmly established laws of nature. Furthermore, these breakthroughs employed new domains of physics, as in the development of nuclear weapons, or fundamentally new technologies, as in the invention of radar.

The current proposals for missile defense, however, do not depend on any fundamentally new physics. Rather, BMD concepts, while displaying great ingenuity, rely on devices that exploit well-known principles of physics and engineering. Should new concepts for total missile defense be put forward, the whole matter will have to be analyzed anew.

Public discussion will help lead the nation to the best path toward national security and will help to avoid the worst pitfalls. Such an open debate will not provide the Soviet Union with information or insights that it does not already possess, because Soviet defense analysts are well versed in the scientific and technological principles under examination.

The remarkable scientific advances of the twentieth century share one essential characteristic: they were not opposed at every step by a resourceful and ingenious adversary. Yet the most important factor in the issue at hand is the certainty that the Soviet Union will react vigorously to a massive U.S. commitment to missile defense. The threat that U.S. defenses would face when they first emerge a decade or more from now would be very different from that posed by current Soviet strategic forces. This dynamic character of the threat is ignored by those who draw a parallel between the quest for a total missile defense and the decision to undertake a landing on the moon. They seem to forget that when the Apollo Program was launched we already knew that the moon was not populated by a powerful nation bent on frustrating our mission.

* * * * * * * *

Part II 7: Systemic Problems and Net Assessment

Our analysis makes clear that total ballistic missile defense—the protection of American society against the full weight of a Soviet nuclear attack—is unattainable if the Soviet Union exploits the many vulnerabilities intrinsic to all the schemes that have been proposed thus far. In none of the three phases of attack can one expect a success rate that would reduce the number of warheads arriving on U.S. territory sufficiently to prevent unprecedented death and destruction. Instead, each phase presents intractable problems, and the resulting failure of the system compounds from one phase to the next.

A highly efficient boost-phase interception is a prerequisite of total BMD, but is doomed by the inherent limitations of the weapons, insoluble basing dilemmas, and an array of offensive countermeasures. As a result, the failure of midcourse systems is preordained. Midcourse BMD is plagued not so much by the laws of physics and geometry as by the sheer unmanageability of its task in the absence of a ruthless thinning out of the attack in boost phase.

Terminal phase BMD remains fundamentally unsuitable for area defense of population centers as opposed to hard-point targets. There seems to be no way of defending soft targets on a continent-wide basis against the broad variety of attacks that could be tailored to circumvent and overwhelm terminal defenses.

* * * * * * * *

Part II 8: Political and Strategic Implications: Summary

The superficial attractions of a strategy of nuclear defense disappear when the overall consequences of BMD deployments are considered. More than any foreseeable offensive arms breakthrough, defenses would radically transform the context of U.S.-Soviet nuclear relations, setting in motion a chain of events and reactions that would leave both superpowers much less secure and could even lead to nuclear war. Deterrence would be weakened and crisis instability increased. Damage limitation would be undermined by a great emphasis on the targeting of cities and the increased vulnerability of command and control systems. And virtually the entire arms control process would be swept away by the abrogation of the ABM Treaty, the launching of a new offensive round of the arms race, and the extension of the arms race into space. Finally, a commitment to BMD would provoke a serious crisis in the Atlantic Alliance.

The late Republican Senator Arthur Vandenberg was fond of attacking schemes he opposed by declaring "the end unattainable, the means harebrained, and the cost staggering." For Vandenberg, this was a politically useful form of exaggeration. For total ballistic missile defense, it is an entirely fitting description.

Accidental Nuclear War

If nuclear war breaks out in the coming decade or two, it will probably be by accident. The threat of a cold-blooded, calculated first strike is vanishing, but beneath the calm surface of constructive diplomacy among the traditional nuclear rivals lurks the danger of unpremeditated use of nuclear weapons. The accidental, unauthorized or inadvertent use of these weapons has become the most plausible path to nuclear war.

Both superpowers, as well as France, Great Britain and China—long-standing members of the nuclear club—are potential sources of accidental missile launch. The emergence of fledgling nuclear powers such as India, Pakistan and Israel—some armed with ballistic missiles—pushes nuclear safeguards even closer to the top of the international security agenda.

The political stability of some nuclear newcomers is questionable, and so any physical and procedural safeguards they place on their weapons might readily be overriden. It is unlikely, however, that a nuclear attack by one of these nations—on either one of the superpowers or a client state—could trigger a massive nuclear response. U.S. and Soviet arsenals still pose the greatest controllable threat of unintended war, and so changes to those arsenals offer the greatest hope of alleviating that risk.

The chances of unwanted nuclear war would be reduced significantly if tamperproof, coded locks were installed on all nuclear weapons and if methods were put in place to disarm nuclear forces even after launch. In addition, the U.S. and the Soviet Union should reduce their reliance on the dangerous policy of launch on warning and reduce the launch readiness of their nuclear forces.

The social and political upheavals in the Soviet Union underscore fears of unintended nuclear war. Civil turmoil raises the possibility that rebellious ethnic groups or splinter organizations could capture nuclear weapons. Other, deeper fault lines run through the whole of Soviet society and may be capable of cracking the foundations of its nuclear command system. Although the U.S. faces no such civil un-

---

By Bruce G. Blair and Henry W. Kendall, 1990. Reprinted with permission from *Scientific American* **263** (6), 53–205 (December 1990).

rest, the country's system of nuclear command carries some risk that nuclear weapons might be used contrary to the intentions of legitimate authorities.

The organization of missile forces in the U.S. and the Soviet Union makes it just as likely that many missiles could be launched without authorization as easily as a single one. A breakdown of control at the apex of the command chain or at lower levels (perhaps resulting from a violent rupture of Soviet political institutions) could lead to an attack causing vast destruction—and possibly trigger a larger nuclear exchange.

Were an unauthorized attack to occur from the Soviet Union, the minimum launch could involve a battalion (6 to 10 missiles) or even a regiment (18 to 30 missiles). Each missile carries up to 10 warheads; the salvo could result in as many as 300 nuclear explosions. In the U.S. a minimum launch could involve either a flight of 10 missiles or a squadron of 50, carrying as many as 500 warheads, each more than 25 times as powerful as the bomb that destroyed Hiroshima. Even if no retaliation ensued, the resulting destruction and loss of life would dwarf any previous calamity in human experience.

Both U.S. and Soviet nuclear commanders face an unavoidable dilemma: they must exert negative control over nuclear weapons to prevent unwanted use, but they must exert positive control to ensure that weapons are used when duly authorized. Measures that reduce the chance of unwanted launch may increase the chance that legitimate launch orders will not be carried out. Military commanders have thus resisted improved safeguards on the grounds that those safeguards would weaken nuclear deterrence. Deficiencies in negative control have been tolerated, and although some remedial measures have gradually been implemented, a completely satisfactory trade-off has yet to be found.

An alarmist position is unwarranted; the dominant peacetime requirement of both the U.S. and the Soviet nuclear command systems is to prevent reliably the illicit or accidental release of even a single nuclear weapon. The nuclear hierarchy is well aware that it would probably not survive the political repercussions of any major failure to perform this function. Both sides have developed sophisticated weapon-design principles and operational procedures to preserve effective negative control over tens of thousands of widely dispersed warheads; their record has been perfect to date.

Complete confidence, however, is equally unwarranted. Even the most thorough study of ways that a disaster might occur cannot exhaust the perverse possibilities, as the explosion of the *Challenger* space shuttle and the nuclear accident at Three Mile Island attest. Furthermore, weaknesses in current safeguards are most likely to surface in a crisis—circumstances under which neither superpower has much experience in preserving negative control.

Crises shift the priority of nuclear command systems toward positive control at the expense of safeguards. When the Soviets invaded Czechoslovakia in 1968, they placed at least one army of their strategic rocket forces one notch below maximum alert. Nuclear warheads were removed from storage depots and affixed to missiles at the launch pads. These actions compromised a strong safeguard against unintended launch: separation of warheads from their means of delivery.

The U.S. engaged in comparable actions during the 1973 Arab-Israel war. Additional long-range bombers were placed on ground alert, ballistic missile submarines left port and nearly all land-based strategic missiles were readied for launch. Individual weapon commanders removed launch keys and presidential launch codes from their dual-lock safes—increasing the possibility of unauthorized release of weapons.

Current procedures since added to this alert level include the activation of special radio communications links that enable a military command aircraft to fire all 1,000 land-based American intercontinental ballistic missiles (ICBMs) by remote control. This decreases the power of ground crews to veto an illicit launch command.

The actions taken in such alerts cannot be completely governed by political leaders. The vulnerability of nuclear forces and the command system itself to nuclear attack generates pressure to delegate authority for nuclear alerts and weapon release down the chain of command.

The U.S. and the Soviet Union, however, appear to differ substantially on the extent to which positive control is delegated in a crisis. The U.S. command system is decentralized and allows individual weapon commands to take virtually all alert steps short of firing weapons [Figure 1]. Military commanders can send bombers to their holding stations near Soviet territory. They also launch the airborne command posts, which issue launch orders in case of an attack that destroys ground-based centers.

Orders affecting the disposition of nuclear forces flow through strictly military channels, with marginal civilian oversight. Furthermore, historical documents leave little doubt that past presidents have delegated to key military commanders the authority to execute nuclear war plans in the event of a communications failure and verification of a nuclear strike against the U.S. There is strong evidence that such arrangements are still in effect. Numerous military installations possess all the codes needed to authorize launch. The portion of the U.S. arsenal that is restrained by hardware locks can be readied for use by many sources within the military chain of command.

In contrast, no Soviet military commander has significant authority to alert or maneuver nuclear forces, much less order an attack. Changes in alert status require the explicit approval of the highest political leaders. Furthermore, nuclear orders are apparently processed in parallel by several separate control channels to ensure strict conformity to political plans. Units of the KGB—the Soviet political secret police—have custody of tactical nuclear weapons and, it is believed, disseminate weapon unlock codes to tactical and most strategic forces, ensuring central political control. The scope for unauthorized release would expand, however, if codes were distributed as part of the preparation for war.

A further weakness in protection against unwanted war stems from launch-on-warning strategies, which call for commanders to fire retaliatory missiles after an

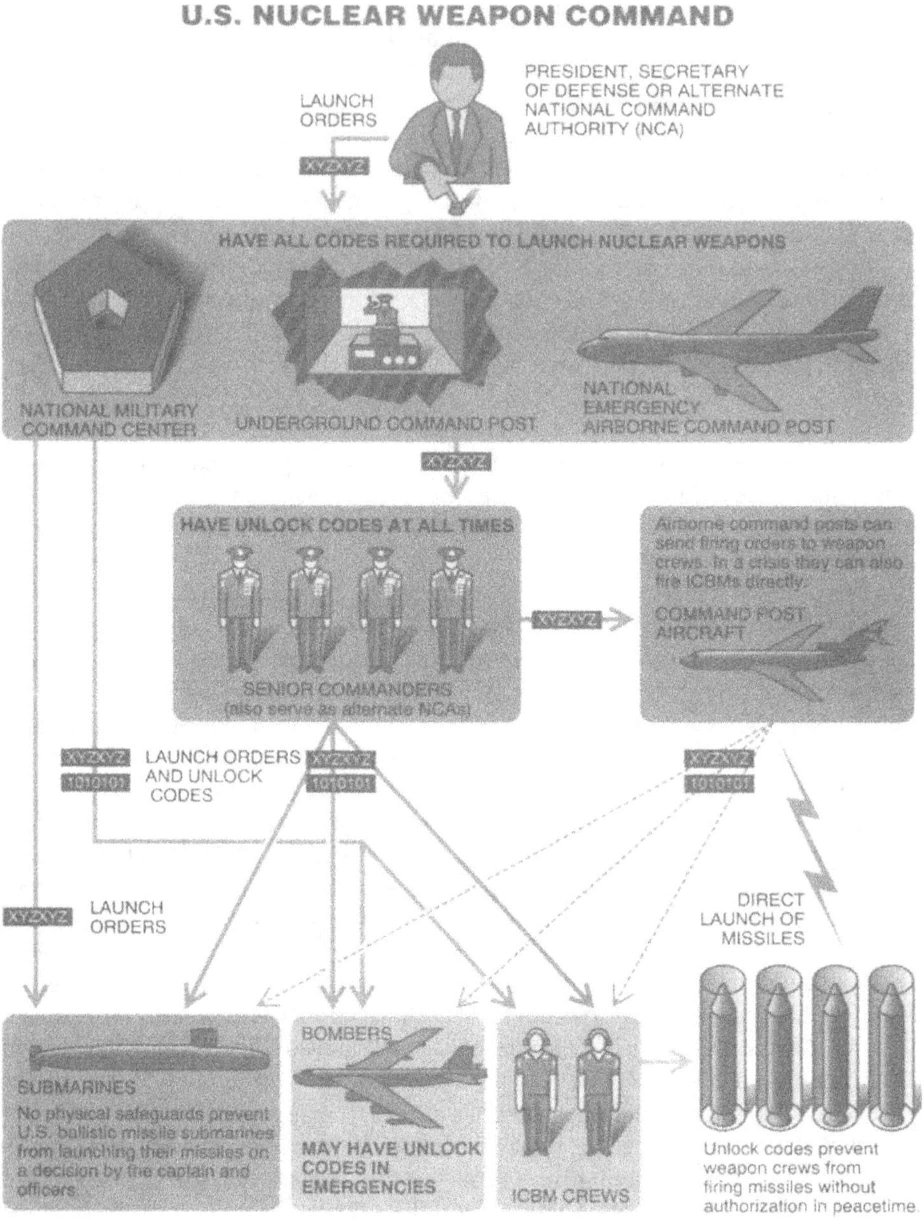

Figure 1. U.S. nuclear weapon command

attack is confirmed but before incoming warheads detonate. Both the U.S. and Soviet Union rely heavily on this strategy. It requires flawless performance from satellite and ground-based sensors and from human beings.

Launch on warning compels authorities to decide whether to fire—and against which targets—in a short time and without a clear picture of the attack supposedly under way. They must respond with no definitive warhead count, no clear idea of the intended targets, no prediction of expected damage and casualties, no knowledge of the objectives of the attack and possibly no way to tell whether the attack is deliberate, accidental or unauthorized. Even if this information were available, commanders could not easily comprehend it and react in measured ways in the time allowed by launch on warning.

The commander of the North American Air Defense Command (NORAD), for example, would have only three minutes from the time of an initial attack indication to pass judgment on whether the continent is under fire or not. Clearly, this decision—and the subsequent ones that must be made during the 10-minute flight time of submarine-launched missiles or the 30-minute flight of ICBMs—entails major risks of premature release of weapons based on false alarms, miscalculations or confusion.

In the U.S. a so-called missile event—indication of a potential attack—typically occurs several times a day. When each event occurs, the command director at NORAD must establish a special conference with the Strategic Air Command and the Pentagon and declare his assessment of the threat to North America. In addition, thousands of anomalous sensor signals annually require urgent attention and evaluation. Each year between 1979 and 1984, the only period for which official information is available, NORAD assessed about 2,600 unusual warning indications. One in 20 required further evaluation because it appeared to pose a threat.

Most false alarms, whether generated by incorrect data, defective computer chips or other malfunctions, are quickly recognized, but perhaps once or twice a year an alarm persists long enough to trigger a nuclear alert. The last such incident to be publicly disclosed occurred in 1980, when a faulty computer chip generated indications of a massive Soviet attack.

In the ensuing confusion, a nuclear alert was declared, and the command director failed to issue a proper evaluation on time. (He was dismissed the next day.) The nuclear alert lasted longer and reached a higher level than the situation warranted. In the midst of a superpower crisis, such confusion could have been far more likely to lead commanders to act as if an attack were actually under way.

Similar Soviet procedures for evaluating indications of attack and initiating retaliation are apparently equally vulnerable to false alarms [Figure 2]. A retired Soviet general recently told how he once witnessed signals from space-based sensors warning of the launch of U.S. Minuteman missiles against the U.S.S.R. A "competent operator," the general recalled, determined that the supposed missile exhaust plumes were in fact merely "patches of sunlight."

Thus far, humans have recognized hardware failures that produced false warnings of attack in time to avoid war. The corrective mechanisms have repeatedly passed muster—albeit during peacetime when alert and anxiety levels were low.

Still, in each case, certain safeguards against unwanted launch were discarded. In the U.S., Minuteman launch crews, for example, removed authorization codes and launch keys from safes. Bomber crews were scrambled, and command aircraft were launched without the knowledge of political authorities. Such actions run some indeterminate risk of provoking precautionary Soviet responses, which could in turn

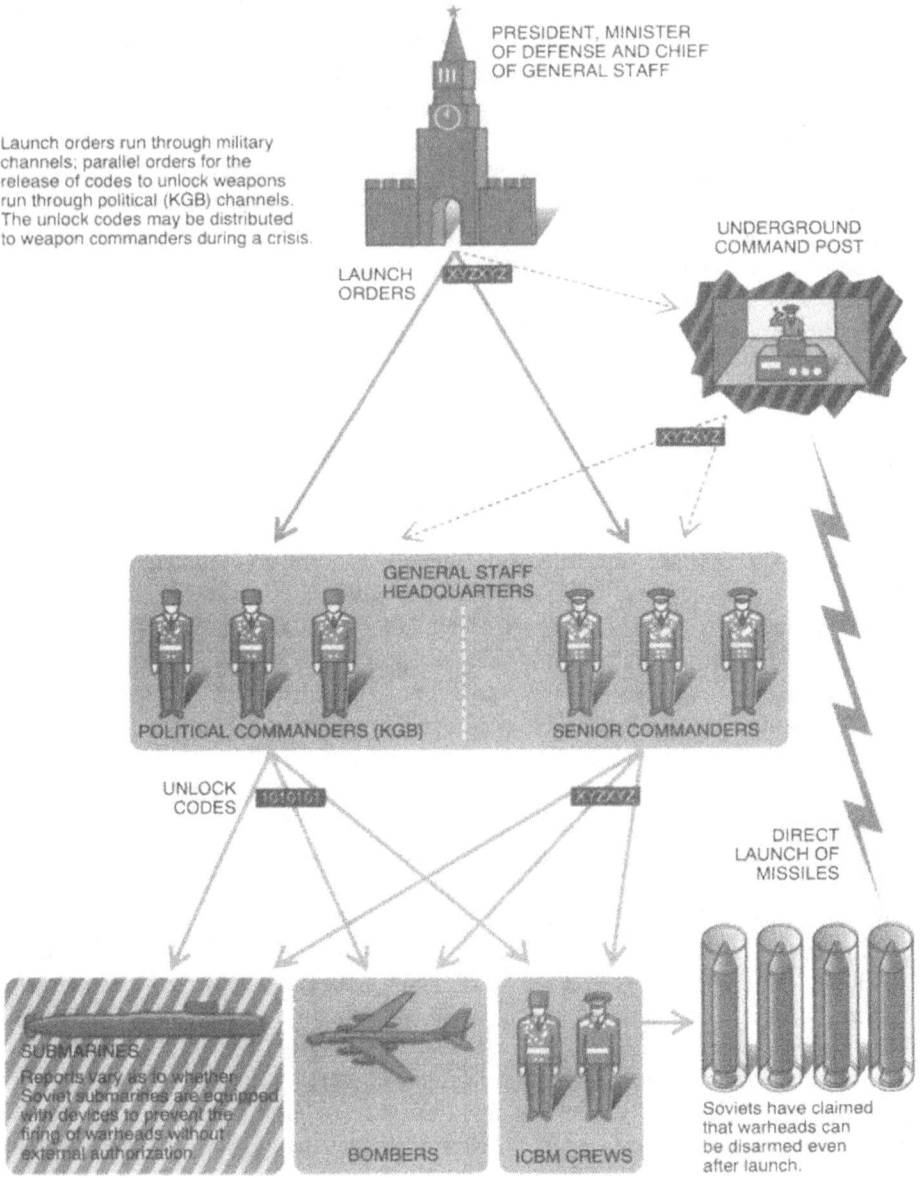

Figure 2. U.S.S.R. nuclear weapon command

have reinforced the U.S. perception of an immediate threat. The question of whether such interactions based on a false alarm could trigger a nuclear attack is an open one.

A number of technical and procedural changes would reduce the chance of unintended war. Heading the list is the comprehensive use of so-called permissive action links (PALs). These electromechanical locks prevent warheads from detonating unless an essentially unbreakable code is inserted. Codes are disseminated to individual weapon commanders by higher authorities only when launch has been properly authorized. PALs prevent unauthorized release by a weapon's own crew or by enemy soldiers or terrorists who might seize a warhead. Similar devices, called coded switch systems, can be employed to prevent bomb bays from opening and missiles from firing.

Such devices were first installed by the U.S. in the early 1960s on tactical warheads assigned to allied forces stationed overseas; today all land-based U.S. tactical weapons are protected by PALs. By the end of the 1970s all Strategic Air Command nuclear warheads were equipped with either PALs or coded switch systems. We believe that Soviet land-based missiles and bombers are similarly equipped.

Naval nuclear forces on both sides, however, are generally not equipped with PALs. The resulting danger of accidental launch is particularly significant in the case of sea-launched cruise missiles. These weapons have a long range, and they are carried by surface vessels and attack submarines that would be in combat during the earliest phases of a conflict. Some British and French warheads are in similarly exposed forward positions.

Another way to reduce the risk of unintended nuclear war is to lower the levels of nuclear readiness. We estimate that 50 to 80 percent of the Soviet ICBM force is routinely maintained in a launch-ready configuration. In peacetime 15 percent of their submarine-launched ballistic missile force is deployed at sea, and none of the long-range bombers are on alert or even loaded with nuclear weapons [Figure 3].

The U.S., meanwhile, maintains about 90 percent of its ICBMs in launch-ready status, capable of firing within three minutes [Figure 4]. Half of the ballistic missile submarine force is at sea at any time, and half of those vessels can fire their missiles within 15 minutes. A quarter of the strategic bomber force is on a five minute ground alert.

This high state of launch readiness is an anachronism in the new era of U.S.-Soviet relations and poses an unnecessary danger. The percentage of both arsenals that is on alert should be cut down to a fraction of current levels. A threefold reduction could be carried out unilaterally without eroding deterence, and deeper cuts could be made in stages under a verifiable agreement.

Furthermore, the warheads for units no longer on ready alert should be placed in the custody of civilian agencies, as they were during the 1950s when the Atomic Energy Commission held custody of nuclear weapons during peacetime. The civilian role in managing warhead stockpiles should be strengthened to maintain tight political control over the arming of nuclear-capable units in a crisis. Although some analysts argue that the risk of misperception in a crisis is lower if nuclear forces maintain high readiness at all times, such a posture runs contrary to current trends in superpower relations.

The adoption of lower alert levels would permit removing warheads or some other necessary component from some portion of the strategic arsenal, thus absolutely preventing an unwanted launch from causing a nuclear detonation. This danger is quite real. The Soviets recently disclosed that a nuclear-tipped ballistic missile had

NATIONAL SECURITY AND NUCLEAR WEAPONS

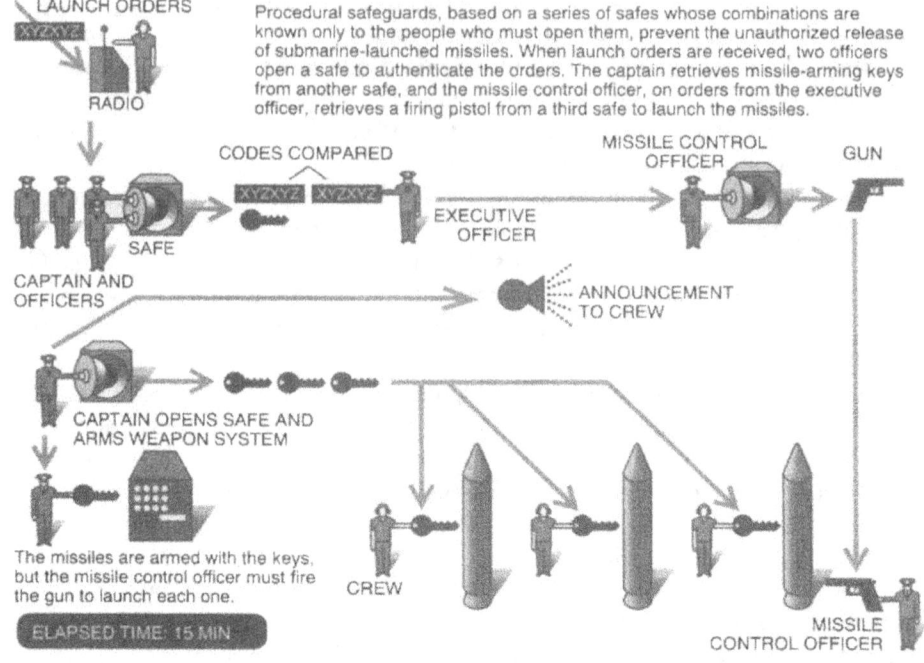

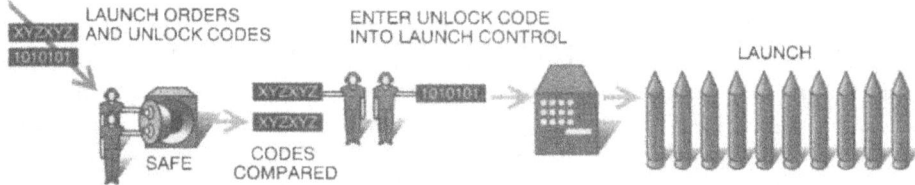

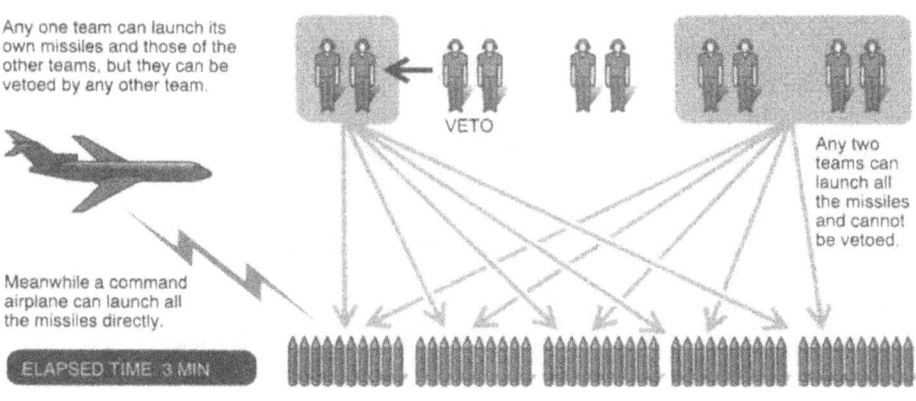

Figure 3. U.S.S.R. ICBM launch-ready status

ACCIDENTAL NUCLEAR WAR 173

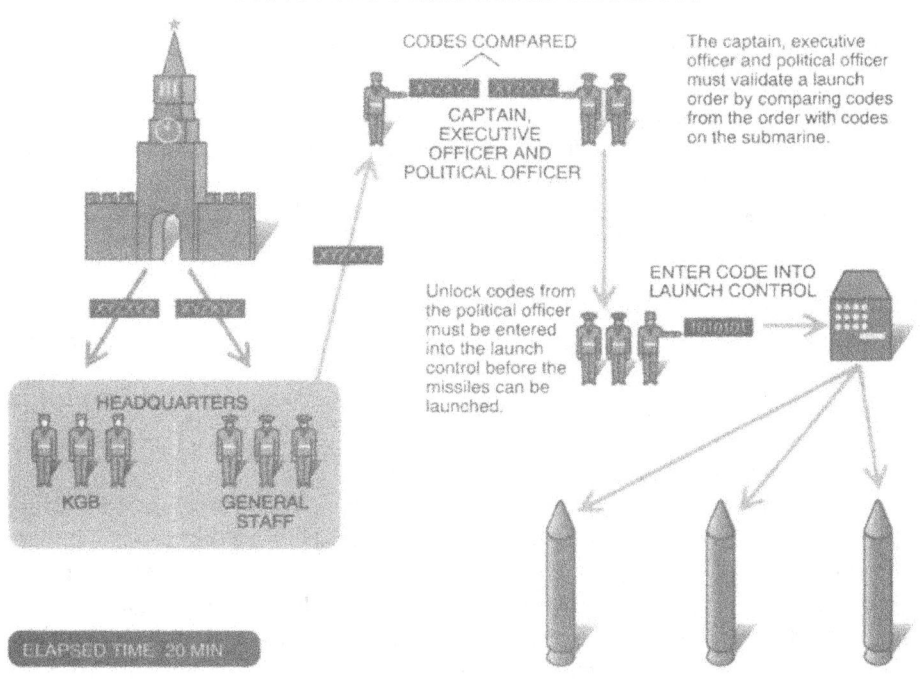

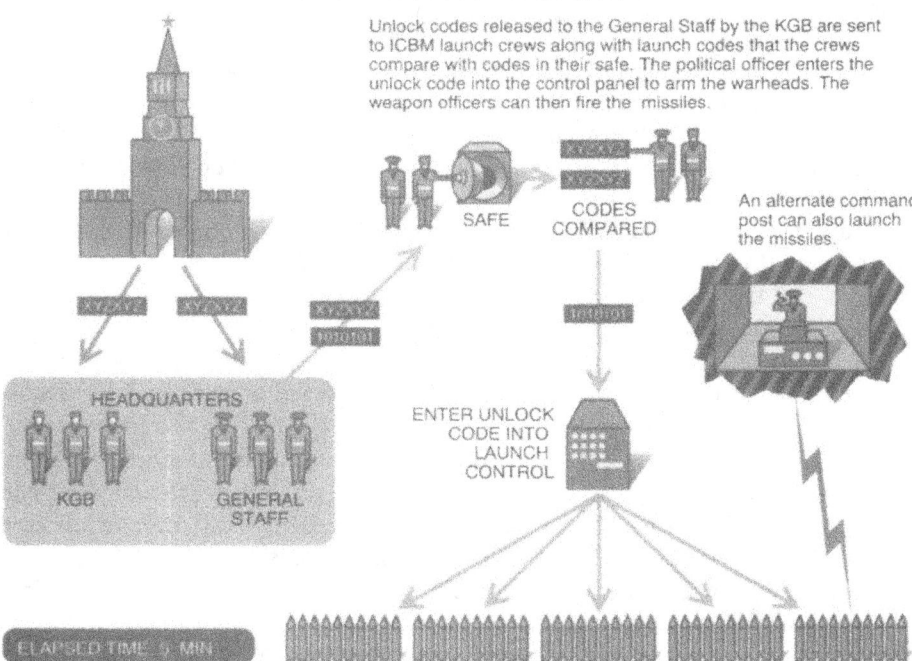

Figure 4. U.S. ICBM launch-ready status

been launched by accident during routine maintenance. Fortunately, it fell a short distance from the launch pad.

Most missiles (including virtually all active strategic missiles belonging to the major nuclear powers) now have warheads attached to them. Launch-on-warning strategies and current alert levels preclude any other configuration. A verifiable agreement to move toward separable warheads and missiles—a common Soviet practice during the 1960s—would reduce the scope for unintended launch. Furthermore, converting weapons to this configuration would be a less drastic, and therefore more palatable, form of arms control.

Restoring the missing parts in time of crisis could be construed as a preparation for attack, thus increasing the possibility of unintended war. On balance, however, the reduced chance of unwanted launch and the easing of tensions resulting from fewer missiles poised for immediate strikes outweigh the possible danger of misperception in a crisis. Furthermore, that risk could be mitigated by agreements that define the conditions for reassembling weapons and by nuclear alert procedures that ensure firm civilian control.

Whatever the alert level, human factors play a key role in the risk of unwanted launch. All those involved in the nuclear weapon chain, from top decision makers to launch officers, are subject to human frailties and instabilities. These frailties are aggravated by work conditions that are boring and isolated. (Duty on missile submarines adds the further stress of trying to adapt to an unnatural 18-hour "day" for the two months of a typical patrol.) Such conditions can sometimes lead to severe behavioral problems, including drug or alcohol use and psychological or emotional instability.

In 1989, for example, of the roughly 75,000 members of the U.S. military with access to nuclear weapons and related components, nearly 2,400 had to be removed from duty. Seven hundred and thirty abused alcohol or drugs, and the rest had psychological or emotional problems, were insubordinate or engaged in criminal behavior. Herbert Abrams of Stanford University has recommended that drug and alcohol use among soldiers with nuclear responsibilities be monitored more closely and that physicians who examine the soldiers be aware of the peculiar nature of their duties. In addition, much more can be done to alleviate stressful working conditions. (All these problems and remedies apply in at least equal measure to the Soviet nuclear apparatus.)

World leaders, meanwhile, are no less subject to stress. They may come to a crisis dependent on alcohol or drugs—whether self-administered or prescribed. Such problems afflicted Winston Churchill and Anthony Eden. Richard M. Nixon was too distraught over Watergate to participate in crucial discussions that resulted in a global U.S. nuclear alert in 1973; senior government officials later took precautions against the possibility that he might act irrationally in his capacity as commander in chief. Even a leader as ruthless as Stalin experienced severe stress during periods of crisis. Such infirmities can lead to behavioral changes, impaired judgment and even irrational actions. Some psychologists have suggested leaders be monitored.

If all these measures fail to prevent an unwanted launch, steps can still be taken to mitigate the consequences. One obvious step is to reduce the number of warheads each missile carries; another is to develop methods for destroying missiles after launch. A third step is to implement launch-detection systems to provide warnings of unintended launches to both sides. Some combination of acoustic and optical sensors would serve for ICBMs; sea-launched ballistic missiles and cruise missiles would require small transmitters that would send signals to relay satellites. Such a detection system might contain provisions for disabling it at the onset of a crisis.

Improvements in the "hot line" (first put in place in 1963) would reduce risks of misperception related to unintended launches, reassembly of missiles and warheads, and other apparently hostile acts. Currently the link between the U.S. and the Soviet Union runs through a Soviet and an American satellite primarily used for civilian communications, with alternate routes provided by cable and radio links. This configuration, however, cannot ensure reliable communication under the adverse conditions in which it is most needed. All parts of the system are vulnerable to nuclear attack. Moreover, the cable routes have been cut inadvertently a handful of times already. Dedicated, radiation-resistant satellites operating in the extremely high frequency band would prevent nuclear static from interrupting signals and would make it possible to continue hot-line conversations through mobile terminals that could accompany relocated command centers.

The critical action, however, is to implement mechanisms for disarming missiles once they have been launched. Costly attempts to develop antiballistic missile systems have led to a dead end. Continuing compliance with the Antiballistic Missile Treaty would limit deployment to 100 interceptors based at Grand Forks, ND. Such a system (at a cost of about $10 billion) could destroy only about 50 reentry vehicles and would be ineffective against cruise missiles or submarine-launched ballistic missiles. A more capable system, exceeding treaty restrictions, would cost far more. Furthermore, there is currently no realistic prospect of pursuing such a project.

The only practical method for stopping a missile after accidental launch is for the country of origin to destroy it or to allow the target country to destroy the warheads prior to impact—a "command destruct system." Indeed, in 1971, the Soviet Union and the U.S. signed a little-known agreement that specifies what each is to do in the event of an unwanted launch. That agreement includes the requirement that the nation "whose nuclear weapon is involved will immediately make every effort to take *necessary* measures to *render harmless or destroy* such weapon without its causing damage."

A typical U.S. system might be based on a coded key automatically generated at launch time to prevent its theft. After the unwanted launch had been verified, the key would be transmitted to the warheads by dedicated relay satellites from a special command and control system located in the National Military Command Center. It would also be sent by hot line to the target country, along with data on warhead trajectories, so that destruct attempts could be made up to the point of reentry. A destruct system could be disabled—by special command units not involved in the launch process—just prior to an authorized launch, to preclude the remote possibility that an adversary would be able to disarm an intentional attack.

Destruct devices should be placed on the warheads themselves to maximize the time available to decision makers. The best place for destruction would be in the midcourse part of the flight, where it would do the least damage.

It would be particularly important, especially during a crisis, that destruction or disarming be verifiable by both sides. ICBMs might emit a coded radio signal or emit a burst of chaff visible to radar. Alternatively, warheads could be fired with minimum yield by draining their tritium before detonation. Cruise missiles might disarm their warheads, then climb to high altitude, transmit a coded signal and then fly into the ocean or the Arctic ice cap, where they could crash or be destroyed by an explosive charge.

Although such a system would add weight and complexity to nuclear reentry vehicles, this penalty is more than balanced by the reduction in the danger of unwanted launch. Nevertheless, the U.S. military and the Defense Department have shown no interest in implementing command destruct systems. This is particularly puzzling in view of attitudes in the Soviet Union: Deputy Foreign Minister Viktor Karpov has told us that Soviet ICBMs already carry command destruct systems.

The risk of accidental nuclear explosions and the worse risk of nuclear war, however low they may be, should be lowered further. The U.S. and the Soviet Union should move promptly to bring about acceptable compromises. There is ample latitude for each country to act independently in this area—most improvements in safeguards can be accomplished unilaterally. The Soviets, for example, have recently moved nuclear weapons from areas of ethnic unrest to storage depots in the Russian Republic. They plan to dismantle ICBM forces in Kazakhstan, the only ethnic region that "hosts" Soviet strategic forces.

It would also be worthwhile for the two governments to exchange views on particular risks of unwanted launch and measures to reduce those risks. Those discussions should be broadened to include other nuclear states. Relations have never been more conducive to fruitful talks on the issue.

Dismantle Armageddon

On May 30, the United States and Russia will no longer aim their strategic missiles at each other. A missile fired accidentally and flying under its own normal motor power and guidance would fall short of its cold-war target, landing instead in the Arctic Ocean.

But the risk of such an accidental launch is low compared to other safety hazards. And this symbolic agreement does nothing to alleviate them.

The real hazards are unauthorized firings, or firings induced by false warning of inbound enemy missiles. These dangers stem from keeping large numbers of strategic missiles in a high state of combat readiness so they can be fired within the 30-minute flight time of opposing missiles. The new plan does nothing to relax this hair-trigger stance.

Under the plan, Russia becomes a reserve target instead of the primary target for U.S. silo-based missiles. Switching back to the original Russian targets is a simple procedure that most missiles need only seconds to complete. This rapid reversibility is supposed to insure that the U.S. missiles can still be fired before Russian missiles could attack them.

But why is this a virtue if rapid reaction is inherently unsafe? The time needed to detect an attack, retarget the missiles and perform the launching procedure would allow a President only a few minutes to consult with advisers and decide whether or not to launch. Under this hasty process, a nuclear war could begin in error. On the other hand, if the attack is real, the slightest delay could result in the destruction of most of our weapons seconds before or after liftoff.

By risking an inadvertent launching while making the survival of land-based missiles precarious, quick-launch tactics put enormous pressure on commanders at all levels. Yet the U.S. and Russia have geared their nuclear strategies to beat a 30-minute deadline. For the U.S., such rapid reaction is necessary to destroy the excessive number of targets required by Presidential directive.

President Clinton could put his good intentions to better work with an agreement that obliges both parties to strip out all the cold-war targets from every missile's

---

By Bruce G. Blair and Henry W. Kendall, 1994. Copyright © 1994 by *The New York Times*. Reprinted with permission.

memory. Eliminating reserve as well as primary targets would strengthen safeguards against unauthorized launchings and make land-based missiles unable not only to respond quickly to attack, but also to initiate a sudden massive attack.

Other steps should be taken. In 1991 President George Bush ordered a unilateral reduction in the readiness of the older Minuteman missiles. Maintenance crews went into each silo and inserted a pin that blocks the motor's ignition. The same could be done with other missiles. Another step is to shut off the power to the missiles. A minimum of several days would be needed to reverse these procedures.

These mutual obligations could be verified by U.S.-Russian crews at designated command centers in both countries. Such monitoring would insure that any illicit retargeting of weapons would be detected in time to prevent the cheater from gaining a decisive advantage.

Farsighted leaders would also take the next logical step: removing all warheads from land-based missiles and removing critical components from sea-based strategic missiles. This would solve the problem of inadvertent or unauthorized launchings and add a safety cushion against coups or irrationality in Moscow.

Keeping missiles on a hair trigger is inherently dangerous. World leaders need to confront the danger with a comprehensive remedy that eliminates the chance of a catastrophic failure of nuclear control. Pointing missiles at the oceans is a timid and feeble beginning.

PART FOUR
GLOBAL ENVIRONMENTAL AND RESOURCE ISSUES

Energy and Greenhouse Gas Issues

By the late 1970s, following the ECCS hearings and the demise of the AEC, cascading problems besetting nuclear power led me and UCS colleagues to ask, If not nuclear power, what else? We set in motion a study to evaluate the potential role of alternative, renewable energy sources, energy conservation and improved end-use efficiency in U.S. and international energy supplies.[1] It was the first move in what remains a major UCS program in energy policy. The program involves comprehensive assessments of renewable energy alternatives and production of a wide range of public educational material, coupled with regional and national lobbying efforts to advance energy efficiency and development of renewable energy technologies.

In the late 1980s, I drew UCS's attention to the issues of carbon dioxide-induced global warming and the resulting prospect of climatic disruption. Energy-related activities are the largest source of greenhouse gases, and seizing this issue was a direct outgrowth of our energy program. We had become aware that unusually warm weather patterns were of considerable concern to the scientific community. Moreover, as a consequence of record temperatures and humidity in the summer of 1988, accompanied by numerous heat-related injuries and fatalities, they were becoming a concern to the American public. With the aim of strengthening the administration's response to global warming concerns and to coincide with a Washington meeting of the Intergovernmental Panel on Climate Change, UCS prepared the "Appeal by American Scientists to Prevent Global Warming." Signed by 52 Nobel laureates and over 700 members of the National Academy of Sciences, it was released to the press on February 1, 1990, along with a letter to President George Bush. The *appeal* is reprinted here along with a partial list of signers. As has been true of many UCS declarations, the signers represented a virtual blue book of the science community.

Two years later, with the Rio Earth Summit a few months away, another opportunity arose from UCS's role as a principal spokesman for the nongovernmental organizations (NGOs) that were monitoring the negotiations prior to the meeting. I was asked to present to the UN Intergovernmental Negotiating Committee, on behalf of UCS and the environmental NGOs, the group's views on the important objectives and priorities for the Rio conference. This speech is included here. Along with other global warming activities by UCS, it amounts to "chipping away at the rock." One usually cannot discern the benefit of such activity, but, in the right circumstances, a long series of similar actions, initiated by numerous groups, can have impact. UCS shortly thereafter launched a major sustained program and effort in this field. As this book is being written, UCS is in the midst of yet another public outreach program on global warming.

The Scientists' Warning

A chain of events started in 1991 that opened up a number of new activities for me and as well as for the Union of Concerned Scientists. In the late spring of that year, the Nobel Foundation, whose responsibilities are the administration of the Nobel prizes and the activities related to their award, posed a question. Would I be interested in participating in a debate, in Stockholm, being planned as part of a celebration of the ninetieth anniversary of the first Nobel awards, the Nobel Jubilee, to which all living laureates would be invited? It was an invitation I found hard to refuse, for the subject was whether the human race was using its collective intelligence to build a better world, a subject that was clearly very close to my heart. Discussions concerning the organization of the debate, eventually referred to as the Great Nobel Debate, finally concluded with my being selected as the spokesman for the negative side of the proposition to be debated. Archbishop Desmond Tutu was selected as the spokesman for the positive side.

I selected, as the general approach a synoptic review of the environmental pressures, resource management and population issues and how a continuation of current practices would affect succeeding generations. I spent several months of reasonably intense effort collecting material and talking to a wide variety of scientists, especially from the life sciences.

Two unexpected matters surfaced in the course of this effort. The first, well known to specialists but of which I had been only dimly aware, was how stressed the global food-producing sector was. A great mass of problems has come to bear on agriculture that will make it quite difficult to provide food for the coming expansion of population. The second was the great distress expressed by biologists whose interests were in ecosystems studies, population dynamics, and other fields that touched on the consequences of the destruction of species and resulting loss of global biodiversity. This made a deep impression on me. I concluded that this scientific community appeared ready to speak out publicly. Indeed, in the light of its great unhappiness, it was puzzling that it had not spoken out powerfully, as a community, before.

Little need be said of the Great Debate itself. It turned out that the Archbishop and I did not disagree with each other's position; his "Yes" was related to the defeat of apartheid in his own country, the peaceful dissolution of the Soviet Union, and the discrediting of Communism, which had just taken place. My "No" was based on my analysis of environmental issues. This agreement left little room for argument among the eight members of the debate panels, so the debate was far from contentious and allowed a fuller discussion of the issues by panelists and from the floor.

My preparation for the debate set the stage for two continuing activities. In the months after the event, I searched out a potential collaborator to help in preparing a synoptic analysis of the global food problems. There is, as is commonly known, no better way to get a reasonably thorough start on a new and contentious subject than to write a scientific paper. Both friends and opponents can be depended on to provide searching and critical comments and suggestions. By midwinter 1992, David Pimentel, a Cornell University entomologist and agricultural expert, and I were at work on "Constraints on the Expansion of the Global Food Supply" reprinted here.

Also in that winter, I began preparation of a declaration, the "World Scientists' Warning to Humanity," having concluded that not only the biologists were distressed and that there were no doubt many other scientists who would support a broad statement of concern. The document required many months to prepare, in soliciting sponsors and sending the mailing to National Academy-level scientists throughout the world. It was signed by over 1700 scientists from 71 countries, including 104 Nobel laureates, and was released in November 1992 by the Union of Concerned Scientists. It is reprinted here, accompanied by the news release that launched it as well as a summary of its principal signers.

The warning, as one of my colleagues remarked, has had "legs," real lasting power. It has, over some five years, become UCS's most widely requested publication, and an estimated 150,000 copies have been distributed. It has been quoted and cited extensively in both scientific and popular publications and has become the document of record in setting forth the consensus of the scientific community on some of the most vexing problems facing humanity. It continues to be cited and reprinted.

As one consequence of the "Constraints" paper, I was invited to join the steering and organizing committee for an international conference on problems of food and agriculture by Per Pinstrup-Andersen, the Executive Director of the International Food Policy Research Institute. This was an opportunity to meet international food and agriculture specialists and so was an invitation important to accept. More opportunities stemmed from a Stockholm meeting of the committee. I was approached by Ismail Serageldin, Vice President of the World Bank for Environmentally Sustainable Development, who asked if I would be willing to organize a panel of experts on global problems of interest to the Bank for the Bank's October 1995 annual meeting in Washington. It was an interesting opportunity, for the audience would be many of the finance and environmental ministers of the developing nations that the Bank's funding helped support. This project took on something of a life of its own, expanding to two panels and a publication. The two chapters in this publication that I prepared are reprinted here.

Evidently the Bank perceived the panels as useful, for in early 1996 Mr. Serageldin issued another invitation: Would I organize and run a panel to study and prepare a report on transgenic crops to help provide some guidance for the Bank in the array of issues raised by the powerful new technology of genetic engineering? I again agreed; the result, *The Bioengineering of Crops: Report of the World Bank Panel on Transgenic Crops* was completed and submitted in the summer of 1997 and is also reprinted here.

A conference organized by the Bank in October 1997[2] was devoted to the bioengineering issue. At the conference, Mr. Serageldin announced that all of the report's recommendations had been accepted by the Bank and that the organization had already started to implement them. It was a pleasing reward to all of the authors for the time and energy they had devoted to the effort.

References and Notes

1. *Energy Strategies—Toward a Solar Future*, edited by Henry W. Kendall and Steven J. Nadis. Contributors: Daniel F. Ford, Carl J. Hocevar, David J. Jhirad, Henry W. Kendall, Ronnie D. Lipschutz and Steven J. Nadis (Ballinger Publishing Company, Cambridge, MA, 1980).
2. *Biotechnology and Biosafety*, Proceedings of an Associated Event of the Fifth Annual World Bank Conference on Environmentally and Socially Sustainable Development, "Partnerships for Global Ecosystems Management: Science, Economics and Law," Ismail Serageldin and Wanda Collins, Editors (World Bank, Washington, July 1998).

Appeal by American Scientists to Prevent Global Warming

Global warming has emerged as the most serious environmental threat of the 21st century. There is broad agreement within the scientific community that amplification of the earth's natural greenhouse effect by the buildup of various gases introduced by human activity has the potential to produce dramatic changes in climate. The severity and rate of climate change cannot yet be confidently predicted, but the impacts of changes in surface temperature, sea level, precipitation, and other components of climate could be substantial and irreversible on a time scale of centuries. Such changes could result in severe disruption of natural and economic systems throughout the world.

More research on global warming is necessary to provide a steadily improving data base and better predictive capabilities. But uncertainty is no excuse for complacency. In view of the potential consequences, actions to curb the introduction of greenhouse gases, including carbon dioxide, chlorofluorocarbons, methane, nitrogen oxides, and tropospheric ozone, must be initiated immediately. Only by taking action now can we insure that future generations will not be put at risk.

The United States bears a special responsibility to provide leadership in the prevention of global warming. It is the world's largest producer of greenhouse gases, and it has the resources to make a great contribution. A thoughtful and vigorous U.S. policy can have a direct beneficial effect and set an important example for other nations.

The United States should develop and implement a new National Energy Policy, based on the need to substantially reduce the emission of carbon dioxide, while sustaining economic growth. The cornerstones of this policy should be energy efficiency and the expansion of clean energy sources.

This policy should include:

1. A steady increase in motor vehicle fuel economy standards, while the search continues for fuels and other technologies that mitigate carbon dioxide impact

---

Union of Concerned Scientists, 1990.

2. A substantial increase in federal funding for research on energy-efficiency technologies, as well as federal activities to enhance the adoption of more efficient energy use
3. Development, demonstration, and commercialization of renewable-energy technologies on a massive scale
4. A nuclear energy program that emphasizes protection of public health and safety, resolution of the problem of radioactive waste disposal, and stringent safeguards against the proliferation of nuclear material and technology that can be applied to weapons construction
5. Full consideration of environmental, social, and economic impacts in the establishment of federal subsidies and regulatory standards for development of energy sources.

These measures, along with others designed to curtail the use of chlorofluorocarbons and promote prudent agricultural and reforestation practices, can form the basis for the lowering of greenhouse gas emissions in the United States and other nations. They will provide other, worthwhile benefits to the nation as well, such as more diverse and flexible energy supplies, reduced dependency on imported oil, and the creation of new energy technologies for export and sale in the international marketplace.

Partial List of Signatures

‡ Christian B. Anfinsen, Department of Biology, Johns Hopkins University
‡ Julius Axelrod, National Institute of Mental Health
■ Howard L. Bachrach, Plum Island Animal Disease Center
■ John Backus, IBM Almaden Research Center; developer of FORTRAN
‡ David Baltimore, Director, Whitehead Institute of Biomedical Research
■ Paul Bartlett, Department of Chemistry, Harvard University
‡■ Paul Berg, Wilson Professor of Biochemistry, Stanford University
‡■ Hans A. Bethe, Newman Laboratory for Nuclear Studies
‡ J. Michael Bishop, Microbiology and Immunology Department, University of California San Francisco
‡■ Konrad E. Bloch, Department of Chemistry, Harvard University
‡■ Nicolaas Bloembergen, Department of Physics, Harvard University
   Norris Edwin Bradbury, former Director, Los Alamos Science Lab
■ E. Margaret Burbidge, University Professor, Astrophysics and Space Science, Univ. of Calif. San Diego

---

‡ Nobel Laureate
■ National Medal of Science

- ‡ Melvin Calvin, Dept. of Chemistry, University of California Berkeley
- ‡ Thomas R. Cech, Chem. and Biochem. Dept., University of Colorado
- Joseph Chamberlain, Space Physics and Astronomy, Rice University
- ‡ Owen Chamberlain, Lawrence Berkeley Laboratory, Univ. of Calif.
- ‡■ Subrahmanyan Chandrasekhar, Lab. for Astrophysics and Space Research, University of Chicago
- ■ Erwin Chargaff, Prof. of Biochemistry, Emeritus, Columbia University
- ■ Morris Cohen, Institute Prof., Emer., Materials Sci. and Engineering, MIT
- ■ Mildred Cohn, Biochem. & Biophysics, Univ. of Pennsylvania Medical School
- ■ George B. Dantzig, Operations Research, Stanford University
- ■ Hallowell Davis, Central Institute for the Deaf
- Margaret B. Davis, Ecology, Evolution, & Behavior Dept., Univ. of Minnesota
- ‡ Gerard Debreu, Dept. of Economics, University of California Berkeley
- ‡ Hans G. Dehmelt, Department of Physics, University of Washington
- ■ Robert H. Dicke, Joseph Henry Laboratory, Princeton University
- ■ Theodore O. Diener, Plant Virology Laboratory, Agricultural Research Ctr.
- ■ Carl Djerssel, Department of Chemistry, Stanford University
- ■ Joseph L. Doob, Department of Mathematics, University of Illinois
- ■ Harold E. Edgerton (deceased), Dept. of Electrical Engineering, MIT; EG&G Inc.
- Paul Ehrlich, Dept. of Biological Sciences, Stanford University
- ‡ Gertrude Ellon, Burroughs-Wellcome Company
- ■ Walter Elsasser, Dept. of Earth and Planetary Sci., Johns Hopkins Univ.
- Katherine Esau, Dept. of Biological Sciences, Univ. of Calif. Santa Barbara
- ■ Herman Feshbach, Department of Physics, MIT
- ‡ Val L. Fitch, Joseph Henry Laboratory, Princeton University
- ‡■ William A. Fowler, Institute Professor of Physics, Emeritus, Cal. Tech.
- ■ Herbert Friedman, Hulburt Ctr. Space Research, UCN Research Laboratory
- Robert C. Gallo, National Cancer Institute, National Inst. of Health
- ‡ Donald A. Glaser, Prof. of Physics and Microbiology, Univ. of Calif. Berkeley
- Marvin L. Goldberger, Director, Institute for Advanced Study
- Gertrude Goldhaber, Dept. of Physics, Brookhaven National Laboratory
- ■ Maurice Goldhaber, Dept. of Physics, Brookhaven National Laboratory

- Herman H. Goldstine, American Philosophical Society
  Stephen Jay Gould, Department of Zoology, Harvard University; columnist, *Natural History*
- ‡■ Roger Guillemin, The Whittler Institute
  James Hansen, Atmospheric Physicist, Adjunct Professor, Columbia Univ.
  Elizabeth Hay, Dept. of Anatomy and Cell Biology, Harvard Medical School
- ■ Michael Heidelberger, Department of Pathology, NYU School of Medicine
- ■ Donald A. Henderson, Sch. of Hygiene & Public Health, Johns Hopkins Univ.
- ‡ Dudley Herschbach, Department of Chemistry, Harvard University
- ‡■ Roald Hoffmann, Department of Chemistry, Cornell University
- ‡■ Robert Hofstadter, Max Stein Prof. of Physics, Emeritus, Stanford University
  John P. Holdren, Professor of Energy and Resources, Univ. of Calif. Berkeley
- ‡ Robert W. Holley, Salk Institute for Biological Studies
- ‡ David H. Hubel, Department of Neurobiology, Harvard Medical School
- ■ William S. Johnson, Jackson-Wood Prof. of Chemistry, Emer., Stanford U.
- ‡■ Jerome Karle, Lab. for the Structure of Matter, USN Research Laboratory
  Isabella Karle, Lab. for the Structure of Matter, USN Research Laboratory
  Carl Kaysen, Science, Technology, and Society Program, MIT
- ■ Joseph B. Keller, Department of Mathematics, Stanford University
  Donald Kennedy, President, Stanford University
- ‡■ H. Gobind Khorana, Department of Chemistry, MIT
- ■ E. F. Knipling, Research Entomologist and Agricultural Administrator
- ■ Walter Kohn, Dept. of Physics, University of California Santa Barbara
- ■ Edwin H. Land, Rowland Inst. of Sci.; Chairman, Polaroid Corporation
- ■ Paul C. Lauterbur, Department of Chemistry, University of Illinois
- ■ Peter D. Lax, Courant Institute of Mathematics, New York University
  Stephen P. Leatherman, Center for Global Change, Univ. of Maryland
- ‡■ Leon M. Lederman, Department of Physics, University of Chicago
- ‡■ Yuan T. Lee, Dept. of Chemistry, University of California Berkeley
- ‡ Wasslly Leontief, University Professor of Economics, New York University
- ‡■ Rita Levi-Montalcini, Institute of Neurobiology, CNR
- ‡ William N. Lipscomb, Department of Chemistry, Harvard University
  Amory B. Lovins, Rocky Mountain Institute
  Lynn Margulis, Dept. of Biology, University of Massachusetts Amherst

- Ernst Mayr, Museum of Comparative Zoology, Harvard University
‡ R. Bruce Merrifield, Professor of Biochemistry, Rockefeller Univ.
- Neal E. Miller, Department of Psychology, Yale University
  Irving Mintzer, Center for Global Change, University of Maryland
  William Moomaw, Center for Environ. Management, Tufts University
- Walter H. Munk, Inst. of Geology and Planetary Physics, U. Cal. San Diego
- James V. Neal, Human Genetics Department, University of Michigan
‡- Severo Ochos, Centro de Biologica Molecular, Universidad Autonoma
- George E. Pake, Institute for Research on Learning
‡- George E. Palade, Sterling Professor, Emer., Cell Biology, Yale University
- Wolfgang K. H. Panofsky, Dir. Emer., Stanford Linear Accelerator Center
  Mary Lou Pardue, Department of Biology, MIT
‡- Linus Pauling, President, Linus Pauling Institute
‡ Arno A. Penzlas, AT&T Bell Laboratories
  Veerabhadran Ramanathan, Dept. of Geological Sciences, Univ. of Chicago
‡- Norman F. Ramsey, Higgins Professor of Physics, Harvard University
- Frederick Reines, Distinguished Prof. of Physics, Emer., Univ. of Calif. Irvine
  Roger Revelle, Prof. of Sci., Tech., and Public Policy, Univ. of Calif. San Diego
‡ Burton Richter, Stanford Linear Accelerator Center
  S. Dillon Ripley, Secretary Emeritus, Smithsonian Institution
‡ Frederick C. Robbins, Dean Emeritus, Case Western Res. Univ. Med. Sch.
- Wendell L. Roelofs, Department of Entomology, NY St. Agr. Exp. Stn.
  Arthur Rosenfeld, Department of Physics, Univ. of California Berkeley
- Bruno Rossi, Center for Space Research, MIT
  Elizabeth S. Russell, Jackson Laboratory
  Carl Sagan, Dept. of Astronomy, Cornell Univ.; author, *Cosmos*
  Ruth Sager, Dana-Farber Cancer Institute
  Stephen Schneider, National Center for Atmospheric Research
‡ Melvin Schwartz, Digital Pathways
‡- Julian Schwinger, University Professor of Physics, Emeritus, UCLA
‡ Glenn T. Seaborg, Lawrence Berkeley Laboratory, University of California
  Phillip A. Sharp, Director, Center for Cancer Research, MIT
‡ George D. Snell, Senior Staff Scientist, Emeritus, Jackson Laboratory
‡ Roger W. Sperry, Trustee Professor, Emeritus, Calif. Inst. of Technology

James Gustave Speth, World Resources Institute
- ■ Earl R. Stadtman, Laboratory of Biochemistry, Nat'l. Inst. of Health
- ■ G. Ledyard Stebbins, Department of Genetics, Univ. of Calif. Davis
- ‡■ Jack Steinberger, European Center for Nuclear Research
- ‡ Joan A. Steitz, Molecular Biophysics and Biochem. Dept., Yale University
- ■ Walter H. Stockmayer, Emeritus Prof. of Chemistry, Dartmouth College
- ‡■ Henry Taube, Department of Chemistry, Stanford University

  Lewis Thomas, President Emeritus, Sloan-Kettering Cancer Institute; author, *The Medusa and the Snail*
- ‡ James Tobin, Department of Economics, Yale University
- ‡■ Charles H. Townes, Department of Physics, Univ. of Calif. Berkeley

  Karen K. Uhlenbeck, Department of Mathematics, University of Texas
- ‡ Harold E. Varmus, Microbiology and Immunology Dept., Univ. of Cal. SF
- ‡ George Wald, Biological Laboratories, Harvard University
- ■ Ernst Weber, Emeritus President, Polytechnic Institute of Brooklyn
- ■ Victor Welsskopf, Institute Prof., Emer., MIT; former Director-General CERN
- ‡ Thomas Weller, Professor Emeritus, Harvard School of Public Health

  Jerome B. Wiesner, President Emeritus, MIT
- ■ Edward O. Wilson, Museum of Comparative Zoology, Harvard Univ.
- ‡ Robert W. Wilson, Radio Physics Department, AT&T Bell Laboratories

This list includes a few of the better-known Signatories to the Appeal. The full list includes 52 Nobel laureates, 67 National Science Medalists, and over 700 members of the National Academy of Sciences. The Appeal was delivered to President George Bush on February 1, 1990, and presented by UCS before the Intergovernmental Panel on Climate Change on February 6.

Affiliations are for the purpose of identification only. Endorsement of this statement does not imply agreement with all positions or activities of UCS.

Speech to the United Nations Intergovernmental Negotiating Committee

On behalf of the numerous environmental organizations from all over the world participating in this fifth session of the Intergovernmental Negotiating Committee, I thank you for this opportunity to share our views on the situation now confronting you—and the world—as this session comes to a close.

Time is running out. As the distinguished delegate from Pakistan told you earlier this week, the environmental clock stands at a quarter to midnight. There is an urgent need to bring an effective climate change treaty to the Rio conference in June, not just an empty framework. The success or failure of these negotiations will also set the tone for the many other environmental and development challenges that confront developing and industrial countries alike. Given the high stakes, failure of these talks would be costly and irresponsible.

It is no surprise that I am greatly disappointed that, despite more than a year of hard work, you have yet to reach agreement on the major issues. I am saddened that the main responsibility for this state of affairs belongs to my own country. The refusal of the United States to make a commitment to reduce, or even to stabilize, its huge emissions of carbon dioxide and other greenhouse gases has been the principal roadblock to progress. Yesterday's announcement of a new "action plan" by the US delegation was useful, but we must have a firm US commitment on emissions limits by the time these negotiations resume later this spring.

In early 1990, I transmitted to President Bush an appeal for action to prevent global warming on behalf of some 55 American Nobel Laureates and more than 700 members of the National Academy of Sciences. It said, in part, that "uncertainty is no excuse for complacency. Only by taking action now can we assure that future generations will not be put at risk."

Henry W. Kendall, 1992.

Other industrial countries must also do more, both to increase their commitments to limit greenhouse gas emissions, and to reach agreement within the OECD [Organization for Economic Cooperation and Development] to stabilize and reduce emissions and to provide financial assistance to developing countries. Industrial nations must not use US intransigence as an excuse for backsliding themselves.

Dr. Bert Bolin's statements to you last week were sharp and to the point, and so I know you are well aware of the broad consensus within the knowledgeable scientific community as to the likely range and consequences of global warming. We scientists are cautious by nature, hesitant to sound the alarm until we are confident of our findings. When a scientist of Dr. Bolin's caliber tells you, on the basis of three years of analysis by several hundred of the world's climate scientists, that "more far-reaching efforts are required than are now being contemplated in order to achieve a major reduction of the rate of carbon dioxide increase in the atmosphere," it is significant. Let me add that in my discussions with a number of the world's leading scientists in recent months, I see an increasing worry on this and other issues.

Dr. Bolin is right: the stabilization of carbon dioxide emissions at 1990 levels by the year 2000, which seems to represent the "upper edge of the possible" in these negotiations, is inadequate to meet the threat of global warming. Based on the IPCC [Intergovernmental Panel on Climate Change] analysis, it is our view that reductions of at least 20 percent by 2000 and 60 percent or more by 2025 are needed if we are to lower the risk of unacceptable damage to natural ecosystems and humanity. Accordingly, the treaty should require further negotiations on emissions reduction targets at the earliest possible date.

Some wish to view the uncertainties in the climate change predictions as a reason to postpone action. This is imprudent. We must ensure against the risk of the expected climate response to greenhouse gas buildup. No military planner would dismiss a potentially grave military threat on the basis that the expectation was somewhat uncertain—neither should we dismiss such a peacetime threat as we now face.

Others argue that an adequate response to climate change would be damaging to national economies. Much of this comes from those whose interests reflect the world's addiction to fossil fuels. The fact is that increased energy efficiency, and a move reliance on renewable energy sources, will enhance, not harm, economic prosperity while doing much to slow the steady buildup of greenhouse gases. Numerous studies by government and nongovernmental groups alike have documented the enormous potential for cost-effective reductions in carbon dioxide emissions. A recent analysis by my own organization, the Union of Concerned Scientists, and several others demonstrates that the United States could cut energy-related carbon dioxide emissions by 25 percent by 2005, and 70 percent by 2030, *at a net savings of $2 trillion to US energy consumers!* The scope of energy savings and the potential for use of renewable resources vary from country to country, but they are substantial in virtually every nation. The climate treaty you negotiate should aggressively exploit the great potential of energy efficiency and renewable energy technologies, for both developed and developing countries.

While all countries must help avert the worst of the global warming threat, the primary responsibility lies with the industrial nations of the world. They are the sources of the bulk of the world's greenhouse gas emissions. They also have an obligation to provide financial resources and access to appropriate technologies (including development of indigenous technologies and capacity) so that the developing nations of the world can meet their legitimate development needs without greatly boosting greenhouse gas emissions. This is by no means charity, or a handout but is a sound investment in a more equitable and environmentally sustainable world for our children and grandchildren and, I would add, very likely a more peaceable world. Of course, these measures must be accompanied by broader reform of international trade, aid, and lending patterns that, in most cases, favor the industrialized countries.

You are meeting in the headquarters of the United Nations. If a nation or group of nations were invading a small country, the Security Council would now be in emergency session, a multinational peacekeeping force would be organized, and the unjustified aggression would be resisted with all the means at the world community's disposal.

The global warming threat demands no less. The greenhouse gas emissions of the rich and powerful nations of the world pose a serious threat to the very survival of many of the world's small island states, along with the well-being of natural ecosystems and human communities over the whole world. Yet because this injury takes place slowly, over years, even decades, and because of the industrialized world's addiction to fossil fuels, no emergency sessions are held, no firm response is organized.

This inactivity must now end. It is unacceptable to the people of the world that these negotiations now teeter at the edge of collapse. Urgent consultations must be held, at the highest levels of government, within and between nations (particularly the OECD nations, whose disunity has brought us to this point), over the next two months. A substantial shift in the US negotiation stance is particularly critical. When these negotiations resume later this spring, it must be with a new spirit of determination and cooperation that will enable agreement to be reached on an *effective* climate change treaty in time for signature in Rio.

A weak treaty will not do. Neither will postponement beyond Rio of these tough decisions so urgently needed.

I thank you for your attention, and wish you well in your efforts over the weeks ahead. Mr. Chairman, distinguished delegates, we are, all of us, quite literally in your hands.

PRESS RELEASE
Union of Concerned Scientists, November 18, 1992

Announcing World Scientists' Warning to Humanity

Many of the world's most prominent scientists, including a majority of the science Nobel laureates, appealed to the peoples of the world to take immediate action to halt the accelerating damage threatening humanity's global life support systems. The World Scientists' Warning to Humanity, signed by 1,575 scientists from 69 countries, has been sent to government leaders of all nations.

The Warning catalogs the damage already inflicted on the atmosphere, oceans, soils, and living species, and stresses that continuation of many destructive human activities "may so alter the living world that it will be unable to sustain life in the manner that we know. . . . A great change in our stewardship of the earth and the life on it, is required, if vast human misery is to be avoided and our global home on this planet is not to be irretrievably mutilated."

Dr. Henry Kendall, a Nobel laureate (1990, Physics), and Chairman of the Union of Concerned Scientists, coordinated the appeal. Signers include 99 out of the 196 living scientists who are Nobel laureates, as well as senior officers from many of the most prestigious scientific academies in Europe, Asia, North America, Africa, and Latin America.

"This kind of consensus is truly unprecedented," remarked Dr. Kendall. "There is an exceptional degree of agreement within the international scientific community that natural systems can no longer absorb the burden of current human practices. The depth and breadth of authoritative support for the Warning should give great pause to those who question the validity of threats to our environment."

The appeal focuses on the environmental and resources damage caused by overconsumption in the industrialized countries—the world's largest polluters—and the pressures on the environment caused by poverty and spiralling populations in the developing world. The scientists emphasize the urgency of the problem. As they note in the appeal, "No more than one or a few decades remain before the chance to avert the threats we now confront will be lost and the prospects for humanity immeasurably diminished."

Calling for a new ethic in caring for ourselves and the earth, the signers of the Warning ask for the help of their scientific colleagues, heads of governments, business and industrial leaders, religious leaders, and citizens throughout the world. Specifically, the Warning identifies five critical tasks:

- Exerting control over environmentally damaging activities to restore and protect the integrity of the earth's systems that we depend on
- More effective management of resources crucial to human welfare
- Improving social and economic conditions in developing nations and expanding access to effective, voluntary family planning in order to stabilize the world's population
- Reducing and eventually eliminating poverty
- Guaranteeing sexual equality, including recognizing women's control over their own reproductive decisions.

These areas are inextricably linked and must be addressed simultaneously. The Warning documents the need for prompt and comprehensive solutions that address the root causes of environmental damage and unrestrained population growth. It notes the obligation of the developed nations to provide aid and support to developing nations.

"We are endangering the ability of our world to support humanity," notes Dr. Kendall. "We simply cannot continue the course we're on; nature won't allow it. We must all pay careful attention to the words of this distinguished group of scientists and act before it is too late."

Signatories

UCS circulated the "Warning" to senior scientists around the world. In an effort to demonstrate consensus among scientists both from developed and developing countries, endorsement was sought from members of the leading scientific academies in North America, Latin America, Europe, Africa, and Asia. The response was very great: more than 1600 signatories from 70 countries. There should be no doubt that a broad consensus is emerging within the senior scientific community regarding the validity of major threats to the future well-being of humanity and the global environment.

As indicated, the quantity and calibre of signatories to the statement is quite unprecedented:

105 scientists who have been awarded the Nobel Prize;

A substantial number of senior officers from national and international science academies (e.g., Third World Academy of Sciences, Brazilian Academy of Sciences, Royal Society of London, Chinese Academy of Sciences),

A majority of the members of the Pontifical Academy of Sciences (the organization that advises Pope John Paul II on scientific issues),

Total Number of Signatories as of February 1, 1993

1,680—all signatories are fellows or members of one or more national or international science academies

Nobel Laureate Scientists

105—a majority of living scientists who have been awarded the Nobel prize

Total Number of Countries Represented

70

Signatories from Leading Scientific Academies or Associations

Regional or Global

Carlos Chagas
  President, Latin American Academy of Sciences
  Former President, Pontifical Academy of Sciences

Mahdi Elmandjra
  Vice President, African Academy of Sciences

Mohammed H. A. Hassan
  Exec. Secretary, Third World Academy of Sciences

Frederico Mayor
  President, International Council of Scientific Unions

Abdus Salam
  President, Third World Academy of Sciences

José Israel Vargas
  President, Third World Academy of Sciences (1995–1996)

National

Carlos Aguirre
President, Bolivian Academy of Sciences

Michael Atiyah
President, Royal Society (London)

Adolphe Butenandt
Former President, Max-Planck Institut, Germany

Ennio Candotti
President, Brazilian Society for the Advancement of Science

Johanna Döbereiner
First Secretary, Brazilian Academy of Sciences

Dagfinn Follesdal
President, Norwegian Academy of Sciences

Konstatin Frolov
Vice President, Russian Academy of Sciences

Carl-Olof Jacobson
Secretary-General, Royal Swedish Academy of Sciences

Torvard Laurent
President, Royal Swedish Academy of Sciences

Leon Lederman
Chairman, American Assoc. for the Advancement of Science

Digby McLaren
Former President, Royal Society of Canada

Gennady Mesiatz
Vice President, Russian Academy of Sciences

Gustavo Rivas Mijares
Former President, Venezuelan Academy of Sciences

Oleg M. Nefedov
Vice President, Russian Academy of Sciences

Cyril Agodi Onwumechili
Former President, Academy of Sciences, Nigeria

Yuri Ossipyan
Vice President, Russian Academy of Sciences

Autar Singh Paintal
Former President, Indian National Academy of Sciences

Sherwood Rowland
   President, American Assoc. for the Advancement of Science

Zhou Guang Zhao
   President, Chinese Academy of Sciences

Other Nobel Laureate Signatories

Camila José Cela
   Spain

Nadine Gordimer
   South Africa

Archbishop Desmond Tutu
   South Africa

Note: Affiliations are listed for identification purposes only.

World Scientists' Warning to Humanity

Introduction

Human beings and the natural world are on a collision course. Human activities inflict harsh and often irreversible damage on the environment and on critical resources. If not checked, many of our current practices put at serious risk the future that we wish for human society and the plant and animal kingdoms, and may so alter the living world that it will be unable to sustain life in the manner that we know. Fundamental changes are urgent if we are to avoid the collision our present course will bring about.

The Environment

The environment is suffering critical stress.

The Atmosphere

Stratospheric ozone depletion threatens us with enhanced ultra-violet radiation at the earth's surface, which can be damaging or lethal to many life forms. Air pollution near ground level, and acid precipitation, are already causing widespread injury to humans, forests and crops.

Water Resources

Heedless exploitation of depletable ground water supplies endangers food production and other essential human systems. Heavy demands on the world's surface waters have resulted in serious shortages in some 80 countries, containing 40% of the world's population. Pollution of rivers, lakes and ground water further limits the supply.

Union of Concerned Scientists, 1992.

Oceans

Destructive pressure on the oceans is severe, particularly in the coastal regions which produce most of the world's food fish. The total marine catch is now at or above the estimated maximum sustainable yield. Some fisheries have already shown signs of collapse. Rivers carrying heavy burdens of eroded soil into the seas also carry industrial, municipal, agricultural, and livestock waste—some of it toxic.

Soil

Loss of soil productivity, which is causing extensive land abandonment, is a widespread byproduct of current practices in agriculture and animal husbandry. Since 1945, 11% of the earth's vegetated surface has been degraded—an area larger than India and China combined—and per capita food production in many parts of the world is decreasing.

Forests

Tropical rain forests, as well as tropical and temperate dry forests, are being destroyed rapidly. At present rates, some critical forest types will be gone in a few years, and most of the tropical rain forest will be gone before the end of the next century. With them will go large numbers of plant and animal species.

Living Species

The irreversible loss of species, which by 2100 may reach one third of all species now living, is especially serious. We are losing the potential they hold for providing medicinal and other benefits, and the contribution that genetic diversity of life forms gives to the robustness of the world's biological systems and to the astonishing beauty of the earth itself.

Much of this damage is irreversible on a scale of centuries or permanent. Other processes appear to pose additional threats. Increasing levels of gases in the atmosphere from human activities, including carbon dioxide released from fossil fuel burning and from deforestation, may alter climate on a global scale. Predictions of global warming are still uncertain—with projected effects ranging from tolerable to very severe—but the potential risks are very great.

Our massive tampering with the world's interdependent web of life—coupled with the environmental damage inflicted by deforestation, species loss, and climate change—could trigger widespread adverse effects, including unpredictable collapses of critical biological systems whose interactions and dynamics we only imperfectly understand.

Uncertainty over the extent of these effects cannot excuse complacency or delay in facing the threats.

Population

The earth is finite. Its ability to absorb wastes and destructive effluent is finite. Its ability to provide food and energy is finite. Its ability to provide for growing numbers of people is finite. And we are fast approaching many of the earth's limits. Current economic practices that damage the environment, in both developed and underdeveloped nations, cannot be continued without the risk that vital global systems will be damaged beyond repair.

Pressures resulting from unrestrained population growth put demands on the natural world that can overwhelm any efforts to achieve a sustainable future. If we are to halt the destruction of our environment, we must accept limits to that growth. A World Bank estimate indicates that world population will not stabilize at less than 12.4 billion, while the United Nations concludes that the eventual total could reach 14 billion, a near tripling of today's 5.4 billion. But, even at this moment, one person in five lives in absolute poverty without enough to eat, and one in ten suffers serious malnutrition.

No more than one or a few decades remain before the chance to avert the threats we now confront will be lost and the prospects for humanity immeasurably diminished.

Warning

We the undersigned, senior members of the world's scientific community, hereby warn all humanity of what lies ahead. A great change in our stewardship of the earth and the life on it, is required, if vast human misery is to be avoided and our global home on this planet is not to be irretrievably mutilated.

What We Must Do

Five inextricably linked areas must be addressed simultaneously:

1. We must bring environmentally damaging activities under control to restore and protect the integrity of the earth's systems we depend on.
   - We must, for example, move away from fossil fuels to more benign, inexhaustible energy sources to cut greenhouse gas emissions and the pollution of our air and water. Priority must be given to the development of energy sources matched to third world needs—small scale and relatively easy to implement.
   - We must halt deforestation, injury to and loss of agricultural land, and the loss of plants, animals, and marine species.
2. We must manage resources crucial to human welfare more effectively.
   - We must give high priority to efficient use of energy, water, and other materials, including expansion of conservation and recycling.
3. We must stabilize population. This will be possible only if all nations recognize that it requires improved social and economic conditions, and the adoption of effective, voluntary family planning.

4. We must reduce and eventually eliminate poverty.

5. We must insure sexual equality, and guarantee women control over their own reproductive decisions.

The developed nations are the largest polluters in the world today. They must greatly reduce their overconsumption, if we are to reduce pressures on resources and the global environment. The developed nations have the obligation to provide aid and support to developing nations, because only the developed nations have the financial resources and the technical skills for these tasks.

Acting on this recognition is not altruism, but enlightened self-interest: whether industrialized or not, we all have but one lifeboat. No nation can escape from injury when global biological systems are damaged. No nation can escape from conflicts over increasingly scarce resources. In addition, environmental and economic instabilities will cause mass migrations with incalculable consequences for developed and undeveloped nations alike.

Developing nations must realize that environmental damage is one of the gravest threats they face, and that attempts to blunt it will be overwhelmed if their populations go unchecked. The greatest peril is to become trapped in spirals of environmental decline, poverty, and unrest, leading to social, economic and environmental collapse.

Success in this global endeavor will require a great reduction in violence and war. Resources now devoted to the preparation and conduct of war—amounting to over $1 trillion annually—will be badly needed in the new tasks and should be diverted to the new challenges.

A new ethic is required—a new attitude toward discharging our responsibility for caring for ourselves and for the earth. We must recognize the earth's limited capacity to provide for us. We must recognize its fragility. We must no longer allow it to be ravaged. This ethic must motivate a great movement, convincing reluctant leaders and reluctant governments and reluctant peoples themselves to effect the needed changes.

The scientists issuing this warning hope that our message will reach and affect people everywhere. We need the help of many.

- We require the help of the world community of scientists—natural, social, economic, political;
- We require the help of the world's business and industrial leaders;
- We require the help of the world's religious leaders; and
- We require the help of the world's peoples.
- We call on all to join us in this task.

Constraints on the Expansion of the Global Food Supply

We examine whether and how global food production may be increased to provide for a world population expected to double by about 2050. Increasing current food production more than proportional to population growth is required so as to provide most humans with an adequate diet. We examine the possible expansion of food supplies to the year 2050, the inventory of presently utilized and potentially available arable land, rates of land degradation, and the limitations of water and biological resources. Serious degradation and loss of the world's arable land is taking place and expansion of irrigation, vital for food production, is becoming more costly and difficult. A business-as-usual scenario points to looming shortages of food. Additional stress from possible climatic alteration and enhanced ultraviolet radiation may make the provision of adequate food supplies extremely difficult to achieve. The nature of the changes that are required to make sufficient food available are identified.

Introduction

World population is projected to continue increasing well into the next century. A central question is whether and how global food production may be increased to provide for the coming population expansion. It would be necessary to increase current levels of food production more than proportional to population growth so as to provide most humans with an adequate diet. There are a number of actions that may be taken to help this food expansion, but there are also a number of constraints that make expansion of food output difficult. In this paper we examine the expansion of per capita food supplies required in the light of the current range of expectations of population growth, the inventory of currently utilized and potentially available arable

---

By Henry W. Kendall and David Pimentel. Reprinted with permission from *Ambio* **23**(3), 198–205 (May 1994). © 1994 Royal Swedish Academy of Sciences.

land, rates of land degradation, and the limitations of water and biological resources. We make assessments of the prospects of achieving the needed growth of the global food supply to the year 2050, when the world's population is projected to have about doubled. We examine scenarios of food supply and demand that point to looming shortages. We do not analyze the problems of providing energy, capital, and other needs to support increasing numbers of people.

Population and Food

Numbers and Growth

The world's population grew slowly over much of the historic past; it was not until after 1900 that growth accelerated.[1] The 1992 population was 5.5 billion. World population is now increasing at about 1.7% yr^{-1}, corresponding to a doubling time of 40 years. Recently, a gradual decrease in the fertility rate has slowed in a number of countries,[2,3] most notably in China and India,[4] which has led to upward revisions in population forecasts. The world population will grow by just under 1 billion people during the decade of the 1990s.

Figure 1 shows three such projections for world population.[5] The United Nations has concluded that if the world's fertility rate were to fall to replacement levels during the period 1990–1995 and remained there, the world population would reach 7.8 billion in 2050. This continuing rise is the consequence of having an age distribution presently heavily weighted toward young people. The populations in many

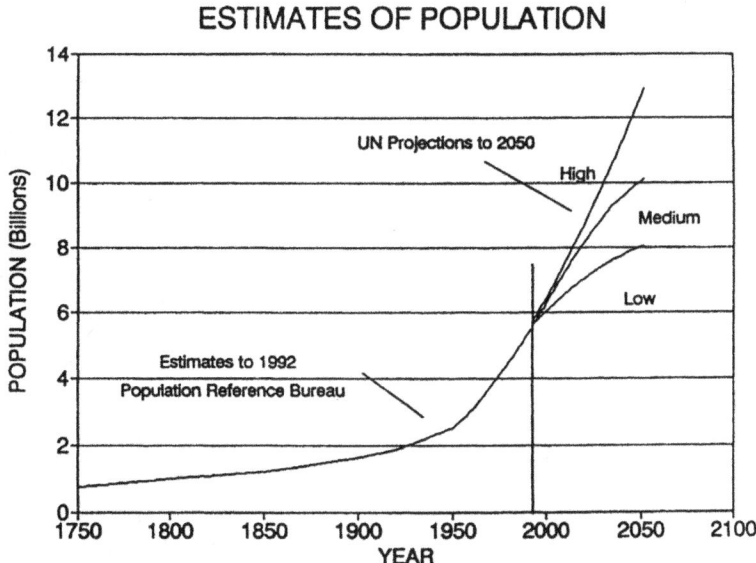

Figure 1. World population to 1992, projections to 2050.

developing countries would double in this case. A population of 7.8 billion, under this implausible assumption, is only slightly less than the low projection for that year, which thus appears unrealistic. For the purposes of this paper we will employ the medium fertility estimate. The 2050 world population in this scenario is expected to be 10 billion.

Food: History and Supply

In the early 1960s, most nations were self-sufficient in food; now only a few are. In the period 1950–1984, the introduction of high-yield crops and energy intensive agriculture ushered in the Green Revolution, leading to increased crop production. World grain output expanded by a factor of 2.6 in this period[6,7] increasing linearly within the fluctuations. Except for parts of Africa, production exceeded population growth throughout the world. Per capita production has now slowed[8] and appears to be declining. Rising growth of population, as shown in Figure 1, and a linearly increasing food production (Fig. 2) have persisted over the recent 40 years. Such circumstances have been of concern since Thomas R. Malthus first called attention, in 1798, to the consequences of their continuation; decreasing per capita food and great human suffering.[9]

The success of the Green Revolution lay primarily in its increased use of fossil energy for fertilizers, pesticides, and irrigation to raise crops as well as in improved

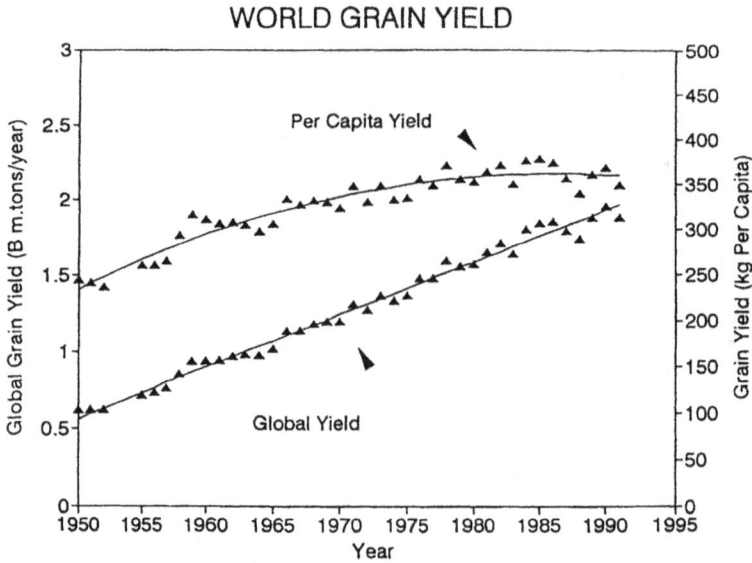

Figure 2. World grain yields, total and per capita, 1950 to 1991. A linear least square fit to the total yield, made to guide the eye, has a slope $+1.5\%$ yr^{-1} in the year 1989 (shown). A quadratic fit to the per capita yield (shown) has a decreasing slope of 1.4% yr^{-1}, in the recent interval, 1990–1991. See text for further discussion.

seed. It greatly increased the energy-intensiveness of agricultural production, in some cases by 100-fold or more. Plant breeding was principally aimed at designing plants that could tolerate high levels of fertilizer use and improving the harvest index.[10] The Green Revolution was technologically suited to special circumstances: relatively level land with adequate water for irrigation and fertilizer, and in nations that could acquire the other needed resources. The Revolution has been implemented in a manner that has not proved to be environmentally sustainable. The technology has enhanced soil erosion, polluted groundwater and surface-water resources, and increased pesticide use has caused serious public health and environmental problems.[11–13] Opportunities exist to reduce these negative environmental and social impacts. Research is underway at most of the International Crop Research Centers to make the Revolution more environmentally and socially sustainable.

Since 1980, there has been some improvement in world crop yields with the rate of increase in total grain production declining slightly. Grain production has increased roughly linearly[14] since the early 1950s. World area planted to grain is down 8% since 1981.[15] However, there are a number of important obstacles to a large further expansion of the energy-intensive practices that underlay the expansion based on the Green Revolution, including economics, technology adoption, and environmental degradation.

At the present time, 2 of 183 nations are major exporters of grain, the United States and Canada.

Food: Availability and Consumption

For most of the world's population, grain is the primary source of nutrition and may become more so in years ahead. It is thus a useful measure in estimating future food needs. The per capita consumption of foods and feed grains supplied per year is shown in Table 1. Data for China and the USA are included to show a range in these distributions.

Per capita grain production in Africa is down 12% since 1981 and down 22% since 1967.[15] Some 20 years ago, Africa produced food equal to what it consumed; today it produces only 80% of what it consumes.[16]

Food from marine sources now provides between 1% and 2% of the world's supply of food[17,18] and the amount, including the contribution from aquaculture, is unlikely to double within the next few decades (John Ryther pers. comm.)

In line with recent studies,[19,20] we estimate that with the world population at 5.5 billion, food production is adequate to feed 7 billion people a vegetarian diet, with ideal distribution and no grain fed to livestock. Yet possibly as many as two billion people are now living in poverty (V. Abernathy, pers. comm.), and over 1 billion in "utter poverty" live with hunger.[7,19,23] Inadequate distribution of food is a substantial contributing factor to this current situation.

It is clear from the above review that current food supplies, with present patterns of distribution and consumption, appear insufficient to provide satisfactory diets to all, although a recent FAO report indicates that chronic undernutrition in developing countries has improved somewhat.[24]

Table 1 Consumption of foods and feed grains

Food/feed	USA[1]	China[2]	World[3]
Food grains	77	239	201
Vegetables	129	163	130
Fruits	46	17	53
Meat and fish	88	36	47
Dairy products	258	4	77
Eggs	14	7	6
Fats and oils	29	6	13
Sugars and sweeteners	70	7	25
Food total	771	479	552
Feed grains	663	100	144
Grand total	1374	605	696
Kcal. cap^1 day^1	3600	2662	2667

[1] J. J. Putnam and J. E. Allshouse, *Food Consumption, Prices, and Expenditures, 1968–1989*, U.S. Department of Agriculture, ERS, Statistical Bulletin 825 (Washington, DC, 1991).
[2] All food item data, exept vegetables, were from Agricultural Statistical Data of China, Agricultural Ministry of PRC (Agricultural Press, Beijing, China, 1990). Feed grains are from D. Wen, Institute of Applied Ecology (personal communication, 1992).
[3] UN Food and Agriculture Organization, *Food Balance Sheets* (Rome, 1991).

It is generally agreed that, among a number of important global changes, economic and social well-being must improve for that large fraction of the world's peoples now in poverty. This includes more and better food. A doubling of the population would necessitate the equivalent of a tripling, or more, of our current food supply to ensure that the undernourished were no longer at risk and to bring population growth stabilization within reach in humane ways, without widespread hunger and deprivation. Improved nutrition may be achieved by dietary shifts and improved distribution as well as by an increased quantity of food, as discussed later in this paper.

Agricultural Productivity

Land Resources

Supply:

The world's land devoted to food production and in forest and savanna is shown in Table 2.

Less than one half of the world's land area is suitable for agriculture, including grazing; total arable (crop) land, in use and potential, is estimated to comprise about 3000 million ha.[25] However, nearly all of the world's productive land, flat and with

Table 2 Land surface (M ha[1])

Total Land Surface of the Globe	13,000
Forest and Savannah	4000
Area Utilized for Food Production	~ 4600
Pasture and Rangeland	3100
Cropland Total[2]	1500

[1]Hectare (ha) = 10,000 m^2 = 2,47 acres.
[2]Table 18.2, Agricultural Inputs, 1975–1989, in *World Resources 1992–1993* (World Resources Institute, Washington, DC, 1992).

water, is already exploited. Most of the unexploited land is either too steep, too wet, too dry, or too cold for agriculture.[26]

There are difficulties in finding new land that could be exploited for agricultural production. Expansion of cropland would have to come at the expense of forest and rangeland, much of which is essential in its present uses. In Asia, for example, nearly 80% of potentially arable land is now under cultivation.[7,27] In the 1970s, there was a net annual gain in world cropland of nearly 0.7%. The rate of gain has slowed and, in 1990, the net annual gain was about 0.35% yr^{-1}, largely as a result of deforestation. As much as 70–80% of ongoing deforestation, both tropical and temperate, is associated with the spread of agriculture.

For these reasons we estimate that the world's arable land could be expanded at most by 500 million ha, or a net expansion of roughly one-third. However the productivity of this new land would be much below present levels in land now being cropped.

At the present time humans either use, coopt or destroy 40% of the estimated 100 billion tons of organic matter produced annually by the terrestrial ecosystem.[28]

Quality and Degradation:

The loss of productive soil has occurred as long as crops have been cultivated. Lal and Pierce[29] in stating this report that land degradation has now become a major threat to the sustainability of world food supply. This loss arises from soil erosion, salinization, waterlogging, and urbanization with its associated highway and road construction. Nutrient depletion, overcultivation, overgrazing, acidification, and soil compaction contribute as well. Many of these processes are caused or are aggravated by poor agricultural management practices. Taken together or in various combinations, these factors decrease the productivity of the soil and substantially reduce annual crop yields,[30,32] and more important, will reduce crop productivity for the long term.[33]

Almost all arable land that is currently in crop production, especially marginal land, is highly susceptible to degradation. We estimate that about one quarter of this land should not be in production.[34] This is depressing food production, as well as requiring increased fossil energy inputs of fertilizers, pesticides, and irrigation in an effort to offset degradation.

Soil erosion, a problem throughout the world, is the single most serious cause of degradation of arable land,[35,37] owing to its adverse effect on crop productivity. The major cause is poor agricultural practices that leave the soil without vegetative cover to protect it against water and wind erosion.

Soil loss by erosion is extremely serious because it takes from 200 to 1000 years, averaging about 500 years, to form 2.5 cm (1 inch) of topsoil[38] under normal agricultural conditions.[39,43] Throughout the world current soil losses range from about 20 to 300 t ha^{-1} yr^{-1}, with substantial amounts of nitrogen and other vital nutrients also lost.[44] Topsoil is being lost at 16 to 300 times faster than it can be replaced.[36]

Worldwide soil erosion has caused farmers to abandon about 430 million ha of arable land during the last 40 years, an area equivalent to about one-third of all present cropland.[6,7] Each year at least 10 million ha are lost to land degradation that includes the spread of urbanization.[45] For example, Tolba[46] reported that the rate of soil loss in Africa has increased 20-fold during the past 30 years.

The estimated rate of world soil erosion in excess of new soil production is 23 billion t yr^{-1}, or about 0.7% loss of the world's soil inventory each year[47] (Table 3). The continuing application of fertilizers[48] has so far masked much of the deterioration and loss of productivity from this process, so that world cropland yield is remaining roughly constant. This appears likely to continue in the next decades. Continued erosion at the current rate will result in the loss of over 30% of the global soil inventory by the year 2050, a truly severe damage and loss, obviously unsustainable over the long run.

Erosion reduces the availability of water[31] as well as nutrients to growing plants and diminishes organic matter and soil biota.[29,49] Reduction of the water available to growing plants is the most harmful effect of erosion.

Soil degradation is affecting 15% of the earth's cropland area.[29] In developing countries, the degradation of soil is growing worse owing to increased burning of crop residues and dung for fuel. This reduces soil nutrients[50,51] and quickly intensifies soil erosion.

Table 3 Gains and losses of arable land (M ha year^{21})

Forest and Savanna Loss[1]	20 (~0.5%· year^{-1})[2]
Cropland Lost or Abandoned	
Erosion	5–7
Urbanization	2–4
Salinization and waterlogging	2–3
Total Cropland Losses[3]	>10
Cropland Gain (from deforestation)	16
Cropland, Net Gain	5–6

[1]N. Myers, *Deforestation Rates in Tropical Forests and Their Climatic Implications* (Friends of Earth Report, London, 1989).
[2]N. Myers. "Mass extinctions: What can the past tell us about the present and future?" *Global and Planetary Change* **2,** 82 (1990).
[3]D. Pimentel, U. Stachow, D. A. Takacs, H. W. Brubaker, A. R. Dumas, J. J. Meaney, J. O'Neil, D. E. Onsi, and D. B. Corzilius, "Conserving biological diversity in agricultural/forestry systems," *Bioscience* **42,** 354–362 (1992).

Water: Resources and Irrigation

Supply and Use:

Water is the major limiting factor for world agricultural production. Crops require and transpire massive amounts of water. For example, a corn crop that produces about 7000 kg ha^{-1} of grain will take up and transpire about 4.2 million L ha^{-1} of water during its growing season.[52] To supply this much water to the crop, assuming no use of irrigation, not only must 10 million liters (1000 mm) of rain fall per ha, but it must be reasonably evenly distributed during the year and especially during the growing season.

Irrigation:

Irrigation is vital to global food production: About 16% of the world's cropland is under irrigation. This area contributes about one-third of crop production, yielding about $2\frac{1}{2}$ times as much per ha as nonirrigated land. In arid lands crops must be irrigated and this requires large quantities of water and energy.[53] For example, the production of 1 kg of the following food and fiber products requires: 1400 liters of irrigation water for corn; 4700 liters for rice, and 17 000 liters for cotton.[54] About 70% of the fresh water used by humans is expended for irrigation.[55]

Much of the world's irrigated land is being damaged by salinization and waterlogging from improper irrigation techniques.[56] It is sufficiently severe over 10% of the area to suppress crop yields.[57] This damage, together with reduced irrigation development and population growth, has led, since 1978, to declining world irrigated area per capita.[58,59] Serious salinization problems already exist in India, Pakistan, Egypt, Mexico, Australia, and the United States. Because salt levels are naturally high in these regions, the problem of salt buildup is particularly severe. Recent research puts the current loss of world farmland due to salinization alone at 1.5 million ha yr^{-1} [60] or almost 1% yr^{-1}, a loss presently being more than made up by expansion of irrigation. If the damage continues, nearly 30% of the world's presently irrigated acreage will be lost by 2025 and nearly 50% lost by 2050, losses increasingly difficult to make up.

Another damaging side effect of irrigation is the pollution of river and stream waters by the addition of salts.

Water Shortages:

Pressures from growing populations have strained water resources in many areas of the world.[59] Worldwide, 214 river or lake basins, containing 40% of the world's population, now compete for water.[55,61]

In many areas of the world, irrigation water is drawn from "fossil" aquifers, underground water resources, at rates much in excess of the natural recharge rates. The average recharge rate for the world's aquifers is 0.007% yr^{-1}.[62] As the aquifers' water levels decline, they become too costly to pump or they become exhausted, forcing abandonment of the irrigated land.[55]

Africa and several countries in the Middle East, especially Israel and Jordan, as well as other countries, are depleting fossil groundwater resources. China has severe agricultural problems.[13] In China, groundwater levels are falling as much as 1 m yr^{-1} in major wheat and corn growing regions of the north China Plain.[64] Tianjin, China, reports a drop in groundwater levels of 4.4 m yr^{-1},[58,59] while in southern India, groundwater levels are falling 2.5 to 3 m yr^{-1}; in the Gujarat aquifer depletion has induced salt contamination.[6,7]

The prospect for future expansion of irrigation to increase food supplies, worldwide and in the US, is not encouraging because per capita irrigated land has declined about 6% since 1978.[57] Greatly expanded irrigation is a difficult, and probably unsustainable solution to the need for expansion of agriculture output[59] because of the rapidly accelerating costs of irrigation.[57]

Greenhouse Effects

The continuing emission of a number of gases into the atmosphere from human activities, including chlorofluorocarbons (CFCs), methane, and, most important, carbon dioxide, is now thought likely to alter the global climate in the years ahead, a consequence arising from the greenhouse effect.[65,66] Worldwide changes in rainfall distribution are expected, including drying of some continental interiors as well as possible increases in climatic variability.

Increased variability in temperature and rainfall can, in many circumstances, be damaging to agricultural productivity. There are expected to be CO_2-induced effects on productivity and growth of plants, including crops and weeds, and collateral effects on plant pathogens and insect pests. There may be decline or loss of exosystems that are unable to accommodate a rapid climate change. The major impact will be caused by changes in rainfall and water availability to crops. Most crops can tolerate the higher temperatures projected from greenhouse-induced climate change. The detailed consequences are difficult to predict, in part because the expected global average temperature rise and changes in weather patterns have substantial uncertainties. The temperature rise expected from a doubling of the atmospheric CO_2 level—which, in the absence of carbon emission controls, will occur a decade or so before the year 2100—is "unlikely to lie outside the range 1.5° to 4.5°C."[67] If the rise were only 2°C (a degree of warming not experienced in the last 8000 years), there could still be pronounced adverse effects.[68]

The 1988 US experience is enlightening. It was the hottest year on record to that time which, accompanied by a midcontinent drought, resulted in a 30% decrease in grain yield, dropping US production below consumption for the first time in some 300 years. Similarly, Canadian production dropped about 37%.[69]

Laboratory studies under favorable conditions indicate that enhanced CO_2 levels can improve growth rates and water utilization of crops significantly.[70] Under field conditions, the estimated increase in yields are projected to be only one-quarter to one-third of that observed in the controlled greenhouse conditions without taking into consideration other deleterious consequences of climate change that also may be present and yields may, in fact, not improve at all.[71]

Ozone Depletion

Ground-level ultraviolet enhancement arising from O_3 loss in the upper atmosphere from the anthropogenic emission of chlorofluorocarbons can affect natural systems' productivity, alter pest balances, as well as affect the health of humans and surface and marine animals. The current ozone loss, as well as its seasonal variation, over equatorial and midlatitude regions is not yet well known but is expected to increase, perhaps greatly.[72] The US Environmental Protection Agency reported in April 1991, a winter-spring O_3 column density depletion of 4.5–5% in midlatitudes. More recently, there is evidence of a slow but steady ozone depletion over most of the globe; between 40° and 50°N the decline is as great as 8% per decade.[73,74] Each percent decrease in O_3 results in about a 3% increase in ground-level ultraviolet intensity. Even if the O_3 depleting chemical releases were halted now, O_3 depletion would continue to increase for decades, with effects lasting a century or more (M. McElroy pers. comm.).

Increased ozone levels may already have decreased phytoplankton yields in the Antarctic ocean.[75] Plant responses to ultraviolet radiation include reduced leaf size, stunted growth, poor seed quality, and increased susceptibility to weeds, disease, and pests. Of some 200 plant species studied, two thirds show sensitivity to ozone damage.[76] A 25% O_3 depletion is expected to reduce yields of soybean, one of civilization's staple crops, 20%.[77] Red Hard disease infection rates in wheat increased from 9% to 20% when experimental ozone loss increased from 8% to 16% above ambient levels.[78] Clearly, the potential exists for a significant decrease in crop yields in the period to 2050 from enhanced surface ultraviolet levels.

Adjusting to modifications of global climate or to altered growing conditions, caused by greenhouse gases or from enhanced ultraviolet, might stress management of agricultural systems greatly, especially if wholly new crops, and new procedures had to be developed for large areas of the world. Important uncertainties in the magnitudes of the effects expected may persist for a decade or so.

Improving the Food Supply

Diet Modification

Currently ruminant livestock like cattle and sheep, graze about half of the earth's total land area.[79] In addition, about one-quarter of world cropland is devoted to producing grains and other feed for livestock. About 38% of the world's grain production is now fed to livestock.[79] In the United States, for example, this amounts about 135 million tons yr^{-1} of grain, of a total production of 312 million tons yr^{-1}, sufficient to feed a population of 400 million on a vegetarian diet. If humans, especially in developed countries, moved toward more vegetable protein diets rather than their present diets, which are high in animal protein foods, a substantial amount of grain would become available for direct human consumption.

Agricultural Technologies

There are numerous ways by which cropland productivity may be raised that do not induce injury over the long term, that is, are "sustainable."[26,80,82] If these technologies were put into common use in agriculture, some of the negative impacts of degradation in the agro-ecosystem could be reduced and the yields of many crops increased. These technologies included:

Energy-Intensive Farming:

While continuation of the rapid increases in yields of the Green Revolution is no longer possible in many regions of the world, increased crop yields are possible by increasing the use of fertilizers and pesticides in some developing countries in Africa, Latin America, and Asia.[83] However, recent reports indicate a possible problem of declining yields in the rice-wheat systems in the high production areas of South Asia. (J. M. Duxbury pers. comm.)

Livestock Management and Fertilizer Sources:

Livestock serve two important functions in agriculture and food production. First, ruminant livestock convert grass and shrub forages, which are unsuitable for human food, into milk, blood, and meat for use by humans. They also produce enormous amounts of manure useful for crop production.

Soil and Water Conservation:

The high rate of soil erosion now typical of world agricultural land emphasizes the urgency of stemming this loss, which in itself is probably the most threatening to sustained levels of food production. Improved conservation of water can enhance rainfed and irrigated crop yields as discussed below.

Crop Varieties and Genetic Engineering:

The application of biotechnology to alter certain crop characteristics is expected to increase yields for some crops, such as developing new crop varieties with better harvest index and crops that have improved resistance to insect and plant pathogen attack.

Maintaining Biodiversity:

Conserving biodiversity of plant and animal species is essential to maintaining a productive and attractive environment for agriculture and other human activities. Greater effort is also needed to conserve the genetic diversity that exists in crops worldwide. This diversity has proven extremely valuable in improving crop productivity and will continue to do so in future.

Improved Pest Control:

Because insects, diseases, and weeds destroy about 35% of potential preharvest crop production in the world,[84] the implementation of appropriate technologies to reduce pest and disease losses would substantially increase crop yields and food supplies.

Irrigation:

Irrigation can be used successfully to increase yields as noted earlier, but only if abundant water and energy resources are available. The problems facing irrigation suggest that its worldwide expansion will be limited.[57] Owing to developing shortages of water, improved irrigation practices that lead to increased water in plant's root zones are urgently needed.

Constraints

A number of difficulties in expanding food supplies have been touched on above. Others are presented below:

There is a need to decrease global fossil-fuel use and to halt deforestation, in order to lessen carbon emissions to the atmosphere.[85] These steps are in direct competition with the need to provide sufficient energy for intensive agriculture and for cooking and heating using firewood. A major decrease in fossil fuel use by the industrial countries would require adoption of new technologies based on new energy sources, with improved conversion and end-use efficiencies, on a scale that would require 40 years at minimum to implement fully, even in favorable circumstances.[86] Yet a three- or fourfold increase in effective energy services to the earth's peoples would be required to yield the improvements needed in the quality of life in a world of roughly doubled population. We do not consider here the considerable challenge that this provides.[87]

Even assuming that sufficient fossil or other fuels were available in the future to support energy-intensive agriculture in developing countries, several constraints appear to make this difficult. These include, the high economic costs of the energy inputs to those countries that already have serious debt problems; the lack of rainfall and/or irrigation water preventing effective use of the inputs; and farmers in developing nations who are not educated in the use of intensive agricultural methods and who change their ways slowly.

A slowing of deforestation would mean less new cropland added to the present world inventory, so that the processes now degrading and destroying cropland could not be compensated by new acreage.

Population growth remains a basic problem. About 0.5 ha capita^{-1} of cropland is needed to provide a balanced plant/animal diet for humans worldwide.[88] For the 1990 population of 5.5 billion, only 0.27 ha capita^{-1} is now available and this is likely to decline further. Moreover, the rate of population growth itself, especially in many developing nations, intensifies the problems of coping with shortages owing to the difficulty of providing the economic expansion required.[89]

A major difficulty arises simply from the rate with which food supplies would have to be expanded to pace or to exceed population growth rates in those countries experiencing high growth rates. In order to stay even with population growth it will be necessary to expand food supplies, globally, by the rate of population increase. For many countries the rate of population expansion is in the range 2–3% yr^{-1}. As an example, in order to achieve an increase of 50% in the per capita food production, by the end of a population doubling, the rate of expansion of agricultural production must be appropriately larger. If the population grows at 2% yr^{-1}, the food production must increase at 3.2% yr^{-1}, if it is 3% yr^{-1}, the food production must grow at 4.8% yr^{-1}.

During the Green Revolution the world grain yield expanded at 2.8% yr^{-1}. As noted earlier, this rate of expansion has slowed and, it appears, is unlikely to be resumed[90] although some countries in Asia and Latin America are still gaining total annual increases in grain yield. In the US, which has one of the best records with corn, the rate of increase from 1945 to 1990 was about 3% yr^{-1}. Since 1980, this rate has slowed. However, with wheat the record is not as good as with corn; the increase in world grain yield is less than 2% yr^{-1}. If the historical record is any guide, no nation with a population growth rate above 2% yr^{-1} has much hope of improving its per capita supply of food unless it receives very substantial external aid and support. Of course these rates of increase for both population and food production, if achieved, cannot be sustained indefinitely.

The Future

Introduction

Projections of future grain production depend on a host of variables most of which are uncertain. It is not possible to make useful forecasts. As an alternative we consider three scenarios, for the period to the year 2050. The first assumes a continuation of present trends, patterns, and activities. This is referred to as Business-As-Usual, or BAU. Population growth is assumed to follow the UN medium projection leading to about 10 billion people by 2050, soil erosion continues to degrade land productivity, salinization and waterlogging of the soil continues, and groundwater overdraft continues with supplies in some aquifers being depleted; there is a modest expansion of cropland at the expense of world forests, and a slight expansion of irrigation. In BAU, the consequences of the greenhouse effect and of ultraviolet injury are ignored, and the developed world does not provide significantly more aid to the developing world than at present, nor does the developing world continue, on balance, its current rate of economic development.[91]

A pessimistic scenario considers qualitatively the possible consequences of climatic changes and ground-level ultraviolet radiation increase that could depress crop yields, coupled with the high UN population growth projection, leading to nearly 13 billion people 2050. The economic debt that many nations face today continues to worsen, especially limiting developing nations in the purchase of fertilizers and other agricultural goods to enhance productivity.

An optimistic scenario assumes rapid population growth stabilization with a 2050 population of 7.8 billion, significant expansion of energy-intensive agriculture and improved soil and water conservation with some reclamation of now-abandoned land. In this scenario, the developed countries provide the developing nations with increased financial resources and technology and a more equitable distribution of food is achieved. There is a shift from high animal protein to more plant protein consumption in the developed countries, freeing more grain for the developing nations.

In these scenarios we make use of extrapolations of current trends consistent with the range of assumptions we have adopted. This procedure is inherently unsatisfactory owing both to the difficulty of determining trends from fluctuating historical data and because few trends continue for periods comparable to the interval of interest to us. Nevertheless, it does, over a number of scenarios, shed light on the range of achievable futures.

Business As Usual (BAU)

Grainland declined from 718 million ha in 1980 to 704 million ha in 1991,[92] a decline we assume continues, leading to 620 million ha in 2050. There is 0.06 ha capita^{-1} available for grain production in that year, or less than half of that available in 1991. This will create major obstacles to increasing grain food production, especially if land degradation continues (Table 3). The rate of loss we assume is about half that projected for the next 25 years in The Netherlands report on the National Environmental Outlook 1990–2010.[93]

In BAU, we make the optimistic assumption that a modest expansion of irrigation will continue as it has recently. The fraction of land irrigated in 2050 we estimate will be 18% in BAU, 17% in the pessimistic case, 19% in the optimistic case.

Estimates suggest that degradation can be expected to depress food production in the developing world between 15% and 30% over the next 25-year period, unless sound conservation practices are instituted now, and that the "total loss in productivity of rainfed cropland would amount to a daunting 29%" due to erosion of topsoil during the same time period.[94]

Despite the increased use of fertilizers, the rate of increase in grain production appears to be slowing.[55,95] Figure 2 shows the world's grain yield from 1950 to 1991 as well as the per capita grain yield.[96] In recent years, 1985 to 1991, the total growth rate has dropped below 1.4% yr^{-1}, less than the current rate of world population growth. Based on past trends we estimate a 300% increase in the use of nitrogen and other fertilizers by the year 2050 and about 12% expansion of irrigated land, consistent with BAU.

In view of the constraints we have identified we conclude that an expansion of 0.7% yr^{-1} in grain production is achievable in the decades ahead. With this rate of expansion, there would be a 50% increase in annual grain production by 2050 compared to 1991, with the world per capita grain production decreasing about 20%. These projections are shown in Figure 3. The 2050 per capita number is about the same as it was in 1960. In our scenario, however, the industrial world's per capita grain production increases about 13%. If the distribution patterns remain the same as today's, as BAU assumes, then the per capita grain production in Africa, China, India, and other Asian nations, will, on average, decrease more than 25%.

In BAU, most developing nations suffer reductions in per capita grain production. Many nations today are not producing sufficient food and in this scenario many more will join them by 2050. This conclusion is consistent with other assessments.[15]

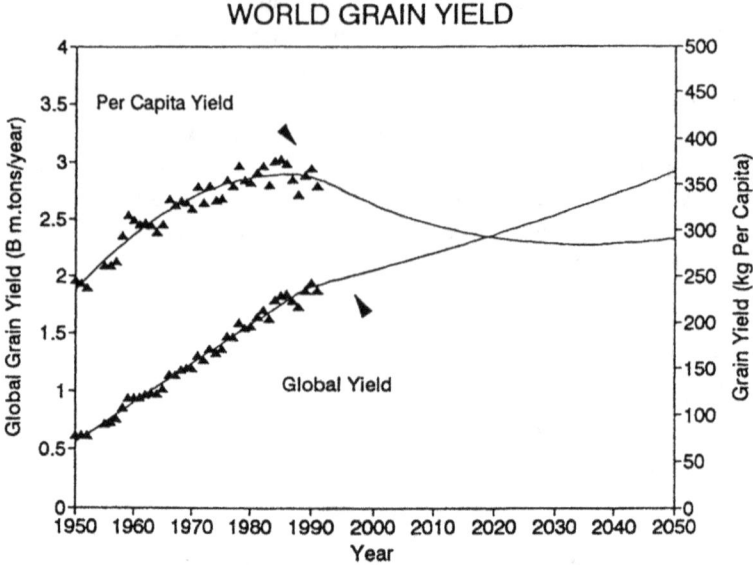

Figure 3. Data and projections of world grain yields, 1950–2050. After 1992 the global production rises at 0.7% yr^{-1}, according to the business-as-usual scenario discussed in the text. The per capita yield employs the UN medium growth case.

One study concluded that if the African population continues to grow, and agricultural production remains roughly constant, the food produced would, in 25 years, support about 50% of its year 2000 population for the Middle East about 60% of its population. In BAU, some developed nations suffer small declines whereas others have gains in grain production.

In general, it appears that Africa, as noted earlier, as well as China and India, will face particularly severe problems in expanding their food supplies in the coming decades. The people of these regions are likely to comprise almost two thirds of the developing countries', and over half of the world's, population, both in 2025 and 2050.

The US appears to have the potential of generating food surpluses for some years, a potential that it shares with parts of Europe, including Eastern Europe, Canada, and possibly other regions. The longer term prospects are unknown in view of difficulties which may appear later.

Pessimistic Scenario (PS)

Scenario PS adopts most of the assumptions in BAU, but includes several other factors which may decrease the rate of grain production in the years ahead. If the population growth rate continues only slightly lower than it is today to the year 2050, the world population will rise to about 13 billion (Fig. 1), more than double the present population. A recent analysis[97,98] of the consequences of climatic change on world food production, not including problems arising from the availability of irrigation water, concluded that decreases in global food production would be small, but with developing countries suffering decreases of as much as 10%. We believe that, in the period to 2050, the greenhouse effect and ozone loss could together depress grain yields on a global basis by 10% to 20%. We base our estimates (Table 4) on current rates of cropland loss (Table 3), continued decline in per capita irrigation,[59] degradation of irrigated land (Table 3), and continued decline on the rate of fertilizer use by some farmers in developing countries.[26,99] A moderate combination of these adverse factors leads to grain production in 2050 about 15% below BAU. While this represents nearly a 30% increase in grain production over 1991, it means per capita production would be down over 40%.

There is, in this scenario, little hope of providing adequate food for the majority of humanity by the middle or later decades of the period we consider.

Optimistic Scenario (OS)

If rapid population growth stabilization can be effected, leading to a world population of 7.8 billion instead of 13 billion by the year 2050, then grain production adequate for the population might be achievable. This would require a near doubling of today's production (Table 4). Soil and water conservation programs would have to be implemented to halt soil degradation and the abandonment of cropland. Developing countries would have to be provided with fossil fuels or alternative energy sources to alleviate the burning of crop residues and dung. Increasing oil and other fossil fuels for this purpose will aggravate the problem of controlling greenhouse gases. Irrigation would have to be expanded by about 20%. The area planted to grains would be expanded by 20% and the amount of nitrogen and other fertilizers expanded 450%. Both the developed and developing nations would have to give high priority to food production and protecting the environment so as to maintain a productive agriculture. The developed countries would have to help finance these changes and also provide tech-

Table 4 Total gains

	1991[1]		2050					
			PS		BAU		OS	
Region	Total	per. cap.	Total	per. cap.	Total	per. cap.	Total	per. cap.
N. America	336	1200	450	1090	530	1630	600	2320
Europe	286	569	320	558	340	700	420	1040
Oceania	37	1370	50	980	50	1220	70	2060
former USSR	172	589	220	491	300	790	350	1100
Africa	99	146	150	52	200	88	220	131
Lat. Amer.	104	231	150	132	200	217	250	326
India	194	226	270	131	320	189	370	280
China	378	315	480	257	500	330	520	423
Other Asia	276	275	310	101	360	151	400	221
World	1865	345	2400	192	2800	280	3200	410

Units: Total, million metric tons year^{-1}; per cap., kg cap^{-1} year^{-1}
[1]UN Food and Agriculture Organization, *Quarterly Bulletin of Statistics* **4**, 26–27 (1991).

nology to the developing nations. At the same time, with diet shifts in the developed world, the 2050 population of 7.8 billion might be fed an adequate diet.

If efforts were made to triple world food production, compared to today's yield, then all of the above changes would have to be made, plus increasing the level of energy intensiveness in the developing world's agriculture by 50- to 100-fold. This would include a major expansion in world irrigation. Such increases appear to be unrealistic. Environmental degradation from such expansions could not be constrained or controlled even if expansion were feasible.

Summary and Conclusions

The human race now appears to be getting close to the limits of global food productive capacity based on present technologies. Substantial damage already has been done to the biological and physical systems that we depend on for food production. This damage is continuing and in some areas is accelerating. Because of its direct impact on global food production, injury and loss of arable land has become one of the most urgent problems facing humanity. Of these problems, this is perhaps the most neglected.

Controlling these damaging activities and increasing food production must now receive priority and resources commensurate with their importance if humanity is to avoid harsh difficulties in the decades ahead.

Attempts to markedly expand global food production would require massive programs to conserve land, much larger energy inputs than at present, and new sources as well as more efficient use of fresh water, all of which would demand large capital expenditures. The rates of food grain growth required to increase the per capita food available, in the light of present projections of population growth, are greater than have been achieved under any but the most favorable circumstances in developed countries.

Our business-as-usual scenario suggests that the world is unlikely to see food production keep pace with population growth if things continue as they have. If they do continue then the world will experience a declining per capita food production in the decades ahead. This decline would include spreading malnutrition and increased pressure on agricultural, range, and forest resources.

Should climatic alteration from greenhouse warming and enhanced ultraviolet levels impose further stress on agricultural systems, the prospects for increased food production would become even less favorable than they are at present.

In our opinion, a tripling of the world's food production by the year 2050 is such a remote prospect that it cannot be considered a realistic possibility. If present food distribution patterns persist the chance of bettering the lot of the majority of the world's peoples vanishes. The likelihood of a graceful and humane stabilization of world population vanishes as well. Fertility and population growth in numerous developing countries will then be forced downward by severe shortages of food, disease, and by other processes set in motion by shortages of vital resources and irreversible environmental damage.

A major expansion in food supply would require a highly organized global effort—by both the developed and the developing countries—that has no historic precedent. As yet a major commitment from the developed nations to support the needed changes is missing, and inadequate commitment in both developing and developed nations has been made for population stabilization. Governments so far appear to lack the discipline and vision needed to make a major commitment of resources to increase food supplies, while at the same time reducing population growth and protecting land, water, and biological resources. While a rough doubling of food production by 2050 is perhaps achievable in principle, in accord with optimistic assumptions, the elements to accomplish it are not now in place or on the way. A large number of supportive policy initiatives and investments in research and infrastructure as well as socioeconomic and cultural changes would be required for it to become feasible. A major reordering of world priorities is thus a prerequisite for meeting the problems that we now face.

References and Notes

1. Horiuchi, S. 1992. Stagnation in the decline of the world population growth rate during the 1980s. *Science 257*, 761–765.
2. United Nations Population Fund. 1991. *Population, Resources and the Environment*, United Nations, NY.
3. UNFPA, UN Population Fund. 1991. *The State of World Population 1991*. UN, New York.
4. Homer-Dixon, T.F. 1991. India's population growth rate has leveled off at 2.1%/year and China's at 1.3%/year. On the threshold: environmental changes as causes of acute conflict. *Inter. Sec. 16*, 76–116.
5. *Long-range World Population Projections*. Department of International and Economic and Social Affairs. United Nations, New York, 1992, Document Number ST/ESA/SER.A/125.
6. World Watch Institute. 1990. *State of The World 1990*. Washington. The 430 M ha estimated to have been abandoned during a 40 year period is further confirmed by the World Resources Institute's estimation that 10 M ha/yr of cropland is lost by erosion.
7. A Report by The World Resources Institute. 1990. *World Resources 1990–1991*. Oxford University Press, New York. Additional losses and abandonment of agricultural land are caused by water logging, salinization, urbanization and others processes as discussed later in the text.
8. Moffat, A. S. 1992. Does global change threaten the world food supply? *Science 256*, 1140–1141.
9. Malthus, T. R. 1970. *An Essay on the Principle of Population*. Penguin Books, New York.
10. CIMMYT. 1987. *The Future Development of Maize and Wheat in the Third World*. Mexico, DF.
11. Dahlberg, K. A. 1979. *Beyond the Green Revolution: the Ecology and Politics of Global Agricultural Development*. Plenum Press, New York.
12. Dahlberg, K. A. 1985. *Environment and the Global Arena: Actors, Values, Policies*. Duke University Press, Durham, NC.
13. World Health Organization/United Nations Environment Programme. 1989. *Public Health Impact of Pesticides Used in Agriculture*. WHO/UNEP, Geneva.
14. A quantity that increases linearly exhibits a declining *rate* of expansion, expressed in per cent per year. A constant *rate* of growth implies exponential increase.
15. *Global Ecology Handbook*. Beacon Press, Boston, 1990.
16. Cherfas, J. 1990. FAO proposes a new plan for feeding Africa. *Science 250*, 748.

17. The 1989 world marine catch was 99.5 metric tons. *Current Fisheries Statistics #9000, Fisheries of the United States, 1990*, Supplemental May 1991, National Marine Fisheries Service, NOAA. The estimate in text based on weight. Less than 1% of world food, based on calories, is obtained from the aquatic environment.
18. *Food Balance Sheets*, Food and Agriculture Organization of the United Nations, 1991. The total yield is considered by the UN Food and Agriculture Organization as at the maximum sustainable yield from the oceans. A number of important world fisheries have started to collapse.
19. Kates, R. W. et al. 1988. *The Hunger Report*. Brown University Hunger Project, Brown University, Providence, Rhode Island.
20. Kates, R. W. et al. 1989. *The Hunger Report: Update 1989*. Brown University Hunger Project, Brown University, Providence, Rhode Island.
21. Ehrlich, P. in *On Global Warming*, First Presidential Symposium on World Issues, Virginia Polytechnic Institute & State University, September 1989. In the early 1980s the World Bank and UNFAO estimated that from 700 mill. to 1 bill. persons lived in "absolute poverty." In 1989 the number was 1.225B, or 23% of world population. In that year poverty increased dramatically in sub-Saharan Africa, Latin America, and parts of Asia, swamping reductions in China and India.
22. McNamara, R. S. 1991. *Rafael M. Salas Memorial Lecture*. United Nations, New York.
23. Administrative Committee on Coordination-Subcommittee on Nutrition. 1992. *Second Report on the World Nutrition Situation, (Vol 1)*. U.N. United Nations, NY.
24. Statistical Analysis Service, Statistical Division, Economic and Social Policy Dept., UN Food and Agriculture Organization. 1992. *World Food Supplies and Prevalence of Chronic Undernutrition in Developing Regions as Assessed in 1992*. United Nations, New York.
25. Lal, R. 1990. Soil Erosion and Land Degradation: The Global Risks. In: *Advances in Soil Science, Volume 11. Soil Degradation*. Lal, R. and Stewart, B. A. (eds). Springer-Verlag, New York.
26. Buringh, P. 1989. Availability of agricultural land for crop and livestock production. In: *Food and Natural Resources*. Pimentel, D. and Hall, C.W. (eds). Academic Press, San Diego. p. 69–83.
27. Sehgal, J. L. et al. 1990. Agro-ecological regions of India. *Technical Bulletin 24*, NBSS&LUP, New Delhi, 1–76.
28. Vitousek, P. M. et al. 1986. Human appropriation of the products of photosynthesis. *Bioscience 36*, 368–373.
29. Lal, R. and Pierce, F. J. 1991. Soil Management for Sustainability. Ankeny, Iowa: Soil and Water Conservation Soc. In: *Coop. with World Assoc. of Soil and Water Conservation and Soil Sci. Soc. of Amer.*
30. McDaniel, T. A. and Hajek, B. F. 1985. Soil erosion effects on crop productivity and soil properties in Alabama. 48–58. In: *Erosion and Soil Productivity*. ASAE Publication 8-85. St. Joseph, MI.
31. Follett, R. F. and Stewart, B. A. 1985. *Soil Erosion and Crop Productivity*. Madison, WI: American Society of Agronomy, Crop Science Society of America.
32. Pimentel, D. 1990. Environmental and social implications of waste in agriculture. *J. Agr. Environ. Ethics 3*, 5–20.
33. Troeth, F. R., Hobbs, J. A. and Donahue, R. L. 1991. *Soil and Water Conservation, 2nd Edition*. Prentice Hall, Englewood Cliffs, NJ.
34. Dregne, H. E. 1982. *Historical Perspective of Accelerated Erosion and Effect on World Civilization*. Amer. Soc. Agron., Madison, Wisc.

35. Economic Research Service, AR-23, 1991. *Agricultural Resources: Cropland, Water, and Conservation Situation and Outlook Report.* US Department of Agriculture, Washington, DC.
36. Barrow, C. J. 1991. *Land Degradation.* Cambridge University Press, Cambridge.
37. Pimentel, D. 1993. *World Soil Erosion and Conservation.* Cambridge University Press, Cambridge.
38. 1 cm of soil over one hectare weighs 112 tons and 2.5 cm weighs about 280 tons. Thus a soil loss of 16 t ha^{-1} yr^{-1}, as in the US, will result in the loss of an inch of soil in 17.5 years.
39. Office of Technology Assessment. 1982. *Impacts of Technology on U.S. Cropland and Rangeland Productivity.* US Congress, Washington, DC.
40. Hudson, N.W. 1981. *Soil Conservation.* 2nd ed. Cornell University Press, Ithaca, NY.
41. Lal, R. 1984. Productivity assessment of tropical soils and the effects of erosion In: *Quantification of the Effect of Erosion on Soil Productivity in an International Context.* Rijsberman, F. R. and Wolman, M. G. (eds). Delft Hydraulics Laboratory, Delft, Netherlands. p. 70–94.
42. Lal, R. 1984. Soil erosion from tropical arable lands and its control. *Adv. Agron. 37,* 183–248.
43. Elwell, H. A. 1984. An assessment of soil erosion in Zimbabwe. *Zimbabwe Sci. News 19,* 27–31.
44. Environment Department, Policy and Research Division, Oct. 1980. *Asia Region: Review of Watershed Development Strategies and Technologies.* Asia Region, Technical Development, Agricultural Division, Washington, DC.
45. Pimentel, D., Stachow, U., Takacs, D. A., Brubaker, H. W., Dumas, A. R., Meaney, J. J., O'Neil, J., Onsi, D. E. and Corzilius, D. B. 1992. Conserving biological diversity in agricultural/forestry systems. *Bioscience 42,* 354–362.
46. Tolba, M. K. 1989. Our biological heritage under siege. *Bioscience 39,* 725–728.
47. Brown, L. 1984. *State of the World 1984.* Worldwatch Institute, Washington, DC.
48. Nitrogen Fertilizers. 1990. *FAO Fertilizer Yearbook 40,* 44–55, UN Food and Agriculture Organization, Rome.
49. Office of Technology Assessment. 1982. *Impacts of Technology on U.S. Cropland and Rangeland Productivity.* US Congress, Washington, DC.
50. McLaughlin, L. 1994. Soil erosion and conservation in northwestern China. In: *World Soil Erosion and Conservation.* Pimentel, D. (ed.). Cambridge University press, Cambridge. (In press).
51. Dunkerley, J. et al. 1990. *Patterns of Energy Use in Developing Countries.* Desai, A. V. *(ed). Wiley Easter, Limited, International Development Centre, Ottawa.*
52. Leyton, L. 1983. Crop water use: principles and some considerations for agroforestry. In: *Plant Research and Agroforestry.* Huxley, P. A. (ed.). International Council for Research in Agroforestry, Nairobi, Kenya.
53. Batty, J. C. and Keller, J. 1980. Energy requirements for irrigation. In: *Handbook of Energy Utilization in Agriculture.* D. Pimentel, D. (ed.). CRC Press, Boca Raton, FL. p. 35–44.
54. Ritschard, R. L. and Tsao, K. 1978. *Energy and Water Use in Irrigated Agriculture During Drought Conditions.* Lawrence Berkeley Lab., University of California, Berkeley.
55. World Resources 1992–93. *A Report by the World Resources Institute.* Oxford University Press, New York.
56. Pimentel, D. 1980. *Ambio.* (Hittar ej Pimentel som förf. 1980).
57. Postel, S. 192. *Last Oasis: Facing Water Scarcity.* W. W. Norton and Co., New York.
58. Postel, S. 1984. *Water: Rethinking Management in an Age of Scarcity.* Worldwatch Paper No. 62, Worldwatch Institute, Washington.
59. Postel, S. 1989. *Water for Agriculture: Facing the Limits.* Worldwatch Institute, Washington, DC.

60. Anonymous 1992. Salt of the earth. *The Economist 323*, 34.
61. Gleick, P. H. (ed). 1993. *Water in Crisis: A Guide to the World's Fresh Water Resources*. Oxford University Press, New York.
62. Global Environment Monitoring System. 1991. *Freshwater Pollution*. United Nations Environment Programme. Nairobi, Kenya.
63. Smil, V. 1993. *China's Environmental Crisis: An Inquiry into the Limits of National Development*. M. E. Sharpe/East Gate Books, New York.
64. Postel, S. 1990. Saving water for agriculture. In: *State of the World*. Worldwatch Institute, Washington, DC.
65. Kerr, R. A. 1992. Greenhouse science survives skeptics. *Science 256*, 1138.
66. Stone, P. H. 1992. *Technology Review*. p 32, Feb/Mar with additional correspondence, ibid, May/Jun, p 10.
67. Houghton, J. T. et al (eds). 1990. *Climate Change—The IPCC Scientific Assessment*. Cambridge University Press, Cambridge. The supplementary report. 1992. *IPCC 1992 Supplement Science Assessment—January 15*.
68. Bruce, J. P. 1992. UN official, chairman of the Science and Technology Committee for the UN International Decade for Natural Disaster Reduction, cited a study projecting significant climatic change over several regions in the next 40 years. Central US and European Mediterranean basin summer temperatures would rise 2–3°C accompanied by a 15–25% reduction in soil moisture. Reported at the MIT Conference "The World at Risk: Natural Hazards and Climate Change," MIT, Cambridge, Mass, Jan 14–16, 1992.
69. US Dept. of Agriculture. 1990. *Agricultural Statistics*. US Gov't Printing Office, Washington.
70. Martin, P., Rosenberg, N. and McKenney, M. 1989. Sensitivity of evapotranspiration in a wheat field, a forest, and a grassland to changes in climatic and direct effects of carbon dioxide. *Climatic Change 14*, 117–152.
71. Bazzaz, F. A. and Fajer, E. D. 1992. Plant life in a CO_2-rich world. *Sci. Am. 266*, 68–74.
72. Kerr, R. A. 1992. New assaults seen on earth's ozone shield. *Science 255*, 797–798.
73. Levi, B. G. 1992. Arctic measurements indicate the chilly prospect of ozone depletion. *Physics Today 45*, 17.
74. Stolarski, R. et al. 1992. Measured trends in stratospheric ozone. *Science 256*, 342.
75. Smith, R. C. et al. 1992. Ozone depletion: Ultraviolet radiation and phytoplankton biology in Antarctic waters. *Science 255*, 952–959.
76. National Academy of Sciences. 1989. *One Earth, One Future*, Washington.
77. Teramura, A. H. and Sullivan, J. H. 1989. How increased solar ultra-violet billion radiation may impact agricultural productivity. *Proc. of the 2nd North American Conf. on Preparing for Climate Change: A Cooperative Approach*. Published by the Climate Institute, p 203–207.
78. Biggs, R. H. and Webb, P. G. 1986. Effects of enhanced UV-B radiation on yield and disease incidence and severity for wheat under field conditions. *NATO ASI Series G-8, Ecological Sciences*.
79. Durning, A. T. and Brough, H. B. 1992. Reforming the livestock economy. In: *State of the World*. Brown, L. R. (ed.). W. W. Norton & Co. New York, p. 66–82.
80. Edwards, C. A. et al. 1989. *Sustainable Agricultural Systems*. Soil and Water Conservation Society, Ankeny, Iowa.
81. Paoletti, M. G. et al. 1989. Agricultural ecology and the environment. *Agr. Ecosyst. Environ. 27*, 630 p.
82. Starr, C. et al. 1992. Energy sources: A realistic outlook. *Science 256*, 981–987, (especially 985–986).
83. Land, Food, and People. 1984. *FAO Economic and Social Development Series, 30*. UN Food and Agriculture Organization, United Nations, NY.

84. Pimentel, D. 1991. Diversification of biological control strategies in agriculture. *Crop Protect 10*, 243–253.
85. Rubin, E. et al. 1992. Realistic mitigation options for global warming. *Science 257*, 148.
86. Union of Concerned Scientists. 1991. *America's Energy Choices: Investing in a Strong Economy and a Clean Environment*. Cambridge.
87. Holdren, J. P. 1991. Population and the energy problem. *Pop. Environ. 12*, 3.
88. Lal, R. 1991. Land degradation and its impact on food and other resources. In: *Food and Natural Resources*. Pimentel, D. (ed.). Academic Press, San Diego, p. 85–104.
89. "Third world countries are building up population at a rate no imaginable economic development could accommodate" Lucien Paye, Head of OECD, quoted in the New York Times, July 14, 1990.
90. Borlaug, N. E. 1985. Making institutions work—a scientist's viewpoint. In: *Water and Water Policy in World Food Supplies*. Proceedings of the Conference, May 26–30 Texas University. Texas A&M University Press, College Station, Texas.
91. Anonymous. 1992. Why the poor don't catch up. *The Economist 323*, 48.
92. UN Food and Agriculture Organization. 1991. Table 15. In: *FAO Quart. Bull. Stat. 4*, 1–103.
93. Stolwijk, H. J. J. 1991. De wereldvoedselvoorziening op de lange termijn: een evaluatie van mogelijkheden en knelpunten. *Onderzoeks Memorandum 83*. Den Haag, The Netherlands: Centraal planbureau. The estimates are that 24 m ha of productive agricultural land will be lost to urbanization and 380 m ha to erosion and desertification, with 320 m ha of new agricultural land added through deforestation.
94. UNFAO Report 1984. *Potential Population-Supporting Capacities of Land in the Developing World*. UN Food and Agriculture Organization, NY.
95. World Resources. 1992. *The Worldwatch Institute*, Washington, DC.
96. UN Food and Agriculture Organization. 1991. *Quart. Bull. Stat. 4*, 26–27.
97. Rosenzwieg, C., Parry, M. L., Fischer, G. and Frohberg, K. 1993. Climate change and world food supply. *Research Report No. 3*. Environmental Change Unit, Univ. of Oxford.
98. Rozenzwieg, C. and Parry, M. L. 1994. Potential impact of climate change on world food supply. *Nature 367*, 133–138.
99. UNFAO Report. 1984. *Potential Population-Supporting Capacities of Land in the Developing World*. UN Food and Agriculture Organization, NY.
100. Putnam, J. J. and Allshouse, J. E. 1991. *Food Consumption, Prices, and Expenditures, 1968–89*. US Dept. of Agriculture, ERS, Statistical Bulletin No. 825, Washington.
101. Agricultural Statistical Data of China in 1990, Agricultural Ministry of PRC, Agricultural Press, Beijing, China.
102. UN Food and Agriculture Organization 1991. *Food Balance Sheets*. Rome.
103. Agricultural Inputs, 1975–89. 1992. Table 18.2. In: *World Resources 1992–93*. World Resources Institute, Washington, DC.
104. Myers, N. 1989. *Deforestation Rates in Tropical Forests and Their Climatic Implications*. Friends of the Earth Report, London.
105. Myers, N. 1990. Mass extinctions: What can the past tell us about the present and future? *Global and Planetary Change 2*, 82.
106. UN Food and Agriculture Organization *Quart. Bull. Stat. 4*, 26–27.
107. The authors wish to express their appreciation to the following people for their help in the preparation of this work: V. Abernethy, M. Brower, S. Chisholm, W. Dazong, M. El-Ashry, P. Faeth, M. Falkenmark, M. Giampietro, R. Goodland, K. Gottfried, S. Harris, D. Hornig, T. Mount, I. Oka, E. Oyer, M. Paoletti, M. Pimentel, P. Pinstrup-Andersen, T. Poleman, S. Postel, P. Raven, K. Robinson, T. Scott, L. Stifel and N. Uphoff.
108. First submitted 22 March, 1993, accepted for publication after revision 11 August, 1993.

Meeting Population, Environment and Resource Challenges: The Costs of Inaction

I Introduction

The task of two panels of senior scientists at the 1995 Third Annual World Bank Conference on Effective Financing of Environmentally Sustainable Development was to set out their views on the major environmental, resource and population issues that constitute obstacles to environmentally sustainable development in the developing nations, to outline the costs of inaction on these pressing matters and to suggest paths for all nations that would blunt the risks we face. Some people believe, and recent media coverage suggests, that the world's senior scientific community is sharply divided on these matters. Let me assure you otherwise, it is not.

In this paper, the seven panel members and one invitee unable to attend the meetings review the most important of the issues facing this and subsequent generations, informally expressing the views of a great number of their scientific colleagues. Their contributions are assembled here in one place, something not possible in the panel presentations.

> Human beings and the natural world are on a collision course. Human activities inflict harsh and often irreversible damage on the environment and on critical resources. If not checked, many of our current practices put at serious risk the future that we wish for human society and the plant and animal kingdoms, and may so alter the living world that it will be unable to sustain life in the manner that we know. Fundamental changes are urgent if we are to avoid the collision our present course will bring about.

Excerpts by Henry W. Kendall from a report of the Senior Scientists' Panels, 1995 Third Annual World Bank Conference on Effective Financing of Environmentally Sustainable Development, Washington, DC, October 4 and 9, 1995. Panel members included: Henry Kendall, Kenneth Arrow, Norman Borlaug, Paul Erlich, Joshua Lederberg, José Vargas, Robert Watson, and Edward Wilson. Proceedings Series no. 14. World Bank: Washington, D.C., 1996.

These sentences are from the 1992 World Scientists' Warning to Humanity[1] a statement supported by over 1600 scientists from 70 countries, members of their respective academies of science, including over one half of the world's Nobel laureates in science, a majority of the Pontifical Academy of Science, as well as officers and former officers of scientific academies from developed and developing countries alike. This is the most prestigious group of scientists to speak out on the threats to global systems, and is a Warning not be disregarded.

The Warning concerns a cluster of grave problems that faces humanity: pressures on vital global systems leading to their injury and possible destruction, systems whose capacity to provide for us is being sorely stressed; resource mismanagement; the critical role that population growth plays as an impediment to gaining control of the threats. It comes from that international community most skilled and knowledgeable about the physical and biological systems of our world. It puts on a firm, authoritative basis the perils that we humans now face and bridges the sharp, often angry split that has developed between representatives of developing nations and those of developed, industrial nations over who is responsible for the increasing pressure on the global environment and for the challenges of population growth, putting these matters in broad perspective. Scientists from developing and developed nations alike are united in their concern for the global environment and the capacity to sustain future generations.

In 1992 and 1993, two groups of national science academies have issued statements on the same issues that are the subject of the Scientists' Warning.[2] Their messages were essentially the same as the Warning's.

There should be no doubt that a strong consensus exists within the scientific community that we are currently headed down a path which is not environmentally sustainable.

The Present

The principal areas of concern among the nest of difficulties that we and our world now face are:

The Atmosphere

Carbon dioxide and other gases released in large quantities into the atmosphere by human activity now appear to have started climate alteration on a global scale, with possible far-reaching effects, lasting for centuries. Stratospheric ozone depletion from other gases having anthropogenic sources is posing the threat of enhanced ultraviolet levels reaching the ground, against which much of the natural world has little protection. Air pollution near ground level, including acid precipitation and enhanced ozone concentrations, is causing widespread injury to humans, forests and crops.

Land and Forests

Human activities have altered between one third to one half of the earth's ice-free surface. Crop and range land is under heavy pressure worldwide and increasingly is suffering from widespread injurious practices: salinization, erosion, overcultivation, and overgrazing. Much cropland is losing its fertility and there is extensive land abandonment. Injury and loss of arable land is one the world's most critical environmental problems, one of the least appreciated and carrying heavy consequences.

Tropical rain forests, as well as tropical and temperate dry forests, are being destroyed rapidly. Deforestation is a worldwide problem, in the US and Canada, South America, Africa, India, SE Asia, indeed, nearly everywhere. Tropical deforestation is essentially irreversible owing to resulting rainfall changes and nutrient-poor soil.

The Oceans and Fisheries

Oceanic regions near the coasts, which produce most of the world's food fish, are being injured by pollution from soil erosion, as well as industrial, municipal, agricultural, and livestock waste.

Pressure on marine fisheries from overexploitation is increasingly severe. Overfishing has already caused some important food fish in some fisheries to become commercially extinct and more are on the way. Of the world's 17 major fisheries, all are either fully exploited, in decline, or collapsing and the world catch of fish has been declining since the early 1990s. The present harvest of fish is estimated to be nearly 25% greater than the sustainable yield that the oceans could provide.

Species Loss and Ecosystem Damage

The combined effects of loss of habitat, overexploitation, spread of alien species, air, soil, and water pollution, and climatic change are already the cause of unprecedented loss of species of living things. In the tropics it has been referred to as "biotic devastation."

Fresh Water

We are exploiting and contaminating depletable ground water supplies heedlessly and heavy pressure on the world's surface waters has become a serious concern. This has led to shortage of water becoming a major global problem. About 40% of the world's population lives in regions where water is in short supply. Water supplies for direct human use and for production of food are coming under stress over large parts of the world; the US, the Middle East, India, and parts of Africa. Water for irrigation is vital for today's food production.

Agriculture and Food Production

Globally, earlier increases in per capita grain production have now leveled off and a decline appears to have started. Per capita food production in many nations is now clearly decreasing. With food producing systems under substantial environmental stress, the task of feeding the world's growing population has become extraordinarily vexing.

Energy

Energy is the lifeblood of modern industrial, as well as developing nations. Yet energy-related activities are top contributors to global environmental pollution. How to manage the side effects and yet provide adequate energy supplies is, along with food production, one of the very top problems our race faces.

Disease

The emergence or reemergence of diseases afflicting humans, as well as plants and animals important to humans, is becoming an international concern. Causative agents and disease vectors are, increasingly, becoming resistant to chemical or biological controls.

People

World population grew slowly up until about 1900, but growth has accelerated since that time. The present population stands at about 5.7 billion. While fertility rates have dropped in recent years, they are still not close to replacement levels and the numbers added annually to the world's total, about 90 million per year, are at a record level. The current rate of increase corresponds to a population doubling time of roughly 40 years. The sheer number of people, coupled with persistent population growth, are linked to numerous environmental and resource concerns, converting many difficult problems into potentially intractable ones.

For a number of decades, there has been an accelerating movement of peoples in developing countries from rural areas to rapidly enlarging cities. A five-fold increase in the number of city dwellers in the years 1950 to 1990 brought the urban population in these nations to 1.5 billion, some 37% of their total population.

Patterns of Economic Activity

Governments as well as private enterprises frequently pursue goals in ways that produce substantial environmental damage, often on a large scale. The patterns of such activities must be scrutinized with a view to reducing short and long term damage they now cause.

IX The Challenges

Business As Usual and Its Costs

The preceding chapters have set out a deeply sobering view of the injury to important global physical and biological systems as well as resources crucial to our well-being that is resulting from reckless human activities. It is made abundantly clear that corrective action will be frustrated by continuing expansion of human numbers and, accordingly, that most *all* aspects of this nest of problems must be promptly and powerfully addressed. The concern of the world scientific community over the future is fully justified.

While it is not possible to make satisfactory predictions of the future course of events, it is possible to outline in broad terms the consequences of humanity's continuing on its present course, that is, business as usual: environmental damage and resource mismanagement remaining uncontrolled while human numbers grow. In what follows, with the material of the earlier chapters as prologue and background, we look at the near future, roughly to the middle of the next century, assuming that the patterns of human activity outlined above remain unchanged.

Population

The United Nations has concluded that the world population could reach 14 billion, a near tripling of today's numbers[3,4] and the momentum of population expansion appears likely to result in about 10 billion persons by the year 2050, or a near doubling of today's numbers. The bulk of the growth, well over 90%, will occur in developing countries. The industrial world's 1 billion people is expected to grow to about 1.2 billion by 2050.

There will likely be a further three-fold expansion of the developing world's population in urban areas to 4.4 billion during the next 30 years, comprising two thirds of the developing world producing numerous "megacities." While such cities have a number of attractive features for developing nations, focussing and easing many productive activities, they can also generate a host of thorny problems, especially so where environmental degradation, poverty, hunger and violence are growing national difficulties. So stressed, these urban areas can become the triggers for civil disruption, if not revolution. Urban expansion typically consumes sorely needed arable land.

Environmental and Ecosystem Damage

The estimated rate of loss of soil from arable land through erosion, in excess of new soil production, is estimated to be 23 billion tons per year globally, or about 0.7% loss of the world's soil inventory annually. Continued erosion at this rate will result in the loss of over 30% of this inventory by the year 2050, an enormous and irreplaceable loss, one which, in the long run, would have an extraordinarily harmful effect on global food production. Other injurious agricultural practices hinder the task of raising crop productivity. Per capita arable acreage is now decreasing and, in business-as-usual, will continue to do so for decades.

If present rates of deforestation persist, or worse, increase, the tropical rain forests may be largely gone by the end of the next century. The irreversible loss of species—which, under business-as-usual, may approach one third of all species now living by the year 2100—is especially serious.

Energy Use and Its Consequences

With world population growing and with no changes in current patterns of energy use, there will almost certainly be a major expansion of fossil fuel use throughout the world. Such an expansion will lead to a doubling of the amount of the "greenhouse gas," carbon dioxide, in the atmosphere sometime about the year 2100, the most severe of the carbon emission scenarios now contemplated. The resulting climatic changes will persist for centuries and, depending on the scale of continuing emissions, may continue to increase.

Food Prospects

With nearly one fifth of the world's population now suffering from inadequate nutrition, the stage is set for a substantial increase in this fraction. Decline in global per capita food production will worsen in coming decades if matters continue as they have. Under business-as-usual, it is possible that even in 25 years, today's number of people living on the edge of starvation would double to some 1 1/2 to 2 billion persons. The expansion of irrigation has slowed for environmental reasons as well as from water shortages. If continued, this will aggravate the food shortages which already loom ahead and further stress numerous impoverished nations. Water shortages have already stimulated a number of conflicts and this will worsen as the demand increases in the years ahead.

Based on historical experience, the 38 nations with population growth rates above 3% per year probably cannot expand their food production to keep pace with this growth. The 63 additional nations whose population growth rates are in the range 2–3% per year may experience great difficulty in keeping pace.

Continued pressure on the world fisheries, as at present, will result in widespread injury and, in some, collapse in the years ahead, all far beyond what has occurred to date. It is quite possible that virtually all of the ocean's major fisheries could be destroyed. Although the global consequences will not be great, because fish constitute no more than a few percent of human food, it will still be a major tragedy. Some nations far more dependant on food from marine sources than the average, will suffer serious consequences from declining catches.

The possibility of providing adequate food for many of the developing world's people is remote if business-as-usual persists much longer into the future. Stress on these people's food supplies arising from climatic change will hasten the troubles.

Linkages and Surprises

With the global systems that we depend on under multiple stresses, an additional array of environmental difficulties appears likely to surface: synergistic effects, aggravated, out-of-proportion damage arising when additional stress reduces a system's tolerance to those already imposed. This can give rise to unexpected disruption. Moreover, there can be known, or unsuspected linkages between activities; for example, climate warming, urbanization, deforestation, and easily accessible transportation all can enhance the spread of disease. Because large scale systems are so complex, they can evolve in ways that cannot be predicted. For all these reasons we are thus subject to surprises. These can be favorable but they may not be, as the CFC injury to the ozone layer proved. Our catalog of the costs of business as usual is necessarily incomplete.

Assessment

It appears to a great many senior scientists that the costs of business-as-usual are very great because the disruption of so much that we depend on is involved. In particular, the following large scale enterprises are currently non-sustainable: present agricultural practices, present energy practices—including dependance on some sources, especially oil, as well as effluent disposal—world fisheries, an important portion of water use and much of the world's logging and timbering and, probably of most importance, procreation.

As the injuries progress and become more acute, many more countries in the third world appear likely to become trapped in spirals of decline, in parts of Africa, Asia, and Latin America: hunger, environmental damage, economic and political instabilities, disease, increasing mass migrations, conflicts over scarce resources. The wealthy nations, although more resilient, will not escape either, for many of the consequences will be visited on all.

An Alternate Future

Controlling Environmental Damage

Great changes, effected very promptly, are now required in our stewardship of the earth if vast human misery is to be avoided and our global home is not to be irretrievably mutilated:

1. Environmentally damaging activities must be brought under control to restore and protect the integrity of the earth's systems we depend on, and we must repair as much of the past damage as we can.

2. We must greatly improve the management of resources crucial to human welfare, including more efficient and frugal use of those in short supply. The necessary goal is to organize human activities on a sustainable basis so that the generations that follow ours may enjoy the benefits that we do. It is far from being a problem solely of the underdeveloped world. There is convincing evidence that most industrial nations are overpopulated as well, given their impact on the environment, and that many of their activities are unsustainable.

Because there is so little time remaining before irreparable injury is visited on critical global systems, the need for a change of course is now urgent. A thoughtful, widespread and effective attack on the problems could still succeed in blunting or averting the most severe consequences. Such a program would provide an alternative future far more humane than business-as-usual. We have only one or a few decades to get remedial actions well started.

Resources and Sustainability

The current rate of increase in food production will need to be enhanced. The bulk of the increases will have to come from productivity improvements, owing to the limited potential for expanding the area of arable land as well as the fraction irrigated. Increases in crop productivity require additional energy for fertilizer and pesticide production and distribution, for tillage, for irrigation, for transportation. The improvements must be effected concurrently with shifts to less damaging agricultural and irrigation practices so that agriculture and animal husbandry become sustainable.

Water shortages will be one of the main constraints on increasing food production and will add to the misery of growing populations in other ways as well. Thus a major priority must be attached to more efficient use of water, in agriculture and elsewhere.

While many of the techniques of the Green Revolution can help to provide increases in food production, a further great expansion of production will still be difficult to achieve. A highly organized global effort involving both the developed and the developing worlds will be required so that food production can be expanded worldwide by even a factor of two by 2050.

Over two thirds of the tropical deforestation now underway is to provide more arable land for food production. Food production has to be increased on deforested, tropic land that is already under cultivation, at the same time introducing sustainable agricultural practices, so as to halt the loss of tropical forest and the resulting loss of animal and plant species.

The urgent need to limit climatic change underlies calls for decreasing reliance on fossil fuels. Thus a move to new energy sources, primarily based on solar, geo-thermal, wind and biomass energy, as well as improved efficiency, can bring major—and badly needed—benefits both to developed and to underdeveloped nations. Indeed, such a move constitutes a vital element in any program to mitigate the risks we face. Its direct importance to industrialized nations such as the United States is generally well understood although not embraced or supported by our political leaders. Development of these sources will be crucially important to the developing nations also, to help meet their aspirations for higher standards of living. Nuclear power could contribute provided its problems of safety, economics, waste disposal and public acceptance can be resolved. Both developing and industrial nations must work together to develop and implement new sources of energy.

No one should underestimate the magnitude of the task of expanding effective energy supplies while moving to new sources of energy of lower environmental impact, with the planning, costs, and commitments that will be required. Nevertheless, it seems an achievable goal. The World Bank should be giving much higher priority to the development of nonfossil sources of energy, especially in forms suited to developing nation's needs, and to improved efficiency of use.

Reducing Conflict

A great diminution of the violence and war that characterize our world is necessary. Many of the resources devoted to the preparation and conduct of war—now amounting to nearly $1 trillion annually—are needed for the new constructive tasks as well as to provide relief from the destruction that war entails.

Stabilizing Population

The destruction of our environment and resources cannot be stemmed unless the growth of the world's population is stemmed and, ultimately, reduced. Continued population growth threatens virtually every step taken to control the challenges we face. This growth is putting demands on resources and pressures on the environment that will overwhelm any efforts to achieve a sustainable future. Almost all scientists who have looked at the problem recognize that without curtailing this growth, virtually all the other major problems of the planet may become wholly intractable. Remedies are overwhelmed. And population growth rates, mainly in the developing world, while declining, are still too large to allow adequate protection of life-sustaining resources and our environment.

The Challenge to Society

The challenge to the present generation is to change the path that we are now on, to start the great programs of remedial action described above so as to greatly lessen environmental damage and stabilize our numbers in decent ways. A major change in our stewardship of the world and the life on it is now required for the barriers and obstacles that have caused so much trouble in the past continue to hinder progress. How can we do this? Can we look to the power of modern science to cope with the dangers?

Barriers and Obstacles

The bleak prospect that business-as-usual provides is in part a consequence of the magnitude and impact of human activities. The human species is now interfering with life and with geophysical processes on a global scale. It amounts to a virtual assault on the environment, with much of the damage permanent or irreversible on a scale of centuries.

The bleak prospect is in part also a consequence of mistaken, shortsighted exploitation, both of renewable and of nonrenewable resources and exceeding the capacity of the globe to absorb effluent without harm. Present patterns of economic development, in both developed and underdeveloped nations, with the environmental and resource damage they now bring, cannot be sustained because limits on the capacities of vital global systems will soon be reached. As this occurs, there is great risk that some of these systems will be damaged beyond repair and trigger catastrophe that would sweep all into its fold.

The earth and its resources are finite, it has limited capacity to provide for us, it has limited capacity to absorb toxic wastes and effluent. There are limits to the number of people that the world can sustain and we are now fast approaching some of these limits.

Governments and their peoples have always had difficulty recognizing concealed costs, one facet of priorities that emphasize dealing with immediate problems over long term ones, immediate benefits over delayed gratification. Depletion of natural resources is unaccounted for in most all national budgets and few, if any, governments coordinate their planning and actions in such areas as environment, health, food production, energy. All too often it appears that nations are unconcerned with their own futures.

The Responsibilities of Nations

Developing nations must realize that environmental damage is one of the gravest threats they face, and that attempts to blunt it will be overwhelmed if their populations go unchecked. The greatest peril is to become trapped in spirals of environmental decline, poverty, and unrest, leading to social, economic and environmental collapse.

The developed nations must greatly reduce their own over-consumption, their own pressures on resources and on the global environment—they are the world's worst polluters—and they must also aid and support developing nations. These are not disinterested responsibilities or altruistic acts. The earth's systems sustain us all and wealthy nations cannot escape if these are hurt.

All nations must recognize that improving social and economic well-being, coupled with effective family planning programs, are at the root of voluntary, successful fertility declines and population stabilization. Reduction of poverty, pervasive in the third world, including improvements in the status and education of women—who have a vital role in development—are irreplaceable needs. Women have important roles in family planning, in family health, in subsistence agriculture and without education the challenges cannot be met. Education, both of men and women, is a wonderful contraceptive.

One cannot overemphasize these needs. If countries in the third world continue to become trapped in decline, the resulting mass migrations will have incalculable consequences for developed and developing nations. Conflicts may develop over scarce resources that none can escape. One of the few humane solutions to such difficulties is help for these nations so that their people do not need, nor wish to leave. It can come only from those nations with the technological skills and the resources to provide it.

Science and Technology

Many among the public and in governments share the mistaken belief that science, with new, ingenious devices and techniques, can rescue us from the troubles we face without our having to mend our ways and change our patterns of activity. This is not so; technology and technological innovation will produce no "Magic Bullets" that can rescue us from the dilemmas and risks that lie ahead for the problems we face are human problems and must be recognized as such.

One should not misunderstand. Scientific advances are absolutely necessary to aid us in meeting the challenges ahead. Science and technology will continue to have most important roles to play in implementing the changes that must come, in agriculture, medicine, public health, energy technologies and a host of other areas. Science and technology have already become a life support system for us all—we cannot do without them. Yet it is a grave error to believe, as many appear to, that science and technology will allow us to avoid the threats that face us all, to continue on as we have been doing, and to evade the deeper problems we now face. Scientists do not believe this, as is clear in the Scientists' Warning and in the Academy statements. No one else should believe it either.

The devastation of global fisheries is close at hand. It has not been halted by environmentalists' pressures. Neither has science or technology helped to spare these marine resources. Quite the contrary; science and technology enabled the destruction. Searchlight sonars, great factory trawlers, monofilament nets tens of miles long "curtains of death," have done what hand fishing could never accomplish: make it possible for fishermen to sweep the sea clean of edible fish. It is a sobering example and a warning, too.

The Role of Governments

Governments now must take lead roles in moving their nations in the right directions. They must now inform themselves, plan programs that will forestall the more distant injuries that earlier would have been set aside, balancing the needs of present with requirements of future. This surely is a major break with the past, where the response of governments to environmental issues was to follow, in some cases needing to be driven to ameliorate distress.

In these new government actions, attention should be given to quelling institutional rivalry between organs of government instead providing motivation for joint attack on the tasks. Disarray is costly, wasting time and money.

To carry out this new array of responsibilities, to coordinate efficient, practical responses, to inform, and gain the agreement of their peoples, will require insight, planning and courage. There will surely be great opposition.

The World Bank can help by supporting powerful programs to mitigate climate change, disease, sustainable energy sources, reform and other actions to increase agricultural productivity and "human development." The last includes education (especially opportunities for women), "capacity building," comprehensive health services and voluntary family planning—all those things that provide the "Human Capital" in the Bank's recent and valuable reassessment of national wealth.

The world scientific community can be of great help. Indeed, it is an under-utilized resource, not usually well coupled to major programs. Generally its members can recognize problems and linkages between problems, in the environmental area, better than any other group and can provide astute views on the balances needed in resource exploitation and on what activities are truly sustainable.

The main responsibility will, in the final analysis, rest with governments. They will have to provide the vision, the discipline and the major commitment of resources necessary for this global effort to succeed.

Conclusions

The magnitude and number of the challenges and speed of their arrival are unlike any that human beings have faced. There is little time remaining. Global limits are so near that is is *our* generation that must face and resolve these challenges.

We can yet avert a grim future. If we move, successfully, to meet these challenges, we can have that increasingly decent and humane world that we all wish for ourselves and for successor generations.

References and Notes

1. Union of Concerned Scientists, "World Scientists' Warning to Humanity" (Cambridge MA, November, 1992).
2. "Population Growth, Resource Consumption, and a Sustainable World," joint statement by the officers of the Royal Society of London and the U.S. National Academy of Sciences, 1992; "Population Summit of the World's Scientific Academies," joint statement issued by 58 of the world's scientific academies, New Delhi, October 1993.
3. *Population, Resources and the Environment*, United Nations Population Fund, United Nations, NY, 1991. See also *The State of World Population 1991*, UNFPA, UN Population Fund, UN, New York, 1991.
4. *Long-range World Population Projections*, Department of International and Economic and Social Affairs, United Nations, New York, 1992, Document Number ST/ESA/SER.A/125.

Bioengineering of Crops

I. Introduction

The primary objectives of the World Bank's international program are the alleviation of poverty, malnutrition and human misery in developing nations while encouraging and supporting a transition to environmentally sustainable activities. The issue of providing adequate food, based on sustainable agricultural practices, looms large in gaining these objectives, for failure in this area virtually guarantees failure to meet other objectives. It would make certain continued misery for many of our fellow human beings. Agricultural systems are already under stress and will become more so as populations continue to swell and the need for increased food supplies swells with them.

The World Bank over many years has made important contributions to the alleviation of hunger and malnutrition through its programs that aid developing world agriculture. Its aid was a major factor in making India self-sufficient in food production in the difficult time after World War II. Its agricultural research coordinating group, the Consultative Group on International Agricultural Research, was a major player in introducing the Green Revolution, "which contributed so much to the economic growth in the developing world."[1] Despite past contributions by the Bank, other organizations and by nations, the need to enhance food security in much of the developing world will remain a critical problem for many years to come.

Among the numerous responses to the need to expand food supplies in the developing world in environmentally benign ways is the bioengineering of crops. Bioengineering has much to contribute but it is a novel system and possible risks need to be evaluated carefully. Opposition to research and application has already arisen, not all of it carefully thought out. As the World Bank has recognized, a considered and technically competent understanding both of the potential and the perceived

---

Excerpts by Henry W. Kendall from a report of the World Bank Panel on Transgenic Crops, 1995 Third Annual World Bank Conference on Effective Financing of Environmentally Sustainable Development, Washington, DC, October 4 and 9, 1995. Panel members included: Henry Kendall, Roger Beachy, Thomas Eisner, Fred Gould, Robert Herdt, Peter Raven, Jozef Schell, and M.S. Swaminathan.

risks of bioengineered crops is a requisite to their successful development and use. Public perceptions that genetically engineered crops and animal products pose specific dangers present an important condition of public acceptance that needs careful consideration and resolution if such products are to reach widespread utilization.

Ismail Serageldin, Vice President of the Bank for Environmentally Sustainable Development, in 1996, initiated a study panel to assess the potential of crop bioengineering as well as the inherent risks, to provide the Bank with guidance in its activities. This is the Panel's report.

In what follows, we review, in Part II, the status of world food supplies and the prospects and needs for the future with emphasis on the developing world. Part III sets out a description of bioengineering technology and the potential contributions that transgenic crops might make to the alleviation of problems of food security. Part IV deals with possible risks from the widespread deployment of genetically altered crops. Finally, Part V includes the Panel's conclusions and recommendations.

Reference

1. Henry Owen, *"The World Bank: Is 50 Years Enough?" Foreign Affairs*, 97–108, September/October 1994.

II. World Food Supplies

Introduction

The current and prospective needs for food and the pressures and stress on the world's agricultural sector[1] generate a need to set priorities among this cluster of problems and the available solutions, including bioengineering of crops. This section of the report sets out and assesses these challenges as they stand today and, so far as one can, what the future may bring.

Current Circumstances

Population

The world's population now stands at 5.8 billion and is currently growing at about 1.5% per year. The industrial, wealthy nations, including those in Europe, North America, Japan, as well as a few elsewhere, comprise about 1.2 billion persons and are growing at a slow rate, roughly 0.1% per year.

Population in the developing world is 4.6 billion, and is expanding at 1.9% per year, a rate that has been decreasing somewhat in the last decade. The least developed nations, with a total population of 560 million, are growing at 2.8% per year, corresponding to a doubling period of 24 years. At present about 87 million people are added to the world's numbers each year.

Foods: Nutrition and Malnutrition

The wealthy nations have high levels of nutrition and little problem supplying all their citizens with adequate food when they wish to do so. Indeed, well over one third of world grain production is fed to livestock to enhance the supply of animal protein, most heavily consumed in the industrial world.

In the developing world, matters are different. Somewhat more than 1 billion people do not get enough to eat on a daily basis, living in what the World Bank terms "utter poverty" and about half that number suffer from serious malnutrition. A minority of nations in the developing world are improving markedly the standard of living of their citizens—in some 15 countries 1.5 billion people have experienced rapidly rising incomes over the past 20 years. In over 100 countries 1.6 billion people have experienced stagnant or falling incomes. Since 1990 incomes have fallen by one fifth in 21 countries of Eastern Europe.

Had the world's food supply been distributed evenly in 1994, it would have provided an adequate diet of about 2350 calories for 6.4 billion people, more than the actual population.[2]

In addition to shortages of food suffered by many in developing countries, there are widespread deficiencies in certain vitamins and minerals. Vitamin A appears to be lacking from many diets, especially in Southeast Asia, and there is deficiency in iron, so that anemia is widespread among women in the developing world.[3]

Food prices have been declining over the past several decades and some observers have argued that this is a sign that food adequate for all is now available. However those in utter poverty do not have the resources to purchase adequate food, even at today's prices. More recently, food prices have risen while grain stocks have fallen to their lowest level in 30 years.[4]

Agriculture

About 12% of the world's total land surface is employed for growth of crops, about 30% is in forest and woodland, and 26% in pasture or meadow. The remainder, about one third, is utilized for other human uses or is unusable for reason of climate, inadequate rainfall or topography. In 1961, the amount of cultivated land supporting food production was 0.44 ha per capita. Today it is about 0.26 ha per capita and based on population projections it is suggested that by 2050 it will be in the vicinity of 0.15 ha per capita.[5] The rate of expansion of arable land is now below 0.2% per year and continues to fall. Because the bulk of the land best suited to rainfed agriculture is already under cultivation, land that is being brought into cultivation generally has lower productivity than land currently used.

Loss of land to urbanization frequently involves the loss of prime agricultural land as cities have usually been founded near such land. The losses in prime land are often not counterbalanced by bringing other land into production owing to lack of the infrastructure that is generally required for market access.

Irrigation plays an important role in global food production. Of currently exploited arable land, about 16% is irrigated, producing over one third of the world crop. Irrigated land is, on balance, over two and one half times more productive than rainfed land.

The situation in India and China is particularly acute as their people comprise nearly one half of the developing world's population. They share the dual problem of expanding population and diminishing per capita arable land and water resources.

The average size of a farm in both countries is one hectare or less. Agriculture, including crop and animal husbandry, forestry and fisheries, has been both a way of life and means to livelihoods in these two countries for several thousand years. Expansion in population and increase in purchasing power coupled with the diversion of prime farm land for non-farm uses make it essential that these two countries, among others, adopt ecologically sustainable, intensive and integrated farming systems. See Appendix for a further description of these systems.

In China, land is socially owned but individually cultivated under a Household Responsibility System. In India, land is individually owned and agriculture constitutes the largest private sector enterprise in the country. India's population of 950 million is still growing at about 1.9% annually, while China's stands at 1.22 billion and is growing at 1.1% per annum. China has nearly roughly 50% of its cultivated land under irrigation, while less than 30% of India's cultivated area is irrigated.[6]

Agriculture in both these countries has to provide not only more food but also more employment and increased income. Modern industry is frequently associated with economic growth but without adequate expansion of employment, while modern agriculture can foster job-led economic growth. Therefore, farming cannot be viewed in these countries as merely a means of producing more food and other agricultural commodities, but rather as the very foundation of a secure livelihood. The role of new technologies such as biotechnology, information and space technologies and renewable energy is pivotal to building vibrant agricultural sectors, producing more from less land and water while contributing to the local economies.

Pressures on Agricultural Systems

Widespread injurious agricultural practices, in both the industrial and the developing worlds, have damaged the land's productivity, in some cases severely.[7] These include water and wind induced erosion, salination, compaction, waterlogging, overgrazing and a host of others. For example, the estimated loss of topsoil in excess of new soil production is estimated to be about 0.7% of the total each year; this amounts to some 25 billion tons, equivalent to the total of Australia's wheat growing area. An additional 0.7% annual loss occurs from land degradation and the spread of urbanization. Erosion has made a billion hectares of soil unusable for agriculture over past years.[8] Asia has the highest percentage of eroded land, nearly 30%, but in all major regions the percentage exceeds 12%.[9] It is estimated that 17% of all vegetated land has been degraded by human activity between 1945 and 1990.

The effect of erosion in crop yield is not well documented because such research is difficult and expensive and because degradation can be masked for short periods of time by more intensive agricultural practices. However, certain data are available.[10] Erosion can ultimately destroy the land's productive capacity by stripping off all of the soil, as has occurred in Haiti. "Haiti suffers some of the world's most severe erosion, down to bedrock over large parts of some regions, so that even farmers with reasonable amounts of land cannot make a living."[11]

Irrigation practices continue to contribute to salinization and other forms of land damage. For example, over one half of irrigated land is situated in dry areas and 30% of this is moderately to severely degraded. Salinization is a serious problem in India, Pakistan, Egypt, Mexico, Australia and the United States. Some 10% of the world's irrigated land suffers from salinization.

There are also serious problems with supplies of water and in much of the world is in short supply.[12] Worldwide, nations with some 214 river or lake basins, containing 40% of the world's population, now compete for water.[13] Much of irrigation depends on "fossil" underground water supplies, which are being pumped more rapidly than they are recharged.[14] This is a problem in portions of China, India, the United States, portions of Africa and several countries in the Middle East, especially Israel and Jordan. The human race now uses 26% of the total terrestrial evapotranspiration and 54% of the fresh water run-off that is geographically and temporally accessible. Most land suitable for rain-fed agriculture is already in production.[15]

The clear picture that emerges is that agricultural production is currently unsustainable. Indeed, human activities, as they are now conducted, appear to be approaching the limits of the earth's capacity. Like all unsustainable activities, these must, at some point, end. The end will come either from changes in practices, thus providing the basis for a human future, or by partial or complete destruction of the resource base with the misery that this would bring.

The Future

Population

Although fertility has been declining worldwide in recent decades, it is not known when it will decline to replacement level. There is broad agreement among demographers that, given the momentum inherent in present growth and if current trends are maintained, the world's population will increase to about 8 billion by 2020, will likely reach 10 billion by 2050 and, barring unexpected events, might possibly reach a level in the range 12 to 14 billion before the end of the next century. Virtually all of the growth in coming decades will occur in the developing world.

Food Demand

In order to provide increased nutrition for an expanding world population, it will be necessary to expand food production faster than the rate of population growth. Studies forecast a doubling in demand for food by 2025–2030.[16] This demand is larger than the projected increase in population, a result of the dietary changes and increased nutritional intake that accompany increased affluence.

Asia, with 60% of the world's population, contains the largest number of the world's poor. 800 million in Asia live in absolute poverty and 500 million in extreme poverty. Projections by the UN Food and Agriculture Organization, by the World Bank and by the International Food Policy Research Institute all show that the demand for food in Asia will exceed the supply by 2010.[17] China, the world's most populous nation with over 1.2 billion persons and an annual growth of 1.1% per year, faces very considerable challenges in years ahead from stress resulting from major environmental damage, shortages of water and diversion or degradation of arable lands.[18] Animal protein has increased in the Chinese diet from about 7% in 1980 to over 20% today, which substantially aggravates her food challenges. Virtually all the water available in China is used for agriculture, and heavy use of fertilizers has polluted much of the water supply.

Lester Brown and Hal Kane[19,20] have argued that India and China will need to import 45 and 216 million tons annually of food grains, respectively, by the year 2030 to feed their growing populations. This is due to the widening gap between grain production and consumption, caused by increased population and purchasing power. They have pointed out that while demand will grow, production prospects are not bright owing to stagnation in applying yield-enhancing technologies coupled with growing damage to the ecological foundations essential for sustainable advances in farm productivity. It is apparent that without substantial change there will not be enough grain to meet the need, a conclusion that at least some Chinese scholars agree with.[21]

Latin America appears to enjoy a relatively favorable situation with respect both to food supplies and to food security and is economically and demographically advanced in comparison with Asia and Africa.[21] Some regions are, however, under stress arising from economic problems coupled with continuing high rates of population increase. Northeast Brazil, Peru, Bolivia, much of Central America, parts of the Caribbean, especially El Salvador, Guatemala, Haiti and Honduras, face important challenges. Latin America's population is expected to increase from 490 million to nearly 680 million by 2025 and it is possible that over one quarter of the area's annual cereal consumption will be imported by 2020. The major countries in this region, including Argentina, Brazil, Chile, Colombia and Mexico, appear to have the resources necessary to meet their projected food needs[22] but this goal would require a stable population and implementing successful land management programs.

Middle East and northern African countries have seen demand for food outpace domestic production. Differences in oil wealth and agricultural production determine differences among the capacities of these countries to import grains and livestock products.[23] The greatest challenges will be faced by those nations that lack the capacity for substantial oil exports or other sources of wealth with which to purchase food imports. These nations include Afghanistan, Cyprus, Egypt, Jordan, Lebanon, Mauritania, Somalia, Tunisia and Yemen, whose 1994 combined population exceeded 125 million. Food self-sufficiency is unattainable for most of these countries. Oil exporting nations will, as their oil resources dwindle, join this less fortunate group.

Sub-Saharan Africa is the region whose prospective food supplies generate the greatest concern; since 1980, agriculture there has grown at 1.7% per year while population, now at 739 million, has grown at 2.9% per year.[24] Some twenty years ago, Africa produced food equal to what it consumed; today it produces only 80%.[25] With a population growth rate close to 3% per year, sub-Saharan African cannot close its food gap. The gap will likely grow, requiring increased imports of food to prevent growing malnutrition and increased risk of famine. If present world-wide decreases in foreign aid persist, these imports may not be forthcoming.

Agriculture and Irrigation

As described above, the current rates of injury to arable land are troubling. Since 1950, 25% of the world's topsoil has been lost and continued erosion at the present rate will result in the further, irreversible loss of at least 30% of the global topsoil by the middle of the next century. A similar fraction may be lost to land degradation, a loss which can be made up only with the greatest difficulty by conversion of pasture and forest, themselves under pressure. In Asia, 82% of the potentially arable land is already under cultivation. Much of the land classed as potentially arable is not available because it is of low quality or easily damaged.

The UN Food and Agriculture Organization has projected that over the next 20 years arable land in the developing countries could be expanded by 12% at satisfactory economic and environmental costs,[26] although it would inflict major damage to the world's remaining biodiversity. The yields per hectare on this land would be less than the land already in production. This is to be compared with the 61% increase in food demand that is expected to occur in these countries in the same period. . . . The last major land frontiers that potentially can be converted to arable land are the acid-soil areas of the Brazil *cerrado* and *llanos* of Colombia and Venezuela and the acid soil areas of central and southern Africa. Bringing these unexploited, potentially arable lands into agricultural production poses formidable, but not insurmountable, challenges.[26]

The prospects for expanding irrigation, so critical to the intensification of agricultural productivity, are also troubling. The growth of irrigated land has been slowing since the 1970s, owing to problems discussed above as well as "siltation" of reservoirs and from environmental problems and related costs that arise from the construction of large dam systems. This can include the spread of disease.

An important "wildcard" in any assessment of future agricultural productivity is the possible consequences of climatic change resulting from anthropogenic emissions of greenhouse gases. The consequences involve a very wide range of technical issues that will not be summarized here.[27] However, in its *Second Assessment Report*, The Intergovernmental Panel on Climate Change concluded, among other matters, that the balance of evidence suggests that there is now a discernable human influence on climate and that, assuming a mid-range emissions scenario, a global warming of about 2 degrees Celsius, with a range of uncertainty of from 1 to 3.5 degrees Celsius, would occur by 2100. Climate change consequent on a 2 degree warming would include regional and global changes in climate and climate-related parameters such as temperature, precipitation, soil moisture and sea level. These, in turn, could give rise to regional increases in

> the incidence of extreme high temperature events, floods and droughts, with resultant consequences for fires, pest outbreaks and ecosystem composition, structure and functioning, including primary productivity.[28]

. . . . .

> Crop yields and changes in productivity due to climate change will vary considerably across regions and among localities, thus changing the patterns of production. Productivity is projected to increase in some areas and decrease in others, especially the trop-

ics and subtropics.... There may be increased risk of hunger and famine in some locations; many of the world's poorest people—particularly those living in subtropical and tropical areas and dependent on isolated agricultural systems in semi-arid and arid regions—are most at risk of increased hunger. Many of these at-risk populations are found in sub-Saharan Africa; South, East and Southeast Asia; and tropical areas of Latin America, as well as some Pacific island nations."[28,29,30]

Further deleterious changes may occur in livestock production, fisheries and global forest product supplies. Salt intrusion into near-coastal aquifers, many of which supply water for irrigation, can occur as a result of rising sea levels. While there will remain important uncertainties in most aspects of possible climatic change and its consequences for some years, the matter must be included in assessing the prospects for the expansion of developing world nutrition that is required.

Prospects

Today, there are hundreds of millions of people who do not get enough food. Given the circumstances set out above, it appears that over the next quarter century grave problems of food security will almost certainly affect even larger numbers, as others have also noticed.[31]

> Given present knowledge, therefore, maximum realization of potential land, and water, supplies at acceptable economic and environmental costs in the developing countries still would leave them well short of the production increases needed to meet the demand scenarios over the next twenty years.[32]

> The task of meeting world food needs to 2010 by the use of existing technology may prove difficult, not only because of the historically unprecedented increments to world population that seem inevitable during this period but also because problems of resource degradation and mismanagement are emerging. Such problems call into question the sustainability of the key technological paradigms on which much of the expansion of food production since 1960 has depended.[33]

As is the case now, the people who will continue to be affected by shortfall in food production will, almost entirely, be those in the lower tier of the developing countries. The industrial nations and the developing nations whose economies continue to improve will face acceptable costs in providing their citizens with adequate nutrition. The extent of the deprivation and the economic and environmental costs remain the subject of controversy between optimists and pessimists.[34,35]

Meeting the Challenges

The twin challenge is to expand agricultural production to exceed the growth of population in the decades ahead so as to provide food to those hungry new mouths to be fed. This must be accomplished with a fixed or slowly expanding base of arable land offering little expansion and, simultaneously, to replace destructive agricultural practices with more benign ones. Thus the call for agricultural sustainability.[36] Owing to the daunting nature of this challenge, every economically, ecologically and socially feasible improvement will have to be carefully exploited. A list of potential improvements would include:

- Energy-intensive farming, including, in some areas, increased fertilizer use
- Soil and water conservation, with special priority on combatting erosion
- Maintaining biodiversity
- Improved pest control
- Expanded and more efficient irrigation
- Improved livestock management
- Develop new crop strains with increased yield, pest resistance, drought tolerance, etc.
- Reduce dependency on pesticides and herbicides.

A key factor in implementing many of these improvements can be the application of modern techniques of crop bioengineering. These techniques can be a powerful new tool to supplement pathology, agronomy, plant breeding, plant physiology and other approaches that serve us now.

If crop bioengineering techniques are developed and applied in a manner consistent with ecologically sound agriculture they could decrease reliance on broad spectrum insecticides which cause serious health and environmental problems. This could be accomplished by breeding varieties with specific toxicity to target pests that don't effect beneficial insects. Furthermore, these techniques could enable the development of crop varieties that are resistant to currently uncontrollable plant diseases.

At their best bioengineering techniques are highly compatible with the goals of sustainable agriculture because they can offer surgical precision in combatting specific problems without disrupting other functional components of the agricultural system.

While it is feasible to use biotechnology to improve the ecological soundness of agriculture, this will not happen unless well-informed decisions are made regarding which specific biotechnology projects are encouraged and discouraged. For example, ten years ago when crop bioengineering was getting started in the US some of the first projects involved engineering crops for tolerance of a dangerous herbicide. These projects were dropped after protest from environmental groups. Projects targeted for developing countries will have to be scrutinized to make sure that their long-term impacts are beneficial.

Not all challenges to sustainable and productive agriculture can be addressed with biotechnology. For example challenges posed by needed improvements in soil and water conservation, maintaining biodiversity and improving irrigation techniques must be dealt with by other means.

We must emphasize that these and other improvements in agriculture, while badly needed, do not address all of the difficulties faced by the lower tier of developing nations plagued by poverty and inequitable food distribution. There is almost no dispute that careful planning and selection of priorities, coupled with substantial commitments on the parts both of industrial and developing nations, will be required in order to provide the increasing food supplies the future will demand and, at the same time, move to sustainable agricultural practices and alleviate hardship in now-impoverished nations.

References and Notes

1. H. W. Kendall and David Pimentel, "Constraints on the Expansion of the Global Food Supply," *Ambio* vol. 23 no. 3, May 1994.
2. "Feeding the World: The Challenges Ahead," comments by Norman E. Borlaug, Fourth Conference on Sustainable Development, World Bank, Washington, DC, September 25–27, 1996.
3. Robert W. Herdt, "The Potential Role of Biotechnology in Solving Food Production and Environmental Problems in Developing Countries," in Agricultural and Environment: Bridging Food Production and Environmental Protection in Developing Countries, ASA Special Publication no. 60, American Society of Agronomy, Madison, WI, 1995, pp. 33–54.
4. Per Pinstrup-Andersen and James L. Garrett, *Rising Food Prices and Falling Grain Stocks: Short-Run Blips or New Trends?* 2020 Brief 30, International Food Policy Research Institute, Washington DC, Jan 1996.
5. R. Engelman and Pamela LeRoy, *Conserving Land: Population and Sustainable Food Production*, Population Action International, Washington, DC, 1995.
6. *World Resources 1994–1995*, World Resources Institute, Washington, DC, 1995, p. 294.
7. See especially D. Norse, C. James, B. J. Skinner, and Q. Zhao, Agriculture, Land Use and Degradation, Chapter 2 in *An Agenda for Science for Environment and Development into the 21st Century*, based on a Conference held in Vienna, Austria in November 1991. Cambridge University Press.
8. Food and Agriculture Organization of the United Nations, *Agriculture Towards 2010*, 27th Session, Rome, 6–25, November 1993.
9. World Resources Institute, *World Resources 1992–93*, Oxford Univ. Press, New York, 1992.
10. Pimentel, D. et al., *BioScience* 34, 277–283, 1987.
11. *Our Common Future*, World Commission on Environmental and Development, Oxford University Press, New York, 1987.
12. Sandra Postel, Water and Agriculture, in *Water in Crisis: A Guide to the World's Fresh Water Resources*, ed. Peter H. Gleick, Oxford University Press, New York and Oxford, 1993.
13. *World Resources 1992–1993*. A Report by the World Resources Institute, Oxford University Press, New York, 1993.
14. Lester R. Brown, "Future supplies of land and water are fast approaching depletion," in *Population and Food in the Early Twenty-First Century: Meeting Future Food Demand of an Increasing Population*, Nurul Islam (ed.), International Food Policy Research Institute, Washington, DC 1995.
15. "Human Appropriation of Renewable Fresh Water," S. Postel, G. Daily, and P. Ehrlich, Science, 271, 785–788, Feb. 9, 1996.
16. Donald L. Plucknett in *Population and Food in the Early Twenty-First Century: Meeting Future Food Demand of an Increasing Population*, edited by Nurul Islam, International Food Policy Research Institute, Washington, DC, 1995, pp. 207–208.

17. Kirit Parikh and S. M. Dev in *Population and Food in the Early Twenty-First Century: Meeting Future Food Demand of an Increasing Population*, edited by Nurul Islam, International Food Policy Research Institute, Washington, DC, 1995, pp. 117–118.
18. Vaclav Smil, *China's Environmental Crisis—An Inquiry into the Limits of National Development*, M. E. Sharpe, Armonk, New York, 1993.
19. Lester Brown and Hal Kane, *Full House: Reassessing the Earth's Population Carrying Capacity*, W. W. Norton & Co. New York, 1994.
20. Lester R. Brown, *Who Will Feed China?—Wake-Up Call for a Small Planet*, W. W. Norton & Company, New York, 1995.
21. "Malthus Goes East," *The Economist*, August 12, 1995, p. 29.
22. Tim Dyson, *Population and Food* (Routledge, London and New York, 1996).
23. Thomas Nordblom and Farouk Shomo in *Population and Food in the Early Twenty-First Century: Meeting Future Food Demand of an Increasing Population*, edited by Nurul Islam, International Food Policy Research Institute, Washington, DC, 1995), pp. 131, et seq.
24. *From Vision to Action in the Rural Sector*, Prepared by the rural development and agricultural staff of the World Bank Group, World Bank, Washington, August 5, 1996 (Draft).
25. J. Cherfas, *Science* 250, 1140–1141, 1990.
26. *Agriculture: Towards 2010*. United Nations Food and Agriculture Organization, Rome, 1993. See also the discussion by Pierre Crosson in *Population and Food in the Early Twenty-First Century: Meeting Future Food Demand of an Increasing Population*, edited by Nurul Islam, International Food Policy Research Institute, Washington, DC, 1995, pp. 143–159.
27. *Radiative Forcing of Climate Change—The 1994 Report of the Scientific Assessment Working Group of IPCC*. Intergovernmental Panel on Climate Change, World Meterological Organization and United Nations Environment Programme, New York, 1994. See also John Houghton, *Global Warming—The Complete Briefing*, Lion Publishing, Oxford, 1994.
28. Intergovernmental Panel on Climate Change, Working Group II, Second Assessment Report, *Summary for Policymakers: Impacts, Adaptation and Mitigation Options*. Released by the IPCC, October 24, 1995, Washington, DC.
29. For more detailed projections of the effects of climate change for the principal nations of Asia, see *Climate Change in Asia: Executive Summary*, published by the Asian Development Bank, Manila, July 1994.
30. David W. Wolfe, "Potential Impact of Climate Change on Agriculture and Food Supply, at Conference on Sustainable Development and Global Climate Change, December 4–5, 1995, Arlington, VA, sponsored by Center for Environmental Information.
31. Klaus M. Leisinger, *Sociopolitical Effects of New Biotechnologies in Developing Countries*, Food, Agriculture and the Environment, Discussion Paper 2, International Food Policy Research Institute, Washington, DC, March 1995.
32. Pierre Crosson, in *Population and Food in the Early Twenty-First Century: Meeting Future Food Demand of an Increasing Population*, edited by Nurul Islam, International Food Policy Research Institute, Washington, DC, 1995, p. 157.
33. Peter A. Oram and Behjat Hojjati, in *Population and Food in the Early Twenty-First Century: Meeting Future Food Demand on an Increasing Population*, edited by Nurul Islam, International Food Policy Research Institute, Washington, DC, 1995, p. 167.
34. John Bongaarts, "Can the Growing Human Population Feed Itself?" *Scientific American*, March 1994, p. 18.
35. Alex F. McCalla, *Agriculture and Food Needs to 2025: Why We Should Be Concerned*, Consultative Group on International Agricultural Research, Sir John Crawford Memorial Lecture, International Centers Week, October 27, 1994, Washington, DC.

36. D. L. Plucknett and D. L. Winkelmann, "Technology for Sustainable Agriculture," *Scientific American*, September 1995, pp. 182–186.

III. BIOENGINEERING TECHNOLOGY

What Can Be Accomplished Now

Plant scientists can now transfer genes into many crop plants and achieve stable intergenerational expression of new traits. "Promoters" (e.g., DNA sequences that control the expression of genes) can be associated with transferred genes to ensure expression in particular plant tissues or at particular growth stages. Transformation can be achieved with greater efficiency and more routinely in some dicots than in some monocots, but with determined efforts nearly all plants can or will be modified by genetic engineering.

Gene Transformation

Genetic transformation and other modern techniques to assist crop breeding have been used toward four broad goals: to change product characteristics, improve plant resistance to pests and pathogens, increase output, and produce to improve nutritional value of foods.

Genetic modification to alter product characteristics is illustrated by one of the first genetically engineered plants to receive FDA approval and to be made available for general consumption by the public, the FLAVR SAVR™ tomato; the fruit ripening characteristics of this variety were modified so as to provide a longer shelf life.[1] Biotechnology has also been used to change the proportion of fatty acids in soybeans, modify the composition of canola oil,[2] and change the starch content of potatoes.[3]

Natural variability in the capacity of plants to resist damage from insects and diseases has been long exploited by plant breeders. Biotechnology provides new tools to the breeder to expand their capacity. In the past, crop breeders were generally limited to transferring genes from one crop variety to another. In some cases they were able to transfer useful genes to a variety from a closely related crop species or a related native plant. Genetic engineering now gives plant breeders the power to transfer genes to crop varieties independent of the gene's origin. Thus bacterial or even animal genes can be used to improve a crop variety. *Bacillus thuringiensis* (Bt), a bacterium which produces an insect toxin particularly effective against *lepidoptera* (e.g., caterpillars and moths), has been applied to crops by gardeners for decades. It is also effective against mosquitoes and certain beetles. Transformation of tomato and tobacco plants with the gene that produces Bt toxin was one of the first demonstration of how biotechnology could be used to enhance a plant's ability to resist damage from insects.[4] Transgenic cotton that expresses Bt toxin at a level providing protection against cotton bollworm has been developed,[5] and a large number of Bt transformed crops, including corn and rice, are currently being field tested.[6]

Other strategies to prevent insect damage include using genes of plant origin for proteins that retard insect growth, such as lectins, amylase inhibitors, protease inhibitors, and cholesterol oxidase.[7]

Genes that confer resistance to virus diseases have been derived from the viruses themselves, most notably with coat protein mediated resistance (CP-MR). Following extensive field evaluation a yellow squash with CP-MR resistance to two plant viruses was approved for commercial production in the United States.[7] Practical resistance to fungal and bacterial pathogens has been more elusive, although genes encoding enzymes that degrade fungal cell walls or inhibit fungal growth are being evaluated. More recently natural genes for resistance to pathogens have been cloned, modified, and shown to function when transferred to susceptible plants.[8]

While protecting plants against insects and pathogens promises to increase crop yield by saving a higher percentage of present yield, several strategies seek to increase the potential crop yield, including the exploitation of hybrid vigor, delaying plant senescence, inducing plants to flower earlier and to increase starch production.

Several strategies to produce hybrid seed in new ways will likely contribute to increasing yield potential. Cytoplasmic male sterility was widely used long before the age of biotechnology, but strategies to exploit male sterility required biological manipulations that can be carried out only using molecular biology tools, and several of these are quite well advanced.[9] Some of these strategies entail suppressing pollen formation by changing the temperature or day length. Delayed senescence or "stay-green" traits enables a plant to continue producing food beyond the period when a non-transformed plant would, thereby potentially producing higher yield.[10] Potatoes that produce higher starch content than non-transformed controls have been developed.[11]

Plants have been modified to produce a range of lipids, carbohydrates, pharmaceutical polypeptides, and industrial enzymes, leading to the hope that plants can be used in place of microbial fermentation.[12] One of the more ambitious of such applications is the production of vaccines against animal or human diseases. The hepatitis B surface antigen has been expressed in tobacco, and the feasibility of using the purified product to elicit an immune response in mice has been demonstrated.[13]

Gene Markers

Far-reaching possibilities for identifying genes have been made possible through various molecular marker techniques with exotic names like RFLP, RAPD, micro-satellites, and others. These techniques allow scientists to follow genes from one generation to the next, adding powerfully to the tools at the disposal of plant breeders. In particular, it enables plant breeders to combine resistance genes, each of which may have different modes of action, leading to longer-acting or more durable resistance against pathogens. It also makes it possible for the breeder to combine several genes, each of which individually may provide only a weakly expressed desirable trait, but in combination have higher activity.

On-Going Research

Research continues to improve the efficiency and reduce the costs of developing transgenic crops and the use of genetic markers. As this research succeeds, it will be applied to more different plants and genes.

By far the greatest proportion of current crop biotechnology research is being conducted in industrialized countries on the crops of economic interest in those countries. Plant biotechnology research in the 15 countries of the European Union probably is a fair reflection of global current plant biotechnology research. Almost 2000 projects are underway, 1300 actually using plants (as opposed to plant pathogens, theoretical work and so forth). Of those about 210 involve work on wheat, barley, and other cereals, 150 on potato, 125 on oil seed rape, and about 90 on maize.[14,15]

The record of field trials reflects the composition of earlier research activities, and those data show that the work on cereals was started somewhat later than on other plants. Some 1024 field trials had been conducted worldwide through 1993; 88% were in OECD countries, with 38% in the United States, 13% in France, 12% in Canada, with Belgium, The Netherlands and the United Kingdom each hosting about 5% of the total number of field trials. Argentina, Chile, China and Mexico led in numbers of field trials in developing countries, but none had more than 2% of the total.[16]

The largest number of field trials was conducted on potato (19%), with oilseed rape second (18%); tobacco, tomato, and maize each accounting for about 12%, with more than 10 trials each on alfalfa, cantaloupe, cotton, flax, sugar beet, soybean, and poplar. Nine tests were done on rice, and fewer than that on wheat, sorghum, millet, cassava, and sugar cane, the food crops that, aside from maize, provide most of the food to most of the world's people who live in the developing countries.

Herbicide tolerance has been the most widely tested trait, accounting for 40% of the field trials for agronomically useful genes. Twenty-two percent of tests were conducted on 10 different types of modified product quality—including delayed ripening, modified processing characters, starch metabolism and modified oil content.[15] About 40% of field trials in developing countries were for virus resistance, one-quarter of the field trials were for crops modified for herbicide resistance, another one-quarter for insect resistance, with the balance for product quality, fungal resistance or agromatic traits.[17]

Although much of the focus of biotechnology research in agriculture has focused on bioengineering (i.e., gene transfer), the techniques of biotechnology extend beyond this approach. The techniques involved in tissue culture have been advanced and refined over the past decade. These techniques can be used to regenerate plants from single cells and have proven especially useful in producing disease free plants that can then be propagated and distributed to farmers. This has resulted in significant yield improvement in crops as diverse as potato and sugar cane.

Another use for biotechnology is in developing diagnostic techniques. Too often, poorly performing crops have observable symptoms that are so general that the farmer cannot determine the specific cause. For example, Tungro disease in rice produces symptoms that are also those of certain nutrient deficiencies. Biotechnology techniques can be used to develop easy-to-use kits that can alert the farmer to the presence of Tungro virus DNA in his/her rice plants. Knowledge of this type can decrease the frustration and money spent on solving the wrong problem.

Current Focused Efforts

Most biotechnology research in the industrialized world is being conducted on human health issues rather than on agriculture. United States government spending for biotechnology research in the US is around $3.3 billion, with $2.9 billion going to health and $190 million to agricultural issues.[18] It is estimated that between 1985 and 1994 an estimated $260 million was contributed as grants to agricultural biotechnology in the developing world, with another $150 million in the form of loans, with perhaps an average of $50 million per year in the more recent years.[19] At least a third and perhaps half of these funds have been used to establish organization designed to help bring the benefits of biotechnology to developing countries.

Maize is the focus of much crop biotechnology work in the United States, most of which is directed toward making maize better suited for production or more capable of resisting the depredations of pests in the industrialized countries. The International Wheat and Maize Improvement Center (CIMMYT) sponsors the largest concentration of international effort directed at identifying traits of maize that could be improved using biotechnology, but it spends barely $2 million per year on those efforts.

There are, at present, only four coherent, coordinated programs directed specifically at enhancing biotechnology research on developing country crops: one supported by USAID, one by the Dutch government, one by the Rockefeller Foundation and one by the McKnight Foundation.

The USAID supported project on Agricultural Biotechnology for Sustainable Productivity (ABSP), headquartered at Michigan State University, is implemented by a consortium of US Universities and private companies. It is targeted at five crop/pest complexes: potato/potato tuber moth, sweet potato/sweet potato weevil, maize/stem borer, tomato/tomato yellow leaf virus, and cucurbits/several viruses. An outgrowth of the earlier USAID-supported tissue culture for crops project, ABSP builds on the network of scientists associated with the earlier project as well as on others.

The cassava biotechnology network, sponsored by the Netherlands Directorate General for International Cooperation, held its first meeting in August of 1992. Its goals include bringing the tools of biotechnology to modify cassava so as to better meet the needs of small-scale cassava producers, processors and consumers. Over 125 scientists from 28 countries participated in the first network meeting. Funding to date has been about $2 million. An important initial activity is a study of farmers' needs for technical change in cassava, based on a field survey of cassava producers in several locations in Africa.

Another important initiative is being developed at The Scripps Institute, La Jolla, California and is called ILTAB (International Laboratory of Tropical Agricultural

Biotechnology). It is jointly administered by the Institute and ORSTOM, a French governmental agency. Funding for research in the control of diseases of rice, cassava and tomato through applications of biotechnology is provided from grants from ORSTOM, the Rockefeller Foundation, ASBC and USAID. Most of the research is carried out by fellows, students and other trainees from developing countries.

The Rockefeller Foundation support for rice biotechnology in the developing world began in 1984. The program has two objectives: to create biotechnology applicable to rice and with it produce improved rice varieties suited to developing country needs, and to ensure that developing country scientists know how to use biotechnology techniques and be capable of adapting them to their own objectives. Approximately $50 million in grants have been made through the program. A network of about 200 senior scientists and 300 trainee scientists are participating, in all the major rice producing countries of Asia and a number of industrialized countries. Researchers in the network transformed rice in 1988, a first for any cereal. Transformed rice has been field tested in the United States, and a significant number of lines transformed with agronomically useful traits now exist and are being developed for field tests.

RFLP maps, that is "road maps" that allow breeders to follow genes, are being used to assist breeding, and some rice varieties developed by advanced techniques not requiring genetic engineering are now being grown by Chinese farmers.

The McKnight Foundation recently established the Collaborative Crop Research Program, which links researchers in less developed countries with U.S. plant scientists, in order to strengthen research in selected countries while focusing the work of U.S. scientists on food needs in the developing world. It is being funded at $12–15 million for the first 6 years. While crop engineering is not the sole research tool being supported by the program, it plays an extremely important role.

Early in the effort to apply bioengineering to crop improvement, there was great hope placed in the potential to engineer the capacity for nitrogen fixation into crops without it. After the investment of millions of dollars in public and venture capital, and many years of research, it has become apparent that the genetic machinery involved in nitrogen fixation by legumes is extremely complex and beyond our current capacity for gene transfer and expression. Although at some point in the future nitrogen fixation may be transferred to crops such as corn and rice, such an achievement must be seen as a far off goal.

It is unlikely that these four focused crop biotechnology efforts, taken together, entail in excess of $20 million annually. Total agricultural biotechnology research in the developing world may not greatly exceed something on the order of $50 million annually.[18] China, India, Egypt, and Brazil and a few other countries have a reasonable base for biotechnology, but most developing countries will find it difficult to develop useful products of biotechnology without sharply directed assistance. Little attention will be paid to the crops or the pests/diseases/stresses of importance in the developing world unless they are also of importance in the more advanced countries. That is, while the grains in fundamental knowledge that apply to all organisms will, of course, be available, the applications in the form of transformation techniques, probes, gene promoters, etc, will not be generated.

Potential Contributions of Transgenic Crops
Potential Applications to Production and Improved Quality of Food

What will be the likely contribution of the developments of molecular biology to the food production problems in developing countries in the years ahead? Contributions may come through two different pathways: molecular biology specifically directed at food needs in the developing world and spill over innovations that are directed at issues in the more industrialized countries but which happen also to contribute to developing country food production.

The discussion of the preceding section shows that the resources directed at food crop production in developing countries is small, especially when compared to that directed at industrialized world crops. Still, some important contributions should come from the former. Training of developing country scientists under various programs means there is a small cadre of plant molecular biologists in a number of developing countries. The Rockefeller Foundation's support of rice biotechnology should begin to pay off in the form of new varieties available to some Asian farmers within 2 to 5 years. In China, varieties produced through another culture, a form of biotechnology, are now being grown on thousands of hectares by farmers in rural areas near Shanghai. The speed with which varieties get into farmers' hands depends largely on national conditions—the closeness of linkages between biotechnologists and plant breeders, the ability of scientists to identify the constraints, identify genes that overcome those constraints and get those genes into good crop varieties, and the success of plant scientists and others to craft meaningful biosafety regulations.

It is likely that the contributions to increased rise yield from biotechnology in Asia will be on the order of 10–25% over the next ten years. These will come from improved hybrid rice systems largely in China, and in other Asian countries from rice varieties transformed with genes for resistance to pests and diseases. The latter will raise average yields by preventing crop damage, not through increasing yield potential. The reason is simple: few strategies are being pursued that attempt to directly raise yield potential because few strategies have been conceived. The use of hybrid rice is one exception. Potential ways to raise yield potential revolve around increasing the "sink" size and increasing the "source" capacity. The first requires increasing the number of grains or the average grain size, the latter increasing the capacity of the plant to fill those grains with carbohydrate. Both are desired but, certainly for rice, there are only a few investigators thinking about how biotechnology might advance each. While there is a community of scientists working to understand basic plant biochemistry, including photosynthesis, this work as yet offers no hints about which genes can be manipulated to advantage using the tools of molecular biology and genetic engineering.

Maize yields in developing countries may be affected by biotechnology if genes useful in tropical countries are discovered in the course of the great amount of research on maize in the United States. Although most of the maize research is being carried out by private firms, it is not unlikely that some of the discoveries will be made available for applications in developing countries either at no cost or at low enough cost as to make them commercially feasible. Applications of biotechnology on cassava are further in the future, as are those on smallholder bananas and other crops of importance in the developing world.

Herbicide resistance is potentially the simplest of traits to incorporate into a plant because application of the herbicide is an ideal way to select a modified individual cell. Thus, a population of cells exposed to DNA that confers herbicide resistance can quickly be screened. A number of different herbicides are available, and there is a strong self-interest on the part of herbicide manufacturers to encouraging farmers to use herbicides. Thus a number of pressures are at work to ensure that transgenic crops with herbicide resistance are produced. To the extent that herbicide use increases, and to the extent that weeds currently constrain crop yields in developing countries, crop yields may rise. Additionally, proper regulatory activities may lead to the increased use of environmentally less damaging herbicides (e.g., biodegradable herbicides). In impoverished countries, cash-poor farmers typically do not have access to such herbicides, especially the expensive ones such as glyphosphate for which resistance is being engineered. Thus herbicide resistance may not prove to benefit the average farmer in such countries unless the cost of herbicides is reduced. It should be noted that this is already occurring as patent production is lost.

Prospects for incorporating resistance to pests and diseases to benefit developing country crops are, in general, more favorable than for yield increases because this is a much simpler class of problem and is where much of the effort in biotechnology is focused. It is possible that many genes that address insect or disease problems of temperate crops may be effective in tropical crops. If they are, there still are the problems associated with gaining access to the genes, and transforming plants with those genes because most have associated intellectual property rights. In one case, Monsanto has made available, without cost, the genes that confer resistance to important potato viruses to Mexico, and has trained Mexican scientists in plant transformation and other skills needed to make use of the genes. The transformed potatoes are now being field-tested in Mexico. Monsanto has also worked with USAID and KARI to develop and donate a similar virus control technology to Kenya and Indonesia for virus control in sweet potato. These are, however, exceptional cases.

Drought is a major problem for nearly all crop plants, and the prospects of a "drought resistance gene" has excited many scientists. However, plant scientists recognize drought tolerance or resistance as having many dimensions, long, thick roots; thick, waxy leaves; the ability to produce viable pollen when under drought stress; the ability to recover from a dry period, and others. Some of these traits can undoubtedly be controlled genetically, but little progress has thus far been made in identifying the genes that control them. Salt tolerance is often discussed along with drought tolerance because salt conditions and drought cause plants to react in similar ways. Unfortunately, some of the genes that confer drought tolerance may be useless for salty conditions and vice versa. Some early workers held out the prospect that by fusing cells of plants tolerant to drought with non-tolerant plants, a useful combination of the two would result, but that has not been demonstrated, despite considerable effort.

Little attention is paid to the crops or the pests/diseases/stresses of importance in the developing world unless they are also of importance in the more advanced countries. That is, while the gains in fundamental knowledge that apply to all organisms will, of course, be available to developing countries, the applications in the form of transformation techniques, probes, gene promoters, and genes for specific conditions will not be generated. Hence plant biotechnology progress will be slower in developing than in developed countries.

The possibility of increasing the starch content of crops through genetic manipulation that modifies biosynthetic pathways of the plant is enticing. Some success has been demonstrated in the case of potato[11] and holds out the hope that it may be possible to achieve the goal of a significant increase in production potential in this crop and other root and tuber crops, such as cassava, yams and sweet potato.

The extent to which this goal is achieved may depend on two factors: (1) the extent to which there are alternative metabolic routes to the same product and (2) the extent to which control of plant metabolism is shared among the component reactions of individual pathways.[20]

> There may well be short pathways in plant metabolism where control is dominated by one or two steps, but the current evidence suggests that this is not so for the longer pathways. This conclusion has far-reaching effects on our ability to manipulate plant metabolism.[18]

Potential Applications to Environmental Problems

Genetic engineering holds out the potential that plants can be designed to improve human welfare in ways other than by improving their food properties or yields.

For example, a biodegradable plastic can be made from the bacterial storage product polyhydroxbutyrate, and the bacterial enzymes required to convert acetyl-CoA to polyhydroxbutyrate have been expressed in the model plant *Arabidopsis thaliana*. This demonstrates the possibility that a plant can be developed that accumulates appreciable amounts of polyhydroxbutyrate.[21] The optimization of such a process in a plant that will produce the substance in commercial quantities remains to be done.

At present 80% of potato starch is chemically modified after harvest. If starch modification could be tailored in the plant, costs might be lower and the waste disposal problems associated with chemical modification would be reduced.[12]

The observation that certain plants can grow in soils containing high levels of heavy metals like nickel or zinc without apparent damage suggests the potential of a deliberate effort to remove toxic substances using plants. Plants with the capability to remove such substances (hyperaccumulators) typically accumulate only a single element, grow slowly, and most have not been cultivated so that seeds and production techniques are poorly understood. One way around these limitations might be to genetically engineer crop plants to hyperaccumulate toxic substances. Some increased metal tolerance has been obtained in transgenic *Arabidopsis* plants.[22] The use of plants for decontamination of soil, water and air is still at an early stage of research and development. "No soil has been successfully decontaminated yet by either phytoextraction or phytodegradation."[23]

Human Health Applications

As a result of biotechnology, compounds that were previously available only from exotic plant species, other organisms, or in limited quantities, can now be produced in domesticated crops. It has already proved feasible to produce carbohydrates, fatty acids, high-value pharmaceutical polypeptides, industrial enzymes and biodegradable plastics.[12] Production of proteins and peptides has been demonstrated, and it was shown that plants have several potential advantages over microbial fermentation systems. Bacterial fermentation requires significant capital investment, and, in addition, often results in the production of insoluble aggregates of the desired material that require resolubilization prior to use. Plant production of such proteins would avoid the capital investment and in most cases produce soluble materials. However, the cost involved in extracting and purifying proteins from plants may be quite significant and may offset lower production costs, although the economics of purifying proteins from plant biomass has not been evaluated broadly.[12] This disadvantage can to some extent be offset by expressing the protein in the seed at a high level.[24]

Plants can potentially be used as the producers of edible vaccines. The Hepatitis B surface antigen has been expressed in tobacco and the feasibility of oral immunization using transgenic potatoes has been demonstrated.[25] The challenges involved in the design of specific vaccines include optimizing the expression of the antigenic proteins, stabilizing the expression of proteins in the post-harvest process, and enhancing the oral immunogenicity of some antigens.[26] There are even greater challenges in developing effective protocols for immunization.

References and Notes

1. Jach, G., Görnhardt, B., Logemann, J., Pinsdorf, E., Mundy, J., Schell, J. and Maas, C. "Enhanced quantitative resistance against fungal disease by combinatorial expression of different antifungal proteins in transgenic tobacco," *Plant Journal* 8(1), 97–109 (1995).
2. Schlüter, K., Fütterer, J. and Potrykus, I. " 'Horizontal' gene transfer from a transgenic potato line to a bacterial pathogen (*Erwinia chrysanthemi*) occurs—if at all—at an extremely low frequency," *Bio/Technology* 13, 1094–1098 (1995).
3. Zhu, Q., Maher, E. A., Masoud, S., Dixon, R. A. and Lamb, C. J. "Enhanced protection against fungal attack by constitutive coexpression of chitinase and glucanase genes in transgenic tobacco," *Bio/Technology* 12, 807–812 (1994).
4. Perlak, F. J. and Fishoff, D. A. in *Advanced Engineered Pesticides*, Kim, L., (ed.), pp. 199–211 (Marcel Dekker, New York, 1993).
5. Perlak, F. J., et al. *Bio/Technology* 8, 939–943 (1990).
6. The URL for the site is http://www.aphis.usda.gov/bbep/bp.
7. Shah, D. M., Rommens, C. M. T. and Beachy, R. N. *Trends in Biotechnology* 13, 362–368 (1995).
8. Song, W. Y., Wang, G. L., Chen, L. L., Kim, H. S., Pi, L. Y., Holsten, T., Gardner, J., Wang, B., Zhai, W. X., Zhu, L. H., Fauquet, C., and Ronald, P. *Science* 270, 1804–1806 (1995).
9. Williams, M. E., *Trends in Biotechnology* 13, 344–349 (1995).

10. Gan, S. and Amasino, R. M. *Science* 270, 1986–1988 (1995).
11. Stark, D. M., Timmerman, K. P., Barry, G. F., Preiss, J., and Kishore, G. M. *Science* 258, 287–292 (1992).
12. Goddijn, O. J. M. and Pen, J. *Trends in Biotechnology* 13, 379–387 (1995).
13. Thonavala, Y., Yang, Y. F., Lyons, P., Mason, H. S. and Arntzen, C. *Proc. Natl. Acad. Sci. USA* 92, 3358–3361 (1995).
14. Lloyd-Evans and Barfoot, Genetic Engineering News, 1996.
15. Snow, A. A. and Palma, P. M. Commercialization of Transgenic Plants: Potential Ecological Risks, *BioScience* 47 (2) (February 1997).
16. Dale, P. J. *Trends in Biotechnology* 13, 398–403 (1995).
17. Krattinger, A. F. *Biosafety for Sustainable Agriculture*, Stockholm Environmental Institute and International Service for the Acquisition of Agri-Biotechnological Applications (1994).
18. Office of the President, *Budget of the United States*, U.S. Government Printing Office, Washington, (1992).
19. Brenner, C. and J. Komen, OECD Technical Paper 100 (OECD Development Center, 1994).
20. ap Rees, T. *Trends in Biotechnology* 13(9), 375–378 (1995).
21. Poirier, Y., Nawrath, C. and Somerville, C. *Bio/Technology* 13, 142–150 (1995).
22. Meagher, R. B., Rugh, C., Wilde, D., Wallace, M., Merkle, S. and Summers, A. O. Abstract 14th Annual Symposium Current Topics in Plant Biochemistry, Physiology and Molecular Biology, pp. 29–30 (University of Missouri, 1995).
23. Cunningham, Scott D., Berti, William R. and Hang, Jianwei W. *Trends in Biotechnology* 19 (9), 396 (1995).
24. Krebbers E. and van de Kerckhove, J. *Trends in Biotechnology* 8, 1–3 (1990).
25. Haq, T. A., Mason, H. S., Clements, J. D. and Arntzen, C. J. *Science* 268, 714–716 (1995).
26. Mason, Hugh S. and Charles J. Arntzen, *Trends in Biotechnology* 3 (9), 388–392 (1995).

IV. POSSIBLE PROBLEMS

All new technologies must be assessed in terms of both benefits and costs. This section outlines a number of potential costs or problems that may be associated with developing and using the new tools of biotechnology in developing countries. Some of the problems associated with biotechnology for crop improvement are not new. Indeed, some of the problems that must be addressed to safeguard the use of agricultural biotechnology are the same problems that were faced 30 years ago during the "Green Revolution." The new tools of biotechnology give us more power to make positive or negative impacts on the environment that was the case with conventional plant breeding technologies used during the Green Revolution. Thus it is essential that we review critically the potential problems that have been raised by scientists and environmentalists.[1] Our intention here is to present a balanced review of current knowledge concerning risks and problems.

Gene Flow in Plants: Crops Becoming Weeds

In most groups of plants, related species regularly form hybrids in nature or around cultivated fields, and the transfer of genes between differentiated populations that such hybridization makes possible is a regular source of enhancement for the populations involved. Thus all white oaks and all black oaks (the two major subdivisions of the genus, including all but a few of the North American species) are capable of forming fertile hybrids, and some of the species and distinct races that have evolved following such hybridization occupy wide ranges in nature and can be recognized as a result of their distinctive characteristics. The characteristics of corn, wheat and many other crops were enhanced during the course of their evolution as a result of hybridization with nearby weedy or cultivated strains or related species, and those related, infertile, sometimes weedy strains have also been enhanced genetically in some instances following hybridization with the cultivated crop to which they are related.

In view of these well-known principles, studied for well over 50 years, it is also clear that any gene that exists in a cultivated crop or other plant, irrespective of how it got there, can be transferred following hybridization to its wild or semidomesticated relatives. The transfer would occur selectively if the gene or genes being transferred enhanced the competitive abilities of the related strains, and the weedy properties of some kinds of plants might be enhanced in particular instances as a result of this process. If so, those new strains might need special attention in controlling them, just as the many thousands of kinds of weedy strains of various plants that have developed over the history of cultivation need control.

Because most crops such as corn and cotton are so highly domesticated, it is extremely unlikely that any single gene transfer would enable them to become pernicious weeds. Of greater concern is the potential for the less domesticated, self-seeding crops (e.g., alfalfa) and commercial tree varieties (e.g., pines) to become problems. These plants already have the capacity to survive on their own, and transgenes could enhance their fitness in the world. For example, a pine tree engineered for resistance to seed-feeding insects might gain a significant advantage due to decreased seed destruction and could then potentially outcompete other indigenous species. If this happened forest communities could be disrupted.

Gene Flow in Plants: From Transgenic Crops to Wild Plants

Crop varieties are often capable of breeding with the wild species from which they were derived. When the two plant types occur in the same place it is possible for transgenes, like other genes in the domesticated plant, to move into the wild plants. In some cases these crop relatives are serious weeds (e.g., wild rices, Johnson grass). If a wild plant's fitness was enhanced by a transgene, or any other gene, that gave it protection from naturally occurring diseases or pests, the plant could become a worse pest or it could shift the ecological balance in a natural plant community. Wild relatives of crops do suffer from diseases and insect attack, but there are few studies that enable us to predict if resistance to pests would result in significant ecological problems. Weeds often evolve resistance to diseases by natural evolutionary processes. However, in some cases, gene transfer from crops could speed up this process by hundreds of years.

Wild rices are especially important weeds in direct seeded rice (which is an agricultural practice that is becoming more widely used in Asia). It has been shown that genes are often naturally transferred between domesticated rice and weedy wild rices. If a herbicide tolerance gene was engineered into a rice cultivar it would be possible to control the wild rice in commercial rice fields with the herbicide until the wild rice acquired the herbicide tolerance gene from the cultivar. Once the wild rice obtained this gene the herbicide would become useless. This wouldn't make the wild rice a worse weed than it was before genetic engineering. However, this natural gene transfer would make the investment in the engineering effort much less sustainable. Therefore, it is important to consider such gene transfer before investing in specific biotechnology projects. For example, weeds could evolve resistance to some herbicides without gene transfer but the process would take much longer. This would be especially true with herbicides such as Round-Up (glyphosate) from Monsanto, which is difficult for plants to resist with their normally inherited genes. (It should be noted that even in this case, the intensive use of glyphosate has led to weed resistance in Australia.)

Development of New Viruses from Virus-Containing Transgenic Crops

Brief Overview of the Problem

Viral diseases are among the most destructive to plant productivity, especially in the tropics. Consequently, the genetic modification of plants to resist viruses has been an important objective of conventional breeding. Over the past decade, biotechnology has made possible the more rapid and precise production of individual strains resistant to particular viruses as a result of the ability to move particular genes into specific crop strains. One of the goals of genetic engineering has been to identify novel virus resistance genes that can be rapidly transferred to many types of crops, thus easing the problems of the plant breeder and meeting the needs of the farmer. As has always been the case with such efforts, the major challenge is to find virus-resistant genes that cannot be overcome easily by the action of natural selection of the virus. Now, however, we have the potential to react more efficiently to this challenge than before.

One potential advantage of genetic engineering is that it may make possible the transfer of multiple genes for disease resistance that impact the disease organism by different mechanisms. In many cases this would make adaptation by the disease organism more difficult. Engineering multiple resistance genes into crops requires more advanced technical effort and the benefits of such effort will only be seen years after the varieties are commercialized. Therefore, it is important that genetic engineers be given a mandate to develop genes that will protect crops for extended periods of time.

Pathogen Derived Resistance

To date the most widely applied genetic engineering technology for the control of plant viruses is the use of genes derived from the plant viruses themselves. When transferred to plants, a set of genes called viral coat protein genes inhibit replication of the virus once it enters the plant. (Other virus-derived genes can have a similar impact when they are transferred to plants in an appropriate manner.)

Transgenes encoding a variety of viral genes have been tested in transgenic plants over the past 10 years with a range of effects. Plants that produce viral coat proteins have been the most widely tested, and some of these plants have received approval for commercial sale in the United States and in China. In 1995 the U.S. Department of Agriculture proposed a rule that would substitute notification requirements for permit requirements for most field tests of selected genetically engineered crops. If this rule is formalized, researchers will only have to notify the Department before field testing certain genetically engineered plants, including those that express viral coat protein genes, a rule that some have found controversial.[2] The U.S. Environmental Protection Agency also ruled that coat proteins are not pesticidal and are considered safe for environmental release. The U.S. Food and Drug Administration has approved for sale and consumption foods derived from transgenic plants that contain viral coat proteins.

Potential Concerns Regarding the Release of Plants That Contain Genes Encoding Viral Sequences

As research and development of plants that exhibit pathogen-derived resistance moved from the lab to the field several concerns were voiced about the release of plants that encode viral sequences, including the following: (a) Virus proteins may trigger allergic reactions if included in foods. This concern has been largely abandoned, in part because many foods are infected with plant viruses, and have been consumed for many years without known deleterious effects; (b) Virus-resistant plants may have a competitive advantage in the field, and outcrossing with weed species may confer increased competition and weediness. As indicated above, we lack data on how important this problem can be. (c) The presence of transgenic viral sequences in large crop acreage would increase the likelihood of creating novel viruses due to recombination between the transgenes and other viruses that infect the plant. While it is known that many crops are simultaneously infected by multiple plant viruses, there are few examples of confirmed genetic recombination between different viruses. And, while there is evidence for recombination between like viruses or virus strains, there is no evidence that this would occur with greater frequency in transgenic plants than under typical situations of virus infection. In conclusion, there is little evidence for the contention that virus recombination will cause ecological problems; (d) Virus coat proteins produced by transgenic crops could combine with natural viruses and produce more harmful strains. It has been concluded that while this is theoretically possible, the risks of such an occurrence are too low to be considered as a factor in assessing impacts of transgenic crops;

(e) Virus genes other than coat protein genes could elicit greater safety concerns. Genes encoding RNAs that do not produce proteins yet provide resistance are likely to receive approval because there is no scientific expectation of risk. However, it is unclear whether or not other genes will receive approval. Viral genes that have the capacity to decrease infection by one virus but increase the chance of infection by another virus will likely not receive approval unless these genes are mutated and made to act only in a protective manner.

Plant-Produced Drugs/Pesticides Affecting Unintended Targets

In terms of plant-produced insecticides, the only insecticidal compounds that are currently commercialized are proteins that are naturally produced by *Bacillus thuringiensis*. These proteins are highly specific in their toxic effects. One group of these proteins affects only certain species of caterpillars (Lepidoptera), while others affect only a restricted set of beetle species. None of these proteins have been shown to have a significantly disruptive effect on predators of pest species (i.e., beneficial insects). These proteins degrade rapidly when exposed to sunlight and have been shown to degrade even when protected by being inside crop residues. Monsanto presented data to the Environmental Protection Agency that confirm the safety of the protein. Studies with enzymes from the human digestive system indicated that these Bt proteins are quickly digested and are unlikely to cause harmful effects.

Ecosystem Damage

Risks of Genes Spreading from Plants to Soil Organisms

Unfortunately very little is known about the flow of genetic information from plants to microorganisms, making it difficult to assess this type of risk. It is a fact that soil organisms, especially bacteria, are able to take up DNA from their environment and that DNA can persist when bound to soil particles. Although one can speculate about a gene-flow scenario in which plant DNA is released from plant material, bound to soil particles, and subsequently taken up by soil bacteria, this scenario is highly unlikely. Any potential risks of such transfer can be eliminated by making transgenes that bacteria are unable to use (e.g., those with introns). It is even more speculative to consider the possible transfer of genes to soil-dwelling fungi (molds), since gene transfer to the fungi is generally much more difficult than to bacteria.

Assessing the Cost-to-Benefit Ratio of Genetically Engineered Crops

Two major questions that must be addressed before investing in a project to engineer a crop cultivar are: (1) Will the gene being transferred serve an important function in the targeted geographical area(s), and (2) How long will the gene continue to serve its function?

Will the Gene Being Transferred Serve an Important Function in the Targeted Geographical Area(s)?

The pests of a specific crop such as cotton or corn vary from one geographical region to another. For example, the caterpillars of two insect species, the cotton bollworm and the budworm, are major pests of cotton in the southern USA. A variety of cotton developed by the Monsanto Company contains a specific protein derived from *Bacillus thuringiensis* that is highly toxic to these two closely related pests. In Central America the major insect pest species of cotton are the fall armyworm and the boll weevil. As the toxins in the cotton developed by Monsanto have no impact on these pests, investing in the transfer of these seeds to Central America would be futile. Instead of taking this simple but ineffective approach, it would be worth investing resources to find more appropriate genes that would truly control Central American cotton pests.[3]

A number of companies have engineered corn varieties to be tolerant of herbicide sprays. At this point the commercial corn varieties that possess herbicide tolerance are developed by a cross between a parent corn line that contains the transgene for herbicide resistance and another line that does not contain the gene. Therefore, all of the commercially sold corn seeds contain one copy of the herbicide resistance gene and are resistant to the herbicide. In the US, farmers buy hybrid corn seed every year and only plant it once so all of their plants are tolerant of the herbicide spray. While this system will work well in the U.S. and other similar economies, it will not work well in agricultural settings in most developing countries unless changes are made. For example, in El Salvador many farmers buy hybrid corn seed once every three years because it is very expensive. The first year they plant the commercial seed and in the next two years they plant the offspring from these plants. Because of genetic segregation in the offspring, only three quarters of the corn plants in the second and third year would have the herbicide resistance gene, so the farmer would kill one-half of his/her crop by applying the herbicide. It has proven too difficult for seed companies to put the gene in both parents of the plant. Clearly, unless this situation is considered, an agricultural plan that works in a developed country may not work in a developing country.

It is often said that biotechnology is a transferable technology because "the technology is all in the seeds." It is important to recognize that the interface between seed traits and worldwide crop production is not always simple. As indicated above, simply sending herbicide-resistant corn seed to El Salvador without educating farmers about the problem of using the second generation seed could lead to immediate economic losses, and could also lead to rejection of the new technology. Similarly, breeding a corn variety in the U.S. and then sending it to West Africa could be useless if the corn was resistant to U.S. pests but not West African pests. It should be noted that corn pests are not even the same in all West African countries, so varieties must be developed by tailoring them to specific problems.

Once a variety is matched with local pest problems, the technology may be transferrable simply by supplying the seed, although this does not mean that the seed will provide a sustainable solution to the pest problem. There is abundant evidence indicating that pests will overcome genes for pest and disease resistance, regardless of whether they have come through biotechnology or classical plant breeding, unless seeds are used properly. Getting farmers to use seeds properly will require educational efforts.

How Long Will the Gene Continue to Serve Its Function?

Insects, disease-causing organisms and weeds are known to adapt to most pesticides and crop varieties that contain resistant genes. In some cases adaptation occurs within one or two years. Some insect strains have evolved under laboratory and field conditions the ability to tolerate high concentrations of the toxins derived from the *Bacillus thuringiensis* that are produced by transgenic cotton and corn sold commercially in the U.S. Considerable theoretical and empirical research has assessed the possibility that certain insect pests will overcome the effects of transgenic crops that produce these insecticidal proteins. If major crops such as corn, cotton and rice that produce these proteins are planted intensively over wide areas, the chance for insect adaptation is high unless approaches are taken in developing and deploying the insect-resistant varieties.

The EPA has put restrictions on the sale of Bt-containing cotton to ensure that within every US farm some fields are planted to varieties that do not produce the Bt proteins. This serves as an example of the types of strategies that can be employed to "conserve resistance" as developed for cotton in the U.S.A. The fields act as refuges for pest individuals that are susceptible and the insects produced in these refuges will mate with resistant insects emerging from fields where the transgenic varieties are planted. This will serve to dilute the frequency of insects that are resistant to the Bt proteins and lead to more sustainable resistance. Instituting such practices in developing countries would likely be difficult. Furthermore, the refuge strategy works best if the transgenic variety produces enough Bt protein to kill close to 100% of the susceptible insects that feed on it. If a variety is developed to kill 100% of the pest individuals of a species that occurs in Mexican corn, it may only kill 80% in a Nigerian corn field. Therefore attempts to build one transgenic corn type to fit the needs of a number of countries may be misguided.

Similar problems with adaptation described for insects may apply to crops engineered for disease resistance and tolerance of herbicides. Although there are some types of herbicides such as Round-Up (glyphosate) that are considered "immune" to weed adaptation, it is not clear that this immunity will hold up when there is intensive use of the herbicide.[4]

When investing in biotechnology for crop protection it is important to consider what the global effectiveness will be, and for how long. The same is true of agriculture in general. Any improved strains of any kind of crop or domestic animal, regardless of how the genetic modification of its features was attained, needs careful management to be as productive as possible. Integrated systems involving the best and most sustainable practices of soil preparation; the most conservative and appropriate use of water, fertilizers, and pesticides, if the latter are part of the system; and the selection of the best and most appropriate strains of the particular crop are the key to success in agriculture. These considerations are true for all agricultural systems and necessary conditions for improving them, regardless of the exact methods used to genetically modify the crop strains that are being grown.

Investments in new and improved crop strains must also be judged by their global effectiveness, irrespective of how they were produced. The Green Revolution was successful in enhancing productivity in many areas because of the whole system of cultivation that was built up around the new strains of crops, and not solely because of the properties of those strains. The design of plantings, which has a great deal to do with the longevity of resistance to particular diseases and pests, has not always been well considered in early efforts to introduce genetically engineered, or other novel strains, and may need special attention in developing countries with respect to the particular conditions found there.

As mentioned above, biotechnologies other than bioengineering can be used to improve agriculture in developing countries. These techniques can also have their drawbacks. Tissue culture can certainly be used to produce disease-free plants and help increase productivity of farms in developing countries. Tissue culture can also be used to shift the production center of specialty agricultural products from developing to developed countries. Vanilla is typically considered a tropical product but recent work with tissue culture allows its production in the laboratory. If such innovations in tissue culture proliferate, it is possible that other tropical products will be manufactured in the laboratory.

References and Notes

1. Snow, A. A. and Palma, P. M. "Commercialization of Transgenic Plants: Potential Ecological Risks, *BioScience* 47 No. 2, February 1997.
2. R. Goldberg, private communication.
3. N. Strizhov, M. Keller, J. Mathur, Z. Koncz-Kalman, D. Bosch, E. Prudovsky, J. Schell, B. Sneh, C. Koncz, and A. Zilberstein, "A synthetic cryIC gene, encoding a *Bacillus thuringiensis* d-endotoxin, confers *Spodoptera* resistance in alfalfa and tobacco," *Proc. Natl. Acad. USA*, in press 1996.
4. Gressel, J. "Fewer constraints than proclaimed to the evolution of glyphosate-resistant weeds," *Resistant Pest Management*, 8, 2–5 (1996); Sindel, B. Glyphosate resistance discovered in annual ryegrass, *Resistant Pest Management* 8, pp. 5–6, 1996.

V. Conclusions and Recommendations

The Panel's recommendations to the World Bank are based on our belief that very high priorities must be assigned to the expansion of agriculture and to increased production of food in the developing world. It is critically important that increases in food production outpace population growth. Damaging agricultural practices must be replaced with lower impact, sustainable activities so that the global capacity to produce food does not decline. Only by these means will it prove possible to lessen hunger and improve food security in the poorest nations in the years ahead.

Because transgenic technology is so powerful, it has the ability to make significant positive or negative changes in agriculture. Transgenic crops are not in principle more injurious to the environment than traditionally bred crops and in this report we have outlined a number of criteria that can be used to determine if a specific biotechnology program is likely to enhance or detract from ecologically sound crop production. Transgenic crops can be very helpful, and may prove essential, to world food production and the reach for agricultural sustainability if they are developed and used wisely.

Biotechnology can certainly be an ally to those developing integrated pest management (IPM) and integrated crop management (ICM) systems.

Support of Developing World Science

The Bank should direct attention to liaison with and support of the developing world's agricultural science community.

It is of the greatest importance to enhance the capabilities of third world science and third world scientists so as to contribute powerfully to the development of sound agriculture, based on the best environmental principles, in their countries.

A specific and urgent need is the training of developing world scientists in biotechnology methods so each nation has a cadre of scientists to assist it in setting and implementing its own policies on biotechnology research and biosafety.

The education of farmers can be greatly facilitated with the aid of scientists from their own nations. These scientists can contribute to the success of newly introduced crop strains as well as help implement an early warning system, should troubles arise during the introduction of new crops or new agricultural methods.

Research Programs

The Bank should identify and support high quality research programs in support of developing countries whose aim is to exploit the favorable potential of genetic engineering for improving the lot of the developing world.

As noted earlier in this report, not all of the research now in progress in the industrial nations will, even if successful, prove beneficial to the developing world. Research should be planned so that key needs are met. Much of the necessary research will need to be done in advanced laboratories in developed countries in conjunction with those in developing countries. Research priorities should center on directed studies for sustainable agriculture and higher yields in the developing world as well as decreasing the variation in food production arising, for example, from environmental stresses.

Variance in production can result in food shortages with numerous attendant complications. Crops with resistance to insects and diseases, including those developed by genetic modification, can markedly decrease production variance if these crops are developed and deployed in ways that minimize the pest's ability to overcome the impact of the resistance factors in the crop. A poorly conceived strategy for developing and deploying these crops can have the opposite effect if the pests adapt to resistant cultivars. If farmers are taught to depend solely on a crop's ability to ward off pests and diseases, the farmers will not be prepared to use alternative means for control when pests become adapted to the resistance factors in the crop.

Surveillance and Regulation

The Bank should play an important role in supporting the implementation of formal, national regulatory structures in its client nations, seeing that they retain their vigor and effectiveness through the years and by providing scientific and technical support to them as requested.

Effective regulation by deploying nations will prove critical should problems arise during the introduction of transgenic crops or, indeed, by other aspects of industrialized agriculture, including chemical inputs, some of which may be promoted in conjunction with genetically engineered herbicide-tolerant crops, as these all may pose problems for developing countries. The necessary comprehensive regulatory structure to carry this out appears to exist in few if any nations, including the United States. To provide the basis for a strong national regulatory structure in each nation, it is necessary that there be a designated agency with a clear mandate to protect the environment and the economy from risks associated with the uncritical application of new methods, including inappropriate new strains of crops or animals that may pose special risks for the environment. It must have the necessary authority. The agency would need the technical capacity to develop competent risk assessment and the power to enforce its decisions.

The Bank should support, in each developing country, an early warning system both to identify emerging troubles which may arise and to be able to introduce improvements in adapting the new strains.

Such a system would spot unexpected success as well, so that gains could be exploited and duplicated elsewhere, and could provide feedback to speed and optimize the introduction of new plant varieties.

The Agricultural Challenge

The Bank should continue, as it has in the past, to give very high priorities to all aspects of increasing productivity on developing world agriculture while making the necessary transition to sustainable methods.

While genetically engineered crops can play an important role in meeting the goal of improved food security, their contribution alone will not suffice. They must be accompanied by numerous other actions:

• Enhance priority on conventional plant breeding and farming practices

- Ensure that adequate energy and water become available and that procedures for their increasingly efficient use are made known and adopted
- Ensure the introduction of modern means of controlling pests, including integrated pest management, the use of safe chemicals and resistant crops
- Support the transition to sustainable activities and the reduction of waste and loss in all elements of the agriculture enterprise
- Provide the necessary education to farmers so they can best implement the array of new techniques that are needed, as, for example, integrated pest management
- Ensure that the changes in agriculture will provide the employment opportunities that will be needed in the developing world.

The scale and importance of the challenge that the Bank faces in the agricultural sector are formidable. We have concluded that the Bank should establish a permanent technical and scientific advisory group dealing broadly with the goal of improving food security while ensuring the transition to sustainable agricultural practices. It would deal with all the elements that comprise a successful program, and could provide the liaison to the target nations' scientific communities that is required.

Appendix: Integrated Intensive Farming Systems

China and India, as well as numerous other developing nations, must achieve the triple goals of more food, more income and more livelihoods from the available land and water resources. An example approach is the adoption of Integrated Intensive Farming Systems methodology (IIFS). The several pillars of IIFS methodology are the following.

1. Soil Health Care

This is fundamental to sustainable intensification of agriculture, and IIFS affords the opportunity to include stem-nodulating legumes like *Sesbania rostrata* and to incorporate Azolla, blue-green algae, and other sources of symbiotic and nonsymbiotic nitrogen fixation in the farming system. In addition, vermiculture constitutes an essential component of IIFS. IIFS farmers will maintain a soil health card to monitor the impact of farming systems on the physical, chemical, and microbiological components of soil fertility.

2. Water Harvesting, Conservation, and Management

IIFS farm families will include in their agronomic practices measures to harvest and conserve rain water, so that it can be used in a conjunctive manner with other sources of water. Where water is the major constraint, technologies which can help to optimize income and jobs from every liter of water will be chosen and adopted. Maximum emphasis will be placed on on-farm water use efficiency and to the use of techniques such as drip irrigation, which help to optimize the benefits from the available water.

3. Crop and Pest Management

Integrated Nutrient Supply (INS) and Integrated Pest Management (IPM) systems will form important components of IIFS. The precise composition of the INS and IPM systems will be chosen on the basis of the farming system and the agro-ecological and soil conditions of the area. Computer aided extension systems will be developed to provide farm families with timely and precise information on all aspects of land, water, pest and postharvest management.

4. Energy Management

Energy is an important and essential input. In addition to the energy-efficient systems of land, water and pest management described earlier, every effort will be made to harness biogas, biomass, solar and wind energies to the maximum extent possible. Solar and wind energy will be used in hybrid combinations with biogas for farm activities like pumping water and drying grains and other agricultural produce.

5. Post-Harvest Management

IIFS farmers will not only adopt the best available threshing, storage and processing measures, but will also try to produce value-added products from every part of the plant or animal. Post-harvest technology assumes particular importance in the case of perishable commodities like fruits, vegetables, milk, meat, eggs, fish and other animal products. A mismatch between production and post-harvest technologies affects adversely both producers and consumers. As mentioned earlier, growing urbanization leads to a diversification of food habits. Therefore there will be increasing demand for animal products and processed food. Agro-processing industries can be promoted on the basis of an assessment of consumer demand. Such food processing industries should be promoted in villages in order to increase employment opportunities for rural youth.

Investment in sanitary and phyto-sanitary measures is important for providing quality food both for domestic consumers and for export. To assist the spread of IIFS, Government should make a major investment in storage facilities, roads and communication and on sanitary and phyto-sanitary measures.

6. Choice of Crops and Other Components of the Farming System

In IIFS, it is important to give very careful consideration to the composition of the farming system. Soil conditions, water availability, agroclimatic features, home needs and above all, marketing opportunities will have to determine the choice of crops, farm animals and aquaculture systems. Small and large ruminants will have a particular advantage among farm animals since they can live largely on crop biomass. IIFS will have to be based on both land-saving agriculture and grain-saving animal husbandry.

7. Information, Skill, Organization and Management Empowerment

For its success, IIFS system needs a meaningful and effective information and skill empowerment system. Decentralized production systems will have to be supported by a few key centralized services, such as the supply of seeds, biopesticides, and plant and animal disease diagnostic and control methods. Ideally, an Information Shop will have to be set up by trained local youth in order to give farm families timely information on meteorological, management, and marketing factors. Organization and management are key elements, and depending on the area and farming system, steps will have to be taken to provide to small producers the advantages of scale in processing and marketing. IIFS is best developed through participatory research between scientists and farm families. This will help to ensure economic viability, environmental sustainability and social and gender equity in IIFS villages. The starting point is to learn from families who have already developed successful IIFS procedures.

It should be emphasized that IIFS will succeed only if it is a human-centered rather than a mere technology-driven program. The essence of IIFS is the symbiotic partnership between farming families and their natural resource endowments of land, water, forests, flora, fauna and sunlight. Without appropriate public policy support in areas like land reform, security of tenure, rural infrastructure, input and output pricing, and marketing, small farm families will find it difficult to adopt IIFS.

The eco-technologies and public policy measures needed to make IIFS a mass movement should receive concurrent attention. The program will fail if it is based solely on a technological quick-fix approach. On the other hand, the IIFS program can trigger an evergreen revolution if mutually reinforcing packages of technology, training, techno-infrastructure and trade are introduced.

The M. S. Swaminathan Research Foundation has designed a bio-village program to convert IIFS from a concept into a field level reality. By making living organisms both the agents and beneficiaries of development, the bio-village serves as a model for human-centered development.

PART FIVE
CLEANUP OF RADIOACTIVE CONTAMINATION

Nuclear waste had been a minor interest at best for both UCS and me for many years. The nuclear reactor safety controversy would, from time to time, generate questions on the subject, as did nuclear weapons matters. With UCS colleagues, I initiated a study, carried out by Ronnie D. Lipschutz, a UCS research staff specialist, of the problems being encountered in the U.S. civilian waste disposal program. It resulted in a useful volume by Lipschutz containing numerous references and technical background material.[1] We did almost nothing else for many years, believing that the risks of catastrophic nuclear plant accident warranted far higher priority.

This situation changed for me in the spring of 1994 when I was appointed to the Department of Energy's *Task Force on Alternative Futures for the Department of Energy National Laboratories*. This task force was organized by then-Secretary of Energy Hazel O'Leary in response to the great changes in mission made necessary by the demise of the Soviet Union, the end of the Cold War, and the halt to nuclear weapons production and testing activities at the Department of Energy's weapons laboratories at Los Alamos, Livermore, and Sandia. Other issues and seven other laboratories were also included in the task force's charge, and several subpanels focused on them. Extensive radioactive contamination was not on the agenda, however, for, as one department staffer told me, it was too large an issue for the group to take on.

I joined the National Security Subpanel, dealing with the specific problems posed by the laboratories' several weapons design teams and their weapons responsibilities. In some of the initial briefings in the late spring, I became aware that the senior laboratories, cast partially adrift by the slowdown in weapons-related activities, could usefully and, it turned out, would willing turn their talents to help the cleanup of the U.S. weapons production complex, contaminated by decades of dumped and leaked radioactive and chemically toxic materials. The extent of the contamination is enormous, and the cleanup effort, driven by attacks from the environmental community allied with large groups of furious affected citizens, had become the biggest environmental remediation program in the western world. It had been most poorly managed, its current costs—$6 billion per year—were huge, and its prospective costs were out of control.

Despite the earlier warning, I proposed to Robert Galvin, the task force chairman, that I be allowed to investigate the state of the department's cleanup program, with a view to suggesting that portions of the cleanup be included in the laboratories' revised missions. He agreed. Later, as my efforts in the area started to bear fruit, I was made chairman of a subpanel on the matter. Others on the task force were, by the time this was all organized, committed to other subjects, and I was, happily, left as a subpanel of one to pursue the enterprise by myself.

Over approximately nine months, I made some 17 trips in connection with one or another of the two subpanels I was on, occasionally being able to study both national security issues and waste problems. I prepared Chapter IV of the task force report reprinted here. It was quite critical of the department's cleanup practices. On its release it drew specific fire from the secretary herself as well as others in the department who suggested that it was based on "incorrect assumptions or misconceptions about the national laboratories." As I learned privately, it was received better by some department managers and scientists who had long been aware of the department's shortcomings in the area.

More cheering has been the response in the period since the submission of the task force report. In 1995, the report led Congress to direct the department's Office of Environmental Management to spend $50 million to fund basic research and development in support of the cleanup program. This was a response to one of the key findings of the report: that basic research and deeper understanding of the cleanup problems was inadequate, underfunded, and of low quality. Early in 1996 the department announced a new initiative in compliance with the Congressional directive. In the announcement of the initiative, both the director of energy research and the acting department undersecretary, who had been earlier the head of the cleanup activities, cited the task force report.[3] Later in the same year the department announced that it expected to devote $112 million to this effort over three years. Senior personnel at two of the laboratories expressed their gratitude for the task force's efforts in getting these matters moving. Support for larger-scale basic research and demonstration programs at the senior laboratories seems still to be increasing, so on balance, the effort was worth making.

References

1. Ronnie D. Lipschutz, *Radioactive Waste—Politics, Technology and Risk* (Ballinger Publishing Company, Cambridge, MA, 1980).
2. Jocelyn Kaiser, *Science* **2273**, 1165–1166, (August 30, 1996).
3. George Lobsenz, in *Energy Daily*, Wednesday, February 21, 1996, p. 3.

Alternative Futures for the Department of Energy National Laboratories

IV. The Environmental Cleanup Role

A. Introduction

The Department of Energy's management of its program for dealing with the radioactive and hazardous wastes at its former nuclear weapons production sites and the national laboratories has been criticized for its expense and the slow pace of cleanup. The program is of great size and the problems that plague it, developed over decades, are acute and pervasive. Involving the national laboratories in more sweeping ways is an important part of a number of needed improvements.

B. Background

Disposal practices for radioactive and chemically hazardous wastes from the start of the Manhattan Project of World War II, excepting high level waste, consisted of shallow burial, of injection underground using deep or shallow wells, the use of cribs or settling ponds, or direct release to rivers or streams. Some of the Atomic Energy Commission's (AEC) practices resulted in the exposure of uninformed members of the general public to substantial levels of radiation, and in later years there have been well-publicized leaks of toxic and radioactive materials.

Environmental concerns in the United States started rising in the 1950s, initiating a new era of citizen participation and major changes in environmental matters. Over three dozen pieces of federal environmental legislation were enacted by the early 1990s. The AEC, and later DOE, did not move as U.S. industry did, main-

---

Excerpt from report by Henry W. Kendall prepared by the Secretary of Energy Advisory Board, Task Force on Alternative Futures for the Department of Energy National Laboratories, February 1995.

taining that they were exempted from compliance with the bulk of U.S. environmental legislation.

Although beset by increasing discontent and criticism over its practices, DOE was slow to accommodate. It continued its old patterns of behavior until, in 1984, it lost a key lawsuit brought against it. Amendments to major pieces of federal environmental legislation now explicitly require DOE compliance. The result has been to make DOE subject to the same array of Federal environmental standards that U.S. industry had already largely adapted to. The DOE found itself ten years behind in Environmental Protection Agency (EPA) compliance.

In 1989, the department announced that it would have all of its sites cleaned up by 2019. This same year it created the Office of Environmental Restoration and Waste Management (since renamed the Office of Environmental Management, or EM) to have responsibility for cleanup of the complex. The EM annual budget has risen from $1.6 billion in 1989 to $6.2 billion in 1994 and will exceed $7 billion when the Savannah River Site is transferred to it from Defense Programs. It has become the largest single item in the DOE's $19 billion budget. It is the largest environmental restoration and waste management program in the world.

Driven by heightened public and Congressional concern, DOE established, in some haste, greatly enhanced requirements governing its own operations. It initiated major growth in the number and scope of environmental, safety, and health regulations, nuclear safety regulations, and DOE Orders. To ensure compliance, the number of audits, reviews, and appraisals was increased dramatically.

DOE now has had to cope with the series of legal commitments to cleanup performance with milestones and penalties for noncompliance, that it signed with state and federal bodies, for each of its sites. Inadequate attention was given by DOE to the feasibility of these commitments. One example was the Tri-Party Agreement at Hanford, signed by DOE, the EPA and the state of Washington's Department of Public Health. It mandates cleanup of the site by the year 2019.

The Department has been hindered by the press of Federal legislation and regulation by other Federal bodies. A dozen or more pieces of legislation all laid on DOE burdens with which it has been poorly equipped to deal. Moreover, by the 1990s, all the states had their own environmental legislation, much of it binding on the Department and not always consistent with its Federal counterpart.

The Department also is hindered by lack of credibility and mistrust, not only on the part of community stakeholders but by Federal and state legislative and regulatory bodies. Some members of these bodies continue to disbelieve the Department, as well as many of its contractors, even when they are telling the truth.

C. Main Findings

1. Technical Challenges

The large quantities of radioactive and hazardous chemical waste that are at the center of concern exist in a broad variety of forms, toxicity, and storage or placement conditions. For the entire 3365 square miles of the DOE complex now or formerly devoted to weapons-related activities there are, for example:

- 700 contaminated sites, 500 facilities now surplus, 1000 more which will be declared surplus soon, and 5000 peripheral properties with soil contaminated by uranium mine tailings. DOE might declare as many as 7000 facilities surplus in coming decades, most of which would require cleanup prior to decommissioning.
- More than 330 underground storage tanks (including those at Hanford) containing 77 million gallons of high level radioactive waste as sludges or liquids.
- Waste volumes from weapons-related activities:

    —High Level Waste (HLW)   385,000 cubic meters
    —Transuranic Waste (TRU)   250,000 cubic meters
    —Low Level Waste (LLW)   2,500,000 cubic meters

The LLW volume is equivalent to a cube nearly 0.1 mile on an edge, which, if water, would weigh 2.8 million tons, if soil, some 8.4 million tons. The costs of disposal of low level radioactive waste (LLW) are currently in the vicinity of $5800 per cubic meter, for HLW as high as $6 million per cubic meter.

- More than one million 55-gallon drums or boxes of stored, hazardous, radioactive, or mixed (radioactive and chemically toxic) waste. An additional 3 million cubic meters of buried waste, much having breached containers, is mixed with soil underground.
- Over 100,000 gallons of plutonium and transuranic waste liquids having a high likelihood of causing environmental contamination and worker safety problems.

More than 5700 individual "plumes," contaminating soil and groundwater, have been identified on DOE lands. For example, plumes of hazardous chemicals underlie about 150 square miles of the Hanford site.

2. Program Assessment

Two yardsticks are useful in judging the EM program: progress toward cleanup goals and the costs incurred, the latter related to the effectiveness of program management.

The remediation program has accomplished far less than many wish. The Government Accounting Office,[1] in a recent review of management changes needed to improve applications of technology in the program, concluded that while "DOE has received about $23 billion for environmental management since 1989, little cleanup has resulted. Experts agree that many cleanup technologies in use are extremely costly and offer only short-term solutions." A May 1994 Congressional Budget Office (CBO) Study[2] noted that DOE "has been criticized for inefficiency and inaction in its cleanup efforts . . . [and] has been severely criticized because of the small amount of visible cleanup that has been accomplished." These conclusions are shared by many senior DOE personnel, both within and outside the program.

One of the consequences of the troubles has been the enhancement of a syndrome common to large bureaucracies: risk aversion. It has a name: "the Hanford Syndrome." It has become widespread and severe in the EM program. Its symptoms are an unwillingness to alter familiar behavior patterns, to stick with unproductive or failing procedures, to enhance tendencies for excessive resource allocation and regulation, and to oppose innovation. It is an important element in sustaining unproductive patterns of work.

The Tri-Party Agreement at Hanford, and similar ones elsewhere, have proven to constitute major constraints on remediation progress because, in many instances, they are unrealistic, not having had proper input from those experienced in actual cleanup. The milestones they incorporate, along with penalties for noncompliance, force continued activities, some of which are make-work and should be abandoned. Other activities should be delayed or modified so as to await more effective and less costly technologies. Virtually no one believes the timetables are achievable and DOE has already been forced into renegotiations, as at Hanford in January 1994. Elsewhere DOE has been paying fines, owing to the Department's incapacity to meet deadlines, as at Rocky Flats where $27 million is now due for missing cleanup deadlines.

Probably the most important reason behind the slow pace of assessment and cleanup is the low quality of science and technology that is being applied in the field. Many of the methods, such as "pump and treat" for contaminated ground water remediation, cannot provide the claimed benefits. There is a lack of realization that many—and some experts believe most—existing remediation approaches are doomed to technical failure. Others would require unacceptable expenditures and much extended time to reach their stated objectives.

Over time, an increasing proportion of DOE resources has been going into DOE management in an attempt to lower environmental costs. The Congressional Budget Office report concluded that "at least 40% of the cleanup program's funds are devoted to administrative and support activities, a level that many reviewers have considered excessive . . . [they] represent a proportion that is significantly higher than the share spent by some other government agencies that may be performing similar tasks."[3]

DOE provides the most expensive environmental services of any government agency, with costs 40% above the average in the private sector. When DOE first became aware of these high costs, the Department's response was to try to lower them by an increase in management attention: it added between 1200 and 1600 Full Time Equivalents to its management and oversight personnel overseeing the remediation program.

How much the program will cost when and if completed cannot now be assessed with confidence. Estimates in the range $300 billion to $1 trillion have been made by DOE officials, but a lack of specific goals and achievable schedules as well as the absence of some critical remediation technologies make fixing the sum difficult. Some part of the facilities' contamination cannot be wholly cleaned up; portions of the Hanford site, as well as others, will still be radioactive after many thousands of years.

D. *Disconnects*

One useful way of understanding the nature of the problems plaguing the DOE program is to look at "disconnects," potentially discordant sets of activities whose discord the Department has been incapable of harmonizing. There are disconnects in three areas of major importance to the EM program: (1) science/engineering and applications, (2) regulatory oversight and compliance and (3) goals, objectives and means, the last involving the stakeholders affected by the program. These persistent disconnects have had numerous adverse consequences on the program.

1. Science/Engineering and Applications

There is a marked incapacity within the Department's EM program to evaluate current and prospective technologies in a wide-ranging and competent manner based on well-assessed risks. Without the resulting information it is not possible to introduce improved technologies into the applications stream or to modify or eliminate inefficient or ineffective ones. The gap between what might be applied and what is applied is well known within the program; it is called the "Valley of Death." In part it reflects the fact that there is inadequate communication between those attempting to remediate the contaminated sites and the research community that holds the key to identifying and readying advanced and powerful technologies.

One of the injurious consequences of the gap has been the failure to carry out a full program to characterize the waste remediation challenge across the many DOE sites: the nature of the risks presented by the diverse array of problem radioactivity and hazardous materials, the identification of applicable and available technologies to deal with them as well as their limitations and provide schedules, costs and expected effectiveness of reasonable and acceptable programs of remediation. The laboratories have not been tasked to perform such a characterization although they are well aware of its lack and have the technical capacity to carry it out.

The new-technology chain is seriously broken within DOE. There is little basic research being carried out relevant to the problems at hand and there is little rigorous analysis to learn from the experience in the field or from current tests. There is, for example, breakdown in communication and cooperation between organizational units within EM from headquarters to field offices to sites. Technologies are being developed independent of field and site needs that are subsequently not field implemented because of a lack of customer interest or involvement or because they replicate work done elsewhere.

The root deficiency, which makes the science/engineering–applications disconnect a persistent problem, is the absence of a sustained, high-quality, scientific/technical review capability at a high level within DOE as well as a lack of leadership and poor management of the science/engineering—operational interface.

2. Regulatory—Oversight—Compliance: Management Disconnects

The host of self-inflicted, complex and frequently contradictory or redundant regulations and requirements that the laboratories and remediation efforts are subject to has become an enormous obstacle. Compliance can be quite burdensome expensive and frequently fails to improve the affected activities. The influence of this disconnect is not confined to the EM program alone. It affects most every DOE activity, including those in both the multiprogram and the program-dedicated laboratories. Its consequences are greatest in the EM program simply because this program is DOE's largest.

In many circumstances there are harsh non-compliance provisions, and legal, personal and civil penalties for failure. People are intimidated, afraid of going to jail, and this forces an excess conservatism, sometimes bordering on inaction. There is no dispute that this aggravates inherent tendencies toward risk aversion, a problem for other reasons, as noted earlier.

* * * * * * * *

3. Goals—Objectives—Means: Stakeholders' Interests

DOE has not set out to determine, in concert with affected stakeholders, the goals it should pursue or the standards to be met in the EM program. There is a disconnect with the customer base. Are waste-contaminated soils to be removed, remediated, left in place? What exactly is to be done to and with low-level waste? What to do about the large quantity of tritiated groundwater? What site conditions are the activities at Rocky Flats intended to achieve? No one is entirely sure. The January 1994 alterations to the Hanford Tri-Party Agreement were, in part, a consequence of some of these issues surfacing.

One result of the disconnect is too much attention to the immediate, acute problems, such as possible tank leaks, explosions, overheating, with relative neglect of longer range difficulties. The immediate matters can be serious and must be dealt with, but the lack of a systems approach to the problems and their solutions, and thus lack of a synoptic view, means a poor priority list and provides bad choices. All of these elements lead to much ineffectual, albeit expensive activit[y].[4]

E. The Future

1. Within DOE

A well-functioning EM program with clearly defined goals is surely within reach, given a Department commitment to move forward. The model that many refer to was the hugely successful Manhattan Project of World War II, with its exquisite blend of basic and applied science underlying a large production complex, based on previously unknown physical phenomena. From it emerged the testing, production and delivery of the weapons employed just at the end of the conflict. The scientific challenge today is less profound, the managerial ones more so. A crisp, well-defined program, fully utilizing national laboratory skills, could prove a model within the Department and for the nation on how to run a major enterprise. We now have a poignant situation, for technology known to senior scientists and engineers both in the national laboratories and in the country's universities is in the wings that, appropriately applied, could dramatically alter the current prospects.

2. The National Laboratories

Because the EM program so badly needs high-quality science and engineering development, the national laboratories together have a critical role to play, a role very much larger than at present. The laboratories have unique resources and facilities and are accustomed to the complex, interdisciplinary blend of sciences and technologies that are the mark of large, technically driven enterprises. They are really the only organizations that can pursue the large-scale basic research and development so badly needed to replace those conventional approaches that blight much of the current EM program. Industrial site contractors cannot carry out such tasks effectively, for much commitment to basic research puts the meeting of compliance deadlines at risk, dangerous in today's climate.

Most of the national laboratories confront large ranges of environmental problems on their own sites, which, while regrettable, can serve as test beds for the development of a broad spectrum of improved remediation, waste minimization, and cleanup technologies for application on far larger scales.

It may be important to designate lead laboratories for major programs to be established from among the laboratories to provide the synoptic view necessary for implementation of the scientific and technical studies and demonstrations necessary for a swift and efficient program. Most all of the national laboratories have important contributions to make to the EM program; a lead laboratory's role would be one of coordination and overall systems analysis and integration for a particular major effort. This does not mean assuming management responsibilities. The responsibilities fall to DOE management and its contractors and should remain there.

An additional benefit from designation of such lead laboratories is that they could become test beds for improvements in DOE regulatory and management practices and DOE Order compliance as well as for enhanced public participation. In brief, they can act as sites for valuable pilot programs, demonstrating the benefits of positive changes.

Formal institutional connections will be required with a number of other Federal bodies whose skills or whose regulatory authority relate to the tasks of the remediation program. These include the Environmental Protection Agency, the Department of Defense, the Bureau of Mines and others. A lead laboratory is the natural place for much of this linkage to be coordinated. Here is where special regulatory provisions must be hammered out so as not to hobble research and development work unnecessarily. Constraints on environmentally injurious activities necessary to "production" cleanup and remediation efforts are not always appropriate to research, where special relief is often required and typically difficult or impossible to get.

The recommendation to create lead laboratories could well arise naturally, in the wake of other beneficial changes, but it might be well to anticipate its arrival. The first task of one lead laboratory would be to organize the long-missing characterization of the remediation challenge mentioned earlier. This must be carried out with stakeholder participation for reasons discussed. It would be a major program, as it would require the participation of many of the Department's laboratories and EM sites. Thoughtful options would then soon appear.

There are difficulties [in] organizing laboratory participation. One is the need to ensure neutrality or to have a sure mechanism for dealing with real or perceived drift from neutrality. A second is the absolute need for strong leadership of the whole EM program. This lead laboratory cannot provide this leadership; it must come from above. Fortunately, resolving the second difficulty would go a long way to resolving the first.

3. The Nation

One consequence of the activities of the United States environmental movement is the massive environmental cleanup underway at numerous designated cleanup sites as well as at many other places in the nation.[5] There are 60,000 EPA Superfund sites, 2400 Resource Conservation and Recovery Act (RCRA) sites, 22,00 state-funded sites, and 7200 DOD sites. The total U.S. cleanup bill is estimated to be about $1.7 trillion. The program is going slowly. Of the $15 billion that has already been spent on Superfund cleanups (across the nation), roughly 75% has gone to legal fees and related costs. The need for more cost-effective cleanup has already become an urgent matter.

Many of the problems are very similar to those that DOE faces. In particular DOD, EPA, and others are struggling with the same technology and management issues as DOE. They will badly need the technical skills that a well-organized, technically competent DOE effort, with national laboratory help, could provide. For example, volatile organic compounds in arid and non-arid soils and ground water is one of the most common environmental problems in the US. Lawrence Livermore has already made important contributions to the technology of dealing with them.

There is abundant evidence for the beneficial role the national laboratories could play in helping resolve national problems in the numerous advances that they have already made. Ocean–climate interactions are being modeled by Los Alamos in support of Global Climatic Change studies with similar global and regional atmospheric modeling at Lawrence Livermore National Laboratory. Many of the laboratories have made contributions in the areas of environmental damage and resource base assessment and diagnostics.

The Department must take positive steps to encourage this attractive opportunity. It will, among others actions, have to consider reducing its cost-recovery fees levied on all "Work for Others." These fees now signal that contributions to the tasks faced by other agencies of government are not a high priority with the Department. The national laboratories could look forward to being available to the entire government system as a powerful environmental technical resource, a great national need. They should become in fact, as well as in name, national laboratories, saving our nation significant resources and improving cleanup efficiency. If the national laboratories do not fill this role, there will be no satisfactory alternative, and the need will remain substantially unmet. In any event, the experience base and the technological developments arising from the continuing EM program from the laboratories', industry's and research universities' contributions should be shared with the country on a continuing basis.

A broader vision sees the U.S. environmental and resource problems as a subset of many similar ones throughout the world. Science and technology must play a key role in coping with them. A strong DOE program would contribute at all levels. We are the nation best equipped to contribute solutions. Within the US, the Department of Energy marshals the best of these skills through its national laboratories, and they could be put at the world's service.

F. Concluding Remarks

The Atomic Energy Commission, and for many years the Department of Energy, broke the unwritten contract between these arms of government and the people they were to serve. The results, contamination on an enormous scale and a bitter distrust, imply a deep obligation to carry through the cleanup that has now been launched with efficiency, speed, and a decent respect for the opinions and needs of those affected. This cannot be accomplished as things are now. The changes required are clear; marshal the skills high in the DOE to bring about the managerial changes that are required; raise the quality of science and engineering in the program among other things by utilizing adequately the great power available in the national laboratories as well as the power among DOE contractors and in the universities. The changes only need to be set in place and exploited.

G. Recommendations

1. *Sustained improvements in DOE management and leadership are needed both at senior levels in the Department and in positions below the Deputy Assistant Secretary level.* It is clear from the above material that those portions of the problem that DOE can control stem from managerial deficiencies at the top levels in the Department.

2. *A comprehensive remedy to the array of problems plaguing the EM program can only be achieved by a substantial commitment and high priority to address the challenges of this program.* These must originate high in the Department. It seems clear that this must occur at the Under Secretary level. This does not imply disassembling the present EM structure under an Assistant Secretary. It does mean a technically adept, flexible and perceptive management of that and related efforts within DOE that acts with power.

3. *Closing the science/engineering applications disconnect should be dealt with by the establishment of an Environmental Advisory Board (EAB) reporting to the Under Secretary.* This should be a permanent Board and should include mostly scientists and engineers from within and without the Department and the laboratories, as well as stakeholders, to ease public acceptance of its recommendations. A good review capability could be provided by the EAB to identify needs so as to stimulate, with Department support, the required basic research, development and demonstrations. Such advances should then be applied by capable management to improve field remediation activities. The Board must have influence and

visibility in order to fulfill its role as an instrument of the Under Secretary. The High Energy Physics Advisory Panel (HEPAP) and the Nuclear Science Advisory Committee (NSAC) have such visibility, enhanced by their ability to give testimony to the Congress and their access to the Office of Science and Technology Policy. They are both widely believed to be quite successful. With members having a spread of skills, the Board should be able to provide technical oversight, flag management and regulatory disconnects as they arise and provide the synoptic view of the array of problems now lacking.

4. *The national laboratories together have a critical role to play, a role very much larger than at present, in performing high-quality science and engineering for the Environmental Management program.* Their principal contributions would be:

- Help to characterize the waste remediation across the DOE complex as a first step in helping the Department establish priorities for environmental work.
- Help communicate the technical challenges to the appropriate research communities.
- Help close the "Valley of Death" by aiding the construction of a seamless technology development chain.

The level of support necessary to implement major laboratory involvement as recommended here is small compared to the sums currently expended in the program. As an example, an increment of $400 million annually for laboratories with a ramp-up over time to twice that sum is roughly the scale needed to pursue research and development in an improved program. In view of the large fraction of the nearly $7 billion annual EM budget that clearly is misspent now, we see no serious difficulty in redirecting funds that are already flowing. No supplemental money should be required.

5. *The Department must take positive steps to make the national laboratories available to the entire government system as a powerful environmental technical resource.* They should become in fact, as well as in name, national laboratories. The Department must take positive steps to encourage this attractive opportunity. It will, among others actions, have to drop, or greatly lower, its cost-recovery fees levied on "Work for Others."

6. *DOE must address more forcefully the task of renegotiating the unrealistic or unfeasible elements of the cleanup compliance agreements that it has made with State and Federal agencies.* These are now impediments from risk management, technical feasibility, and public perception standpoints as well as forcing large and fruitless expenditures. The Federal government's Superfund legislation also incorporates unrealistic goals; legislation in 1993, which failed to pass, addressed many of the issues which make many current remediation schemes impractical and expensive. The new Congress, as well as DOE, should revisit the issue, benefitting DOE's remediation efforts and other cleanup under Superfund.

7. *Much more comprehensive involvement by members of the affected public in decision making should be employed to reduce the bitterness, distrust, and distress that continue to provide a troublesome element in DOE's conduct of its affairs.*

8. *The bulk of the EM environmental challenges, although presenting no immediate threats to public health or safety, still should be addressed with a heightened sense of urgency.* They have already been changing from acute to chronic problems are becoming calcified, and the vast flow of funds into the program acts as an anesthetic, numbing the Department, State regulatory agencies and affected stakeholders, hindering and delaying beneficial change.

References and Notes

1. "DOE Needs to Expand Use of Cleanup Technologies," GAO/RCED-94—9205.
2. "Cleaning Up The Department of Energy's Nuclear Weapons Complex," The Congress of the United States, Congressional Budget Office, Washington, DC, May 1994. This reference contains an extended discussion of DOE's managerial practices, its approach to risk assessment and to the incorporation of new technologies on remediation efforts.
3. Ibid.
4. For example, after the forced shutdown of Rocky Flats in the fall of 1989, acidic plutonium solutions were left in a half dozen tanks in one building, with concentrations up to 125 grams of plutonium per liter. They remain there to this day, with seals and gaskets deteriorating and occasional leaks occurring. It would have required two weeks to one month to process and eliminate the immediate risk. There [are] 70 miles of piping containing Pu-nitric acid solution with 30 kg of Pu in them.
5. *Business Week*, Aug. 2, 1993.

PART SIX
CURRENT ISSUES

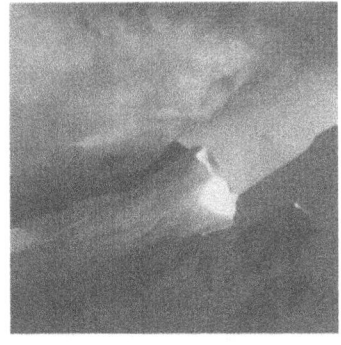

In early 1991, after Iraq had set fire to over 600 oil wells in Kuwait in the course of the Gulf War, the Union of Concerned Scientists held its winter board of directors' meeting in Washington. During the meeting I was handed a fax from Richard Garwin, a fellow experimental physicist and collaborator on several important UCS projects. In the message, he pointed out that the number of such fires was unprecedented, that the tasks of extinguishing them efficiently and rapidly might be advanced by some shrewd technical and scientific attention, and why didn't UCS organize a symposium to help achieve this? It struck a resonance with me. I passed the fax to the president of the John D. and Catherine T. MacArthur Foundation, Adele Simmons, a board member who happened to be sitting on my left with the remark that I wanted money. She read the note and handed it back, saying, "Yes." It was the most rapid and most successful fund raising I have been engaged in, clearly a model for other foundations to adopt, and it made it feasible for us to proceed. The symposium brought together a most interesting mix of national laboratory and academic scientists, explosives experts, fire fighters from Houston and oil operations managers from Kuwait. Some novel proposals emerged from the effort. A summary of the symposium, written in collaboration with Dr. Garwin is reprinted here.

One of the prospective troubles that we all expected to arise in dealing with the flaming wells was extensive booby-trapping with high explosive charges as well as mine fields containing antipersonnel and antivehicular mines. In the end, it turned out that this threat was minimal, and the symposium's attention to mine clearing and the several ingenious ideas that emerged never had to be explored or implemented.

Some of the problems of mine clearing, or demining as it is now called, stuck in my mind after the symposium, and I followed them up with some reading, as well as consulting with a number of experts. In Senate testimony, given in June 1991, I presented my conclusions along with suggestions for a better approach to demining in the aftermath of a conflict. This testimony, too, is included here.

Quenching the Wild Wells of Kuwait

Before the liberation of Kuwait by US forces and their coalition partners, the government of Iraq detailed about 1,000 military explosive experts and 30–40 engineers, together with the enforced assistance of people who worked in the oil fields, to sabotage and set fire to the 1,000 or so oil wells in Kuwait. Thirty to forty pounds of the military explosive C4 were used with sandbag tamping. Sabotage attempts were made on 732 wells and were successful in damaging and starting oil flows in 690 of them, about 640 of which were also set on fire: Data are scarce, but it seems that something near 2–3 million barrels per day were being lost initially, considerably less than the 6 million barrels per day estimated by Kuwaiti authorities, yet still representing a loss of as much as US$50 million per day.

The dollar loss was not the only consequence of this act of sabotage: the heavy flows damaged the underground reservoirs; plumes of smoke and combustion products from the flames created serious health and environmental problems, not only in Kuwait but at much larger distances; and lakes of oil developed, more than a metre thick and many square kilometres in extent, creating additional difficulties.

Iraqi land mines and unexploded coalition munitions in Kuwait were expected to add to the hazard and delay, as were administrative impediments. The scale of the disaster was unique: there had never before been more than five wells simultaneously out of control. Earlier experience had shown that an expert well-control company could typically kill and control a burning well in 2 weeks, although months might occasionally be needed. With an estimated 18 teams potentially available, it was believed that 700 wild wells would take several years to control. Numerous ideas were proposed by specialists to deal with the problems of bringing the oil fields under rapid control. But oil wells are far more complex than many realized and the vast majority of the ideas were poorly suited to the challenge. For example, the big problem is not in extinguishing the fires, but in stopping the flow of oil and gas with low risk to the well-control teams. Moreover, innovative techniques were thought to be required to deal with large numbers of mines in non-combat cir-

---

By Richard L. Garwin and Henry W. Kendall. Reprinted with permission from *Nature* **354** 11–14, (November 1991). Copyright © 1991 Macmillan Magazines Ltd.

cumstances. There was initial concern, later shown to be groundless, that wells would be booby-trapped. Mine fields were later found to have been laid outside, but not within, the oil fields.

We together with Howard Ris and the Union of Concerned Scientists, organized a symposium (later called Desert Quench) in Washington, DC, on 2–3 April this year, to bring together scientists and experts in wells and well control—including several people from the Kuwait Petroleum Company—to address these problems, discuss the usual solutions, and to search for means to expedite control of the wells and deal with challenges posed by the mines.

The problems are interesting, some remaining open despite the speed and efficiency of the capping-in effort so far. By early October, more than 550 wells had been brought under control, with [the] remaining 140 expected to be capped by the end of this week. Nearly 30 crew were used.

It is evident that we are not the most knowledgeable people about well control, but we write to provide some relevant information, and to indicate how people could contribute to the solution of such problems and similar ones.

The Classical Approach

The classical and well-practiced control of a wild well involves delivering cubic meters of water per minute to the equipment involved close to the fire, to keep it at operating temperature at a distance at which it would normally melt, and to cool the corrugated-iron shields behind and below which firefighters work to remove debris by the use of caterpillar-tracked bulldozers, cranes and large water-cooled hooks. Getting the fire jetting straight up is almost always the first order of business. For instance, Red Adair in his successful quenching of the Devil's Cigarette Lighter in Algeria (a blazing plume of 10 million m^3 of gas per day), first drilled water wells to fill polyethylene-lined reservoirs 10-ft deep and the size of three football fields, about 30,000 m^3 of water. (We hope that readers will tolerate a mixture of English and metric units, English being conventional in the oil fields. Conversion is readily achieved as follows: 1 (42-gallon) barrel = 0.159 m^3; 1 inch = 2.54 cm; 1 foot = 30.48 cm; 1 pound per square inch (p.s.i.) = 0.0703 kg cm^{-2} = 6,980 N m^2.) Before he snuffed out the fire, Adair cleared the site to table-top neatness and cooled the vicinity of the fire for days to ensure that there would be no re-ignition when the fire was put out. The flame is snuffed out in the classical method by exploding hundreds of kilograms of high explosive carefully shaped inside a thermally insulated oilfield drum supported by a crane at the base of the flame, and detonated from the safety of a trench or burrow at a distance of hundreds of meters.

Adair's biography, *An American Hero*, by P. Singerman (Little, Brown, 1990) is a highly readable account of the technology and of the remarkable people who have developed and practiced the demanding profession of control of wild wells, often with an unbeatable safety record. Not one of Adair's team has died in fighting wild wells. But the traditional approach typically takes much time in preparation, and much water; it had been practiced at most against five simultaneously burning wells, compared with the hundreds in Kuwait.

The Sabotaged Wells

The nature of the problems of gaining control of a sabotaged well can be understood by reference to Fig. 1. The well has a production casing of 7-inch diameter extending nearly to the total depth of the well, containing a 3.5-inch production tub-

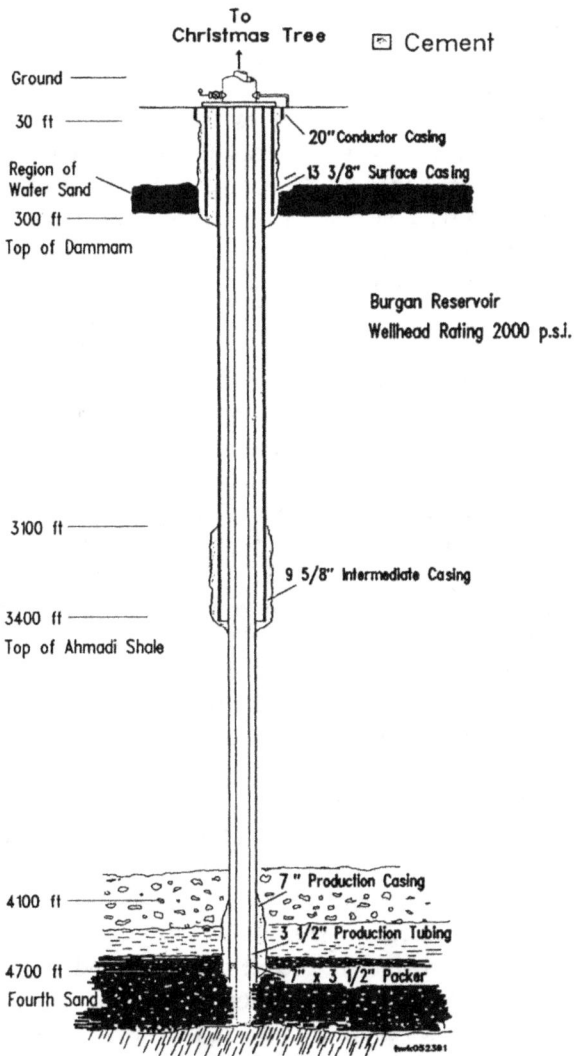

Figure 1. A typical oil well in the greater Burgan field, south of Kuwait city, not to scale.

ing; each tube is of wall thickness adequate for the expected capped pressure. Oil may be produced from the annulus between these two pipes in addition to the production tubing. Figure 1 shows the casing set for a typical well in the Burgan field of Kuwait, indicating the depth of the production zone and other characteristics. The wellhead pressure rating is 2,000 p.s.i., with a test pressure of 4,000 p.s.i. Some wells have wellhead pressures of more than 6,000 p.s.i. The conductor casing of steel tubing 20–30 inches in diameter is drilled or driven about 30 feet into the ground before well drilling begins to stabilize the drill rig.

The well itself consists of four to seven concentric tubes, including the two described above, that terminate at different depths and may have different pressure ratings. Tubes other than the production tubing are called casings and may be up to 20 inches in diameter. The lower end of each casing is cemented to the formation for a hundred feet or more to prevent flow between otherwise hermetic geological strata. The annulus just inside the largest casing may also be cemented. Annuli may also be closed permanently or temporarily by the use of packers near their lower ends, installed along with the next smaller casing, lining or tubing.

The wells in Kuwait are topped by "Christmas trees" of plumbing for delivering the oil into the gathering pipeline and for access for well maintenance and control (Fig. 2). The Christmas tree includes many features for control and safety. Blowout protectors, essential to safe drilling, are removed from completed wells, as are the derricks used for drilling and casing. The production tubing and the casings normally hang from flanges in the Christmas tree and, while cemented in at their lower ends, can be under great tension at the surface, the weight of steel—60 tons or more—corresponding to about 4 p.s.i. stress per foot of depth. The Christmas tree starts above the floor of a small concrete cellar, typically 2 m in diameter, extending 2–3 m below ground, which reduced the required derrick height during the drilling and casing of the well. Some Kuwait wells are "sour," yielding substantial H_2S, a gas lethal at concentrations of 200 parts per million, and one that leads to hydrogen embrittlement of high-strength steels. The petroleum from various wells in Kuwait has a great range of viscosity, and a gas content reportedly in the range 35–70 m^3 of gas at STP per m^3 of oil. This high gas content requires a gas-liquid separator under normal production conditions. Sand entrained in the flow at the well foot becomes more abrasive at the higher linear flow speed existing in the two-phase flow near the wellhead; under normal conditions sand-induced erosion of production tubing is monitored and the production tubing replaced when necessary.

After the Iraqi sabotage, it was reported that two-thirds of the wellheads had been blown off and the rest badly damaged. Some of the uncontrolled and burning wells needed only to have their valves closed, but in most cases substantial work had to be done first. In some the valves and wellhead facilities were removed by explosives above ground level, whereas others had the casing and tubing severed in the well cellar. With pipehangers gone the production casing and the production tubing will drop into the well, damaging it, but not stopping the flow. Many wells are gushing free from unimpeded bores at typical rates of one or two thousand cubic meters per day. Others, whose jets impinge on wreckage or which are venting laterally, are flooding oil at similar rates, which then burns (incompletely) near the ground in

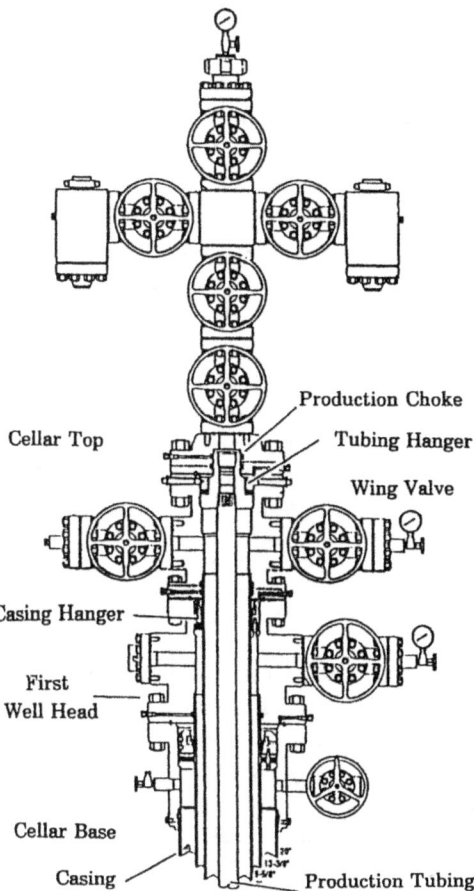

Figure 2. Wellhead valving and piping, known as the Christmas tree, typical of many in Kuwait. The cellar top is at ground level, the base about 5 ft in the ground. The production tubing is 3½ inches in diameter and successive casings are 9⅝, 13¾ and 20 inches in diameter, respectively.

great pool fires. The thermal power from a well flowing at this rate is about 300 megawatts. Conical mounds of hardened coke will have accreted around many of the wellheads. The well-known problem of clearing hundreds of tons of drilling rig from the well is absent on occasions like these, of course.

Flow Control

With destruction only of the gathering pipeline or of a valve on the Christmas tree above the lowest control valves, the well might be brought under control simply by closing one of these lower valves. But if there are no valves left on a production casing, and no flange, a massive fitting must be assembled to be fitted over the casing after preparation, cooling, and usually after snuffing the flame. This phase of the operation is fraught with danger, for example, from ignition by equipment spark, by lighting from dust or rain storms, or from static electricity.

With eroded pipe, a pit is usually dug to give access to the casing set at a depth at which sound casing can be found. Even if there is no erosion, the casing or casings may need to be exposed and then cut cleanly. The fitting (weighing at least a ton) must be pulled over the geyser in a controlled manner and pulled down over the casing, or clamped around the pipe. Grooved, tapered ramps, called slips, can be inserted between fitting and casing to prevent the large forces that will be present after flow is stopped from pushing the fitting off the casing. For a typical casing of 7-inch diameter and a wellhead pressure of 5,000 p.s.i., this amounts to about 100 tons of force. An appropriate packing or seal is used to retain the fluid after the valve is eventually closed. The high-flow Kuwaiti wells pose particular problems to the traditional approach, as the typical well will spill 1,000 tons of oil a day on the activities at the wellhead between the time the flame is snuffed and the flow is stopped, and may flood the area with poisonous and explosive gas.

Snuffing the Flame

Classically, flames are snuffed by the use of high explosive. More recently, as described at the symposium and used simultaneously by a well-control company in Kuwait, liquid nitrogen has been used to quench the flame in a controllable fashion, with clear advantage over any technique that uses explosives that would imperil people and equipment near the wellhead. Nevertheless, liquid nitrogen has been used only on wells of relatively small flow.

Many have proposed the use of fuel-air explosives instead of high explosives, or the use of Halon gas or even ferrocene to terminate the flame-propagating chemical chain-reactions. Although some of these techniques may be feasible and valuable, it remains true that existing techniques of snuffing flames are cheaper.

Snuffing the fire has traditionally been a prerequisite to stopping the flow, and difficult operations frequently have to precede snuffing. It would be highly advantageous to be able to stop the flow without (or before) snuffing the flame to avoid dealing with the flood of oil and gas that would both hinder work and pose a hazard of explosion. Three approaches are immediately apparent and were discussed by participants of Desert Quench.

1. Approaching the wellhead with a gathering culvert system, into which the geyser is suddenly diverted without snuffing the flame. This is being developed by several groups.
2. Tunneling below the surface to reach the casing set, drilling the production casing (and producing tubing, if necessary) by machines similar to those used at relatively low pressure in domestic distribution systems for hot-tapping an existing gas main, and inserting a stopper(s) adequate to stem or divert the flow. This is not viewed with favor by the drilling industry.

3. Inserting and fixing a hollow probe or stinger in the bore, which is then valved off. Such a technique has been in use for some time in undamaged wells and was used successfully, early on, in small, damaged wells in Kuwait. It became the technique of choice for the bulk of the well-capping and, with numerous small improvements, has been responsible for the great success of the effort.

In this method, a stinger equipped with appropriate valving is inserted into the flow and restrained by heavy downward force. Powerful pumps are then used to inject drilling mud into the well at a high flow rate until the weight of the mud column halts the flow from the well. Stingers can also be used in wells with flow both through the production tubing as well as through the production casing: the production tubing hanger is cut off and the tubing drops a few feet down, allowing a stinger to be used in the casing.

Other Proposals

Several proposals were introduced and discussed at Desert Quench for speeding up both access to mined and booby-trapped wells and bringing the unconstrained flows of gas and oil under control. A few were unrealistic or unworkable; we discuss some of these briefly here to report the unexpected obstacles that surfaced.

Mine Clearing by Robust Harrows

Conventional mine-clearing techniques are unsatisfactory for the circumstances in Kuwait, particularly because of the higher safety standards traditionally demanded for operations in peace time. The mines are virtually non-magnetic and could be equipped with anti-tamper fusing. Many land antitank mines consist of 10 kg of high explosive packaged in a disk-shaped plastic container, and are capable of driving a hole through the bottom of a heavy armored tank, and of killing passengers in military or civil vehicles. Such mines can be swept by ploughing, requiring exceptionally well-protected sweep vehicles to prevent injury to their operators. The plough structures rapidly degrade from the detonations. An alternative is to search for the mines on foot by probing with sharp rods. Complex fusing options, which may involve more than one signature, such as pressure or magnetic influence, or which count or use inert periods, make confident clearing uncertain. There have been no satisfactory clearing techniques for mines submerged under a layer or oil.

One potentially valuable approach to mine clearing, proposed by W. H. Wattenburg and tested at Yuma proving grounds by Lawrence Livermore Laboratories, uses a sled of chains that incorporates harrow-like blades that can reach to a depth of a foot or so, clearing a 20-ft path. The device digs up mines which either detonate, degrading the sled only slowly, or are entangled in the chains for later destruction. It is drawn by a helicopter at a safe distance. For area-clearing in Kuwait, the sled could be dragged back and forth among several winches, to reduce the cost below the $3,000 per hour typical of helicopter operations.

Mine Clearing with Compressed Air

Compressed air with sufficient pressure and flow could be used to clear away earth and mines rapidly with a relatively high degree of safety. To illustrate this concept at Desert Quench, S. A. Colgate described a unit with a compressor in the 10,000-hp range; air from a 4-inch nozzle at a pressure in the range 200 to 400 psi will have a flow of about 5000 cubic feet per minute and will develop a reaction force of more than a ton, sufficient to remove desert soil with great speed down to a depth of from 1–4 ft, as desired. Mounted at the end of a movable boom 30–50 ft long on a tracked vehicle with a protected cab, it might move at 100 ft per min. It would either detonate mines or transport them out of the way along with other debris. Used in conjunction with sled mine sweeping and with multiple passes over the ground, areas could be cleared to a reasonable level of confidence. It could be used to construct access roads, staging and work areas, and ditches and ponds for either water or oil. The device might dig a pit around a wellhead 60 ft in diameter and 20 ft deep in less than an hour. Not only could such an approach clear mines and booby traps in an area inaccessible because of radiant heat from the fire, it would also remove ground in the well head area heated by months of exposure to the flame.

Sand added to the air stream would provide an effective abrasive cut-off tool which would rapidly deal with oil well pipes and valving, reinforced concrete and other materials, facilitating debris removal and nondestructive preparation of the well head. Air-blast excavation and high flow air-abrasive cutting would benefit from increased theoretical and practical study, including the scaling laws applicable to currently available information.

Explosive Pipe Closure

Ring or cylindrical shell explosives can be used to pinch shut a pipe or multiple annular pipes. The explosive charge or charges must be selected, placed and detonated with more care than, for example, in pipe cut-off. The technique appears unsuitable for oil-well control because of the difficulty of ensuring no further damage. The production tubing and production casing may have suffered wall thinning from stream erosion that is difficult to assess; some of the wells are decades old and the metal of the pipes may have become embrittled from exposure to H_2S. Hence the task of selecting a charge that will confidently close the pipes without severing them becomes difficult, especially as the pipes are suspended under tension. Squeezing pipes shut inside a cemented casing set poses additional difficulty. In some cases, the task may be complicated by the transient pressures expected from the hydraulic ram effect induced by the near-instantaneous halt of the mixed oil–gas fluid in a couple of kilometers of well tubing and, even more important, shock-induced overpressure of the fluid in the pipe or annuli near the explosion that can induce rupture. With the very gassy oil in Kuwait, these transient overpressures would be mitigated.

Penetrating Bomb

Use of an air-delivered earth-penetrating bomb to seal off an injured well 50–100 ft below the surface suffers all the problems and more of explosive pipe closure, owing to the lack of precision with which the bombs can be delivered at depth, and to the inherent one-sided nature of the explosion. Severing a well below the surface converts a difficult situation into a disastrous one, in which the excess pressure in the sub-surface soil exceeds by far the available lithostatic pressure, and leads to the formation of a crater.

Shallow Relief Well

In cases where the stinger approach is for some reason unworkable, consideration might be given to using a very shallow intercept by a large-diameter, low-pressure, steeply angled well. Connection could be made in conventional manner by explosives at the foot of the relief well, blocking the original well so long as no impediment was posed to the flow of oil out the relief well and into the gathering system. Replaceable lining, resistant to abrasion, would probably be necessary for the relief well. In due time, re-entry could be made through the original well head with several options, including pulling the production tubing from depth and restoring the well.

Tunnelling to the Well

Efficient tunneling techniques could provide a very shallow intercept by a cased, large-diameter tunnel from a point 30–40 m or so from a damaged well to the cemented casing set of the well 30–50 ft below the surface of the ground. The tunnel shaft would be cased and a room mined, shored and cased to provide working space around the casing. To avoid potential explosion hazards from encountering gas or petroleum fumes, or from minor leaks in the apparatus, the tunnel might be kept inerted with CO_2, exhaust gas or nitrogen, with air supplied to masks or hoods for any workers. For the many low-pressure wells, one might bare the outermost casing and attach a saddle to allow drilling under pressure, removal of successive cores, insertion of stoppers and the like—all within the environment of a massive set of pressurized valves and tubes. This technique is essentially foreign to the well-control teams, who emphasize open and quick escape routes, but it might be practiced by those experienced in tunneling, with much help from those experienced in well control.

Flow Diverters

It is possible in principle to place a flow diverter over a well (extinguished or not) and conduct the oil-gas mixture into a long culvert-like pipe fitted with one or more sumps that would provide for separation of the gas from the oil, with the gas flared under control. This scheme temporarily halts the production of soot and other pollutants, and recovering the oil. Care would have to be taken to ensure that an explosion could not develop in the tube and that erosion of the diverter was controlled. The oil could be delivered to plastic-lined holding ponds, dirt tanks, constructed in the desert and using heat-sealed plastic sheets to prevent evaporation and seepage into the sand.

The flow emerging from an uncontrolled well is highly abrasive, in view of the entrained sand and the high velocity to which the gas content carries the fluid on expansion to near-atmospheric pressure. Means are required to retain the integrity of the elbow exposed to this environment; measures such as diamond-coating or massive self-feeding rubber blocks might be useful.

Stove Pipes and Liquid Nitrogen

Cylindrical tubes (3–8 m long) have been used to elevate the flame from a burning well much in the manner of a Bunsen burner. They do not require the cooling screen needed with the burner to prevent flashback; the relative absence of oxygen in the fuel and its high velocity result in the flame base remaining at the top of the tube. A tube hinged along one side, parallel with the tube axis, could prevent its interrupting the fuel stream during insertion but for the lengths so far used this has not been necessary. Liquid nitrogen was used in the first of the damaged Kuwaiti wells, a small one, to have its fire put out. A tube stove pipe was used and the fire extinguished and reignited several times to fine tune the process. Liquid nitrogen can be advantageous where water is in short supply.

Pollution from Burning Oil

Wells should burn more cleanly if there is better mixing with air (see R. Seitz, *Nature* **350,** 183; 1991), and North Sea drilling operations have long practiced the addition of sea water to the oil after the start of free flow of oil to the production platform and into the atmosphere, where the oil is burned to avoid pollution of the sea. The water clearly reduces the atmospheric pollution from the burning oil, and similar approaches could help in Kuwait.

Ducts or driven air can swirl a rising, flaring column of burning fuel. In small flames this produces a narrowing of the column, will raise the level at which burning of the fuel first starts, and appears to promote improved air-fuel mixing and therefore combustion efficiency. If successfully applied to a well fire, such flame control could decrease the radiant heat at the wellhead without the need to extinguish the fire. This would decrease or eliminate the production of pollutants.

Further research into the potential of vortex alteration of large flames would be valuable. On the other hand, if less effort is involved in stopping the flow from a typical well than in providing air or water entrainment, pollution could be reduced sooner by well control than by modification of the burn.

Gaining Control

Shields against the radiant heat of the flame have for decades been made like medieval shields, from corrugated steel, with handles. Spray from water monitors has been used to cool the shields and the firefighters, as well as the hot ground. It would be useful to have a light, thin, robust, multi-layer shield that would not be damaged by contact with water or oil. Reflective shields do not long remain reflective in the rain of combustion debris around a burning well. It would also be desirable to have a shield mat that could be placed on the ground to keep shoes from melting or burning.

For wild wells that feed a pool fire instead of a plume fire, control might most readily be achieved in two steps (without first snuffing the flame). First, a surface culvert could be built to approach near the wellhead, with oil diverted into it from the pool by means of a centrifugal pump placed remotely in the well cellar or in an expedient sump. A large lid could then be placed remotely over the wellhead area, excluding air and serving as a crude gas-liquid separator, with the gas taken through a large elevated culvert to be flared (burned) at a reasonable distance.

Conclusions

We have attempted to provide an analysis of the problems and constraints related to controlling the types of wild wells encountered in Kuwait. We hope that new ways of dealing with such catastrophes will emerge from our discussion. But it is difficult to prepare for a disaster that is likely to happen to somebody else, in a community in which interests are divided and even opposed. In the case of Kuwait, there have certainly been missed opportunities. For instance, if a share in the future production of several wells had been given to one company (and that replicated tenfold), with the proviso that the methods used to control their wells would be available to be practiced by others, progress might have been made more rapidly and with higher confidence. As $50 million a day was being lost in Kuwait, considerable money could have been saved by a fast response to the problem. Not only have important resources been destroyed by the action of Iraq, they have been unnecessarily squandered by an inability to permit technical measures to be implemented to control the wild wells of Kuwait.

Post-conflict Mine Clearing

The widespread mine placement that occurred during the 1900–1991 Gulf war may make access difficult and hazardous to large areas required in gaining control of the damaged oil wells in Kuwait. Portions of the Iraq–Iran and Iraq–Turkey borders, it is said, have also been extensively mined. The mines and booby traps no longer serve a military purpose—if indeed the majority of them ever did. They now must be removed, and the challenge this presents has some features, and requirements, normally not a part of mine clearance during wartime. Satisfactory techniques are not available to meet this challenge, so innovative thinking seems to be required. Moreover, there is no existing organization that can take on responsibility for meeting the need.

Dealing with post-conflict clearance of extensive mine fields is likely to arise, as it has in Iraq, Kuwait, Afghanistan, and Cambodia, with increasing frequency in today's world. The purpose of this note is to lay out the problems, review several candidate approaches, and set out some new ideas.

Mine Clearance in Wartime

Defensive mine fields encountered during combat must usually be penetrated so as to allow men and vehicles access to a mine-free region beyond the field. Relatively narrow passages typically will suffice for this; rarely will it prove necessary to clear large regions. The press of military necessity, including time urgency, permits the use of aerial or artillery bombardment as well as surface-delivered explosives to detonate mines in the selected corridors. World War II technology employing metal-locating electronic sensors [is] no longer satisfactory as mines now contain virtually no metallic parts. During war, mine clearing may frequently have to be carried out in the face of enemy suppressive fire, itself a potential source of casualties. Losses both of men and materiel unacceptable in peacetime may frequently be encountered and accepted in battle.

Testimony by Henry W. Kendall before the Gulf Pollution Task Force, Senate Environment and Public Works Committee, June 11, 1991.

Mine Clearance in Peacetime

After a conflict ends the scenario changes. Time urgency is less and the threat of enemy attack has vanished. Now, however, much larger areas may require clearing to permit normal, risk-free access as in the pre-war period. This implies thorough and credible mine removal. Even the urgency of gaining access to demand oil wells in Kuwait will not justify a residual risk as much as one tenth or one hundredth as great as what might be acceptable in war. Moreover, the clearing itself, very likely to be carried out by civilians, must itself be made far safer than what may be imposed on military personnel. The techniques usable by troops, while a starting point for peacetime operations, are lacking some essential ingredients, most especially safety but also low cost and, in some cases, low collateral impact. No really satisfactory means of clearing post-combat mines is presently available.

Few if any modern mines are fabricated to become inert after a set period, for example, one or two years. Most have tamper-fusing and can remain deadly for many decades.[1] Thus the absence of satisfactory peacetime techniques results in old battlefields frequently remaining semi-permanently forbidden to human presence and posing a continuing source of injury to wildlife.[2]

Because peacetime needs are so different from wartime ones, it is unrealistic to expect that the military will see to the development of the required procedures. An unfortunate situation results, which is the central problem: there is now a requirement for new clearing technologies but there is no existing institution or organization with skills, resources, and appropriate mandate to provide overall management and resources to develop them. This appears to be true both for the US as well as its allies.

Present Mine Clearing Techniques

There are a number of clearing techniques which have been employed in the past and whose level of success is well known. They generally fall into one of two categories:

1. Detect the mines and then remove or explode them. This method requires very sensitive and reliable detection techniques for any appreciable failure rate makes the approach unacceptable.

2. Ignore the detection phase and subject the area to be cleared to enough physical force to detonate or possibly destroy whatever mines may be present. This poses other problems. Anti-tank mines often require heavy overpressure to detonate them, equivalent to the weight of a heavy armored vehicle, and may also have delay or sequenced fuses that will detonate only after a set number of overpressures, for example after the passage of the sixth or seventh tank. It may be necessary to subject the area to enough force to physically destroy or remove the mine, whether or not it detonates. This may require severe overpressure that could cause collateral damage to installations or equipment in the area to be cleared.

In addition to the clearing methods mentioned above, there are:

1. **Vehicles with rakes or plows.** This will surely locate, and may explode, mines. Vehicle operators are at considerable risk and the vehicles and the plow structures suffer a high rate of attrition. This technique is not particularly well suited to peacetime operations, although [it is] occasionally employed.

2. **Explosive detonation of mines.** This is expensive, suited best to lane, rather than area clearance, and the detonations may entail unwanted destruction. Fuel–air explosives apparently do not work well enough (that is, there is not sufficient sustained downward pressure) to defeat "sequence" fuses or even to be certain of detonating anti-tank mines set for heavy pressure vehicles, and they won't produce enough force to physically destroy mines. Explosive foam—used by the South Africans—has the same difficulties. Long strings of detonating cord can be satisfactory if enough explosive is employed to destroy any mines in the treated area, but it is expensive and may cause unwanted damage.

3. **Manual clearing.** This classic technique is simple and hazardous: mines are sought by searching with sharpened probes by men on foot or on their hands and knees. Mines that are located in this way are then gingerly exposed and then detonated in place or even more gingerly removed. It is a procedure frequently employed but, in addition to the risk, it is slow and ill-suited to large area clearance. It does have the feature that it works and may become the means of last resort.

4. **Flails.** Giant flails, using chains to beat the ground ahead of a slowly moving vehicle, have been employed, much in the manner of the medieval flagellants updated by modern technology. These may, but will not surely, detonate mines. It appears poorly suited to peacetime use for it won't defeat sequence fusing or physically destroy mines.

5. **Rollers.** During [World War] II, the Soviets employed large rollers filled with concrete, which, pushed ahead of vehicles, were found to detonate mines in a satisfactory manner. The rollers proved moderately resistant to the explosions. For peacetime use the vehicles would need to be remotely controlled, for the vehicles themselves are at some risk, especially to mines that, intentionally, do not detonate on the first pressure signature. This technique also appears poorly suited to peacetime use, for it will not destroy mines that it doesn't detonate.

Prospective Mine Clearing Techniques

A number of new and innovative approaches to mine clearing have been proposed for postconflict clearing. While some have had limited testing, none is now available in fully operational status.

1. **Chain matrix (robust harrow).** Devised by W. H. Wattenburg, this scheme utilizes a chain mat that includes thin harrow blades that cut beneath the surface of the ground to a foot or so and either entangle mines or detonate them. It is dragged

by helicopter from a safe distance or pulled back and forth by winches. Detonations degrade its performance slowly. It, and its prospective use in Kuwait, are discussed at length by Garwin [and] Kendall[3] and by Wattenburg.[4] Potentially it appears quite attractive, especially for use in the desert.

2. **Compressed air jetting.** Devised by Stirling Colgate,[5] it employs a movable jet nozzle mounted on a vehicle that discharges a powerful blast of air with a reaction force in excess of a ton. This can cut away earth and other obstacles and either remove or detonate mines with low risk to the vehicle or its operator. A model constructed by Dresser-Rand has undergone preliminary and successful tests. Like the harrow, it continues to appear attractive; it too is discussed [by Garwin and Kendall]. Used in conjunction with the chain matrix, it appears to be the best and most reliable proposal.

3. **Ground-penetrating radar.** General Dynamics Corporation as well as the Israeli military have been studying the potential of small, hand-held radar units for detecting mines. The potential for success of this scheme is presently unknown, but it may prove satisfactory for mines that are not buried very deeply.

4. **Explosives detection.** Several approaches to the detection of explosives have been developed in connection with the task of identifying bombs in air transport luggage. They depend on the presence of hydrogen or nitrogen in the explosive material and one or more may have application to mine clearing.

   a. **Sniffing.** It is claimed that chemical "sniffers" can detect $1:10^{13}$ parts of specific nitrogen-bearing organic molecules in air and that this may form the basis of a new clearing technology. One of the potential problems is that it may prove *too* sensitive and unable to function properly for mine clearing. Moreover, many mines are quite well sealed, and it is not yet clear whether the approach is in principle practical.

   Several organizations have been developing such detectors and units are now apparently in commercial production for explosives detection. But sniffing is vulnerable to spoofing, wherein some material, to which the detector is sensitive, is spread about. The resulting false alarm rate could then become unacceptable and paralyze the search. It is said that the South Africans have trained dogs to detect a mine area by a somewhat circuitous method. The suspect area is sprayed with solvent and gas samples are taken from spaced, separate locations. Dogs kept in air-conditioned, filtered caravans are used to "sniff" samples (filter pads from air pumps) from each area in turn. The success rate, that is, identification rate, is reportedly very high. Even if successful, the technique only resolves a portion of the mine clearing problem.

   b. **Bremsstrahlung-induced gamma-*n* reaction in nitrogen.** The proposal, which has been laboratory tested, is to employ a portable 15-MeV electron linac and use it to irradiate land containing mines. The nuclear reaction produces a unique signature: a ten minute half-life positron emitter, N^{13}, an early portion of whose subsequent annihilation radiation may be observed easily in an array of nuclear gamma detectors. This has been carried out by Titan Corporation (Albuquerque, New Mexico). The project has been funded since 1987 from the Defense Advanced Research Projects Agency (DARPA).

c. **Fast neutron irradiation.** Fast neutron irradiation of a mine, which will contain hydrogen, will result in the production of some thermal neutrons, which can be detected by counters insensitive to fast neutrons. Cf^{252} furnishes the fast neutrons. Soil will not, in most circumstances, provide a false signal, but the environment must be dry and free of oil. Development is by Science Applications International Corp (SAIC) with funding [from] DARPA. The company claims that a prototype has been tested (at the U.S. Army's Fort Belvoir) and could be used immediately. A claimed advantage is that the device can be small enough to be hand held.

d. **Slow neutron irradiation.** Slow neutron irradiation will produce N^{15} by neutron capture. About 14% of these neutrons captured will produce a characteristic 10.8-MeV gamma ray. This work is being carried out by another branch of SAIC and has had DARPA funding.

None of the explosive detection schemes has been reduced to practice as a commercial and proven mine detector. Although some question exists that any will prove out, there is a chance that one or another might work.

A Proposed Clearing Technique

Should one of the latter proposals prove successful, then an effective means of deploying it would be desirable. The deployment would profit by having a very low probability of inadvertently detonating a detected mine. A vehicle with an extremely low ground pressure (which in addition could be remotely operated) can be built based on a scheme now employed to shift small boats. This scheme uses a fabric tube, a foot or so in diameter and several feet long, inflated to a low pressure, as a roller. With an axle and internal fabric spider-web supports, it would become a sort of extended, cylindrical tire, several of which could be employed as the "wheels" of a vehicle transporting the detector gear. Such a vehicle would serve as the platform for the mine detecting apparatus. A detected mine would have its position marked—by dye, for example—which would allow subsequent removal. The practicality of a system based on this is naturally dependent on the practicality of one or another of the electronic detectors.

Summary

In this paper we have focused mainly on the problem presented by larger anti-tank mines. Anti-personnel mines and unexpected munitions in Kuwait also are a problem, but development of satisfactory means of dealing with the larger mines represents a big step towards the neutralizing the smaller ones.

That post-war mine clearing presents challenges different from those met in combat clearing is not yet widely recognized and so little effort has gone into dealing with the difficulties. The challenges are now being faced in Kuwait and no doubt will continue to arise elsewhere. We are at a stage where resources should be devoted to the matter, if only to help speed up gaining control of the wild cell catastrophe in Kuwait.

References and Notes

1. It should be *required* by the rules of war that mines should explode or, preferably, expire so as not to be dangerous after a pre-set time.
2. In the aftermath of the Falklands conflict, in which very large numbers of mines were laid by the Argentinians before their defeat, little or no back-country clearance has been effected. The British Army conducts regular helicopter sweeps of parts of the islands to destroy wild horses who have had their legs blown off by mines. J. P. Kendall, private communication.
3. R. L. Garwin and H. W. Kendall, Controlling damaged oil wells in Kuwait, submitted to *Science*, April 29, 1991.
4. W. A. Wattenburg, Lawrence Livermore National Laboratory consultant, unpublished descriptions.
5. Stirling Colgate, Los Alamos National Laboratory, described in Garwin and Kendall 1991.

Epilogue

The panoply of problems that generation faces, the great impact they have—and will have—on human affairs, the likelihood that some may not be adequately resolved and that great sorrow will therefore be inflicted on our descendants raises deeply troubling questions. What should thoughtful people do? Are there means to brighten future prospects in the face of such human myopia and stubbornness?

There are many in the scientific community whose views are bleak. One biologist, speaking to me many years ago, remarked that many of his colleagues thought of the behavior of the human race as similar to a one-celled animal's. Such an animal may extend a pseudopod in search of food; if hurt or blocked, it pulls back, but otherwise its activity continues, mindlessly, without understanding, without end. With larger perils unperceived, it can destroy itself.

Over 40 years ago, the scientist Harrison Brown set out a searching inquiry into the human prospect. He too came to be deeply troubled about the future.

> It [humanity] is behaving as if it were engaged in a contest to test nature's willingness to support humanity and, if it had its way, it would not rest content until the earth is covered completely and to a considerable depth with a writhing mass of human beings.[1]

E. O. Wilson, the distinguished entomologist, has revisited the oft-raised question: Is humanity suicidal?[2] He is troubled, as we all are, by the traits that have led to such destructive impact on our world, our environment-devouring technology, our willingness to dismantle a global support system that is too complex to understand, and he wonders if our species has concluded that it is released from the "iron laws of ecology." His answer to the query as to whether humanity is suicidal is "No, but . . . " He concludes that we are intelligent enough and have enough time to avoid catastrophe but that there is no assurance that we will. He is surely correct.

There have been clear successes among the numerous efforts to control risks and diminish damage to ourselves and to our world. They include the termination of the nuclear arms race between the United States and the Soviet Union; the sweep of powerful legislation, national and international, that environmental damage directly affecting people has spawned; some efforts to preserve resources from destructive harvesting; and a slow but steady spread of understanding of the dangers posed by excessive procreation. Yet as I have set out in this volume, the most intractable problems, those whose consequences remain largely unperceived, remain unresolved.

All we who can gauge the threats can do is soldier on, exploiting what tools we have, gaining as much ground as time permits: seizing issues when they are ripe, remaining patient and careful with facts, even when faced with relentness and reckless opposition, mounting sustained campaigns and avoiding simple shots across the bow, combining solid analysis with persistent outreach and public education, touching people as widely as we can, and, as Winston Churchill emphasized, never giving up.

Many of us in science understand well what the costs of inattention and lack of care will be. The evidence offered in this volume should make this clear. Yet neither we nor others have yet caught the sustained attention of our fellow humans, and, until we do, the world cannot escape from its troubles. Thus the deepest question before us all is: How will our species reach the understanding and gain the political will to alter the prospects on the horizon? No one now has the answer to that need. It is indeed a distant light.

References and Notes

1. Harrison Brown, *The Challenge of Man's Future* (Viking Press, New York, 1954).
2. Edward O. Wilson, "Is humanity suicidal?" *The New York Times Magazine*, May 30, 1993.

Index

A

ABM. *See* Antiballistic missile systems
ABM Treaty of 1972, 140, 141, 158
Abrams, Herbert, 174
Absentee handicap, 147–148, 148
ABSP. *See* Agricultural Biotechnology for Sustainable Productivity
Accidental Missile Launch, 131
Accidental nuclear war, 165–176, 177–178
  factors contributing to possibility of, 165–166
  human weakness and, 174
  prevention by country of origin, 175
  removal of safeguards against, 169
Accuracy, of media reports, 6
Acid rain, 198, 225
Adair, Red, 286
AEC. *See* Atomic Energy Commission
Aerojet Nuclear Company, 52, 54, 55
  criticism of Atomic Energy Commission, 59–60
Africa, food demand in, 241
Agricultural biotechnology, 250
Agricultural Biotechnology for Sustainable Productivity (ABSP), 250
Agricultural productivity
  greenhouse effect and, 210
  land resources and, 106–108
  ozone depletion and, 211
  water resources and, 209–210
Agricultural technologies, 212–213
Agricultural use restrictions, distances for, 27

Agriculture, 238–239
  future of, 242
  pressures on, 226, 239–240
Air Cleaning as an Engineered Safety Feature in Light-Water-Cooled Power Reactors, 28
Air jetting, for mine clearance, 292, 299
Air pollution, 198, 225
Alternative Futures for the Department of Energy National Laboratories, 271–281
An American Hero, 286
Announcing Declaration on the Nuclear Arms Race (press release), 133–134
Antiballistic missile (ABM) systems, 140
Antiballistic Missile Treaty, 175
Antineutrino scattering, 93
Antiquarks, 90
Antisatellite weaponry (ASAT), 130, 160–161
Appeal by American Scientists to Prevent Global Warming, 183–184
  signatures of, 184–188
Arab-Israel war of 1973, 167
Arnold, Thomas B., 50
ASAT. *See* Antisatellite weaponry
Asia, food demand in, 240–241
Asymptotic freedom, 122
Atmosphere, 198, 225
Atom
  Rutherford's experiments on, 73–77, 74f, 75f, 76f
  structure of, 71

Atomic Energy Act of 1954, 66
Atomic Energy Commission (AEC),
    13–17, 66
  breakup of, 15
  criticisms of, 56–60
    from within, 58
    response to, 33–34, 49–50, 57, 60–61
  custody of nuclear weapons by, 171
  design criteria of, 43
  Division of Reactor Development and
    Technology, 55
  Interim Acceptance Criteria of, 54
  maximum accident scenario of, 61–63
  and nuclear waste disposal, 271
  press release on, 19–20
  Regulatory Staff of, 53–54
  Safety Program Plan of, 42
  safety standards of, 42
  Task Force of, 35
  use of mathematical modeling by, 46–47
  WASH-740 of, 25
  Water Reactor Safety Program Plan,
    34, 52–53

B

Bacillus thuringiensis (Bt), bioengineering
  and, 247, 260, 261, 262
Bacterial fermentation, 255
Ballistic missiles
  intercontinental
    flight of, 142–144, 143*f*, 143*t*
    full-scale attack with, 144
    launch-ready status of, 171, 172*f*,
      173*f*
    pop-up interceptors of, 145–148,
      146*f*, 147*f*, 151–153, 152*f*
  submarine-launched, 143*f*, 144
Bank for Environmentally Sustainable
    Development, 237
Baryons, 90
Battelle Institute, testing at, termination
    of, 60–61
BAU. *See* Business-As-Usual scenario
Bethe, Hans, 130
Beyond March 4, 9–11
Biodegradable plastic, 254
Biodiversity
  loss of, 181, 199, 226, 228
  maintenance of, 212

Bioengineering, of crops, 236–268
  agricultural, 250
  application to environmental problems, 254
  challenge of, 265–266
  contributions of, 252–255
  cost-to-benefit ratio of, 260–263
  current focus of, 250–251
  and decontamination, 254
  in developing countries, 250–251,
    252–254, 264
  diagnostic techniques of, 249–250
  and disease resistance, 257–258
  and ecosystem damage, 260
  and food production, 252–254
  and food quality, 252–254
  and gene markers, 248
  and gene transformation, 247–248
  and geography, 261–263
  and herbicide resistance, 253
  and human health issues, 250, 255
  length of effects of, 262–263
  of maize, 250, 252–253
  and nitrogen fixation, 251
  on-going research on, 249–250
  and pest control, 253
  and plant-produced drugs, 260
  problems with, 256–263
  pros and cons of, 244
  recommendations for, 263–266
  regulation of, 265
  research on, 264–265
  of rice, 251, 252
  and starch content increase, 254
  technology of, 247–255
  and tissue culture, 249
  and vaccines, 255
  and viral diseases, 258–260
  and weeds, 256–257
  and wild plants, 257–258
*The Bioengineering of Crops: Report of
    the World Bank Panel on
    Transgenic Crops*, 182
Bio-village program, 268
Bjorken limit, 106
Blair, Bruce G., 131
Boiling-water reactors, 38*f*
  emergency core cooling systems for,
    31, 42
  tests of, 44–45

Bolin, Bert, 190
Booster protection, 156
Boost-phase interception, 141, 142, 144
   countermeasures to, 153–156
      active, 155
      categories of, 155
      passive, 155–156
      threatening, 155
   definition of, 162
   initiation of, 145
   successful, 153, 154f
Boost-phase layer, 144–145
Bootstrap theory, 96
Born approximation, 103
Bremsstrahlung-induced gamma-n reaction in nitrogen, for mine clearance, 299–300
Brown, Harrison, 302
Brown's Ferry nuclear power plant, 16
Bulletin of the Atomic Scientists, 2
Business-As-Usual (BAU) scenario, for food supply, 214, 215–216, 216f, 217t, 227–230
Businessmen in the Public Interest, 13–14

C

Callan, Curt, 122
Callan-Gross relation, 121f, 122
Carbon dioxide
   and greenhouse effect, 210, 225
   stabilization of emissions, 190
Carter, Ashton B., 158
Case, Edson, 59
Cassava biotechnology network, 250
CERN. *See* European Organization for Nuclear Research
Cesium-137, 27
Chain matrix, for mine clearance, 298–299
Chance, 5
Charged particle beam weapons, 153
Chemical explosions, 27
Chernobyl, 67
Cherry, Myron M., 50
China
   agriculture in, 238–239
   food demand in, 241
   groundwater resources of, 209–210
China Accident, 24

Christmas tree tops, of oil wells, 288, 289f
Churchill, Winston, 9, 174, 303
CIMMYT. *See* International Wheat and Maize Improvement Center
Civil turmoil, and accidental nuclear war, 165
Classified information, 161
Clausen, Peter A., 140
Climate change
   limiting, 231
   report on, 242
   speech to Intergovernmental Negotiating Committee on, 189–191
Colgate, S., 292, 299
Collaborative Crop Research Program, 251
Colmar, Robert, 54, 56, 58
Comey, David, 13–14
Command destruct system, 175
Compressed air jetting, for mine clearance, 292, 299
Computer predictions, for emergency core cooling systems, 46–47
   criticisms of, 59
   disclosure to public, 47
Conant, James Bryant, 16
Constraints on the Expansion of the Global Food Supply, 181
Consultative Group on International Agricultural Research, 236
Consumption, of food supply, 205–206, 206t
Continuum, of inelastic scattering, 87
Cooling water, 37–48
   loss of. *See* Loss-of-coolant accident
Cooper, Robert S., 142
Core nuclear reactor
   geometry of, during loss-of-coolant accident, 42–43, 45
   meltdown of
      events preceding, 45
      rate of descent, 29
   schematic of, 51f
Cottrell, William, 56, 61
Coulomb's law, 73, 75f
Crops
   becoming weeds, 256–257
   bioengineering of. *See* Bioengineering, of crops

Crops (*continued*)
  choice of, 267–268
  management of, 267
  and post-harvest management, 267
  varieties of, 212
  and wild plants, 257–258
Current algebra, quarks and, 98–99
Cytoplasmic male sterility, gene transformation and, 248
Czechoslovakia, invasion by Soviet Union, 166–167

D

Declaration on the Nuclear Arms Race, 135–138
Decontamination, bioengineering and, 254
Deep inelastic scattering, 94–126
Deforestation, 213, 228
DeLauer, Richard D., 153, 156
Demining. *See* Mine clearance
Democratic process, 9–11
Department of Energy (DOE), and nuclear waste disposal, 269–281
Desert Quench, 290. *See also* Oil wells
DESY. *See* Deutsches Electronen Synchrotron
Detonation, of land mines, 298
Deuterium, in electron scattering, 81
Deutron studies, 122
Deutsch, Martin, 69
Deutsches Electronen Synchrotron (DESY), 110, 113f
Developed nations
  environmental responsibilities of, 233
  global warming responsibilities of, 191
  pollution and, 201
Developing nations
  bioengineering in, 250–251, 252–254, 264
  environmental responsibilities of, 233
Devil's Cigarette Lighter, 286
Diet modification, 211
Directed Energy Missile Defense in Space, 158
Directed energy weapons, 145
Disease resistance, 227
  bioengineering and, 257–258
  gene transformation and, 248
Dismantle Armageddon, 177–178
DOE. *See* Department of Energy

Drought resistance gene, 253
Drugs, plant-produced, 260
Dyson, Freeman, 16–17

E

EAB. *See* Environmental Advisory Board
ECCS. *See* Emergency core cooling systems
Economics
  and food prices, 238
  of nuclear power, 16–17, 66
  patterns of, 227
  of space-based missile defense system, 148–151
Ecosystem damage, 226, 228, 260
Eden, Anthony, 174
Editorial boards(s), 7
Eightfold Way, 97
Elastic scattering, 75f. *See also* Scattering
  cross-section of, 77, 87, 88f, 104f, 115f
  definition of, 77
  form factors in, 83–84
  radiative corrections and, 110, 111f
  vs. inelastic scattering, 87
Electric form factor, 83–84
Electric power shortage, 21
Electromagnetic force, 71–72
Electron
  spin of, 77
  weak force of, 80
Electron accelerator, at Stanford Linear Accelerator Center, 69, 71, 72f, 81–82, 82f, 99, 100f
Electron scattering. *See* Scattering
Electron-positron annihilation, 91–93
EM. *See* Office of Environmental Management
Emergency core cooling systems (ECCS), 40–42
  for boiling-water reactors, 42
  computer predictions for, 46–47
  effectiveness of, 14, 51–52
  evaluation criteria of, 54
  failure of, 21–22, 39–40
    consequences of, 23–28
    reasons for, 32
    total, 29
  flow blockage and, 32
  flow reduction and, 42
  leidenfrost migration and, 32

in loss-of-coolant accident, 28–34
press release on, 19–20
prima facie problem with, 31, 41
reliability of, 23
  engineering data pertaining to, 33–34
safety standards for, 42–46
schematic of, 38f
spray arrangement of, 31
steam expansion and, 32
tests of
  1970–1971, 29–30, 40–42
  conditions for, 30
  future, 42
  by Idaho Nuclear Corporation, 44–45
  mockup used for, 30, 30f, 41
  at Oak Ridge National Laboratory, 43–44, 52
  results of, 31–32, 41–42, 53
  termination of, 60–61
Emergency Core-Cooling Systems for Light-Water Cooled Power Reactors, 33
Energy issues, 179–180
Energy supply, 227
  expansion of, 231
  management of, 267
  use of, 228–229
Energy-intensive farming, 212, 266–268
Environmental Advisory Board (EAB), 279–280
Environmental damage, 228, 230
Environmental ethic, 201
Environmental problems, 198–199
  bioengineering and, 254
Ergen, W.K., 52
European Organization for Nuclear Research (CERN), 93
Evacuation, distances for, 27
Excimer laser defense, 148–150, 149f–150f
Explosive pipe closure, for oil well control, 292
Explosives detection, for mine clearance, 299

F

Facts, 5
Failure, 5
The Fallacy of Star Wars, 159–164
False alarms, of nuclear launch, 169, 170f

Farming
  energy-intensive, 212
  intensive, 266–268
Fast neutron irradiation, for mine clearance, 300
Fay, J., 3, 29
Federal Register, 35
Federation of American Scientists, 2
Fertilizer, sources of, 212
Feshbach, Herman, 2
Feynman, Richard P., 89–93, 104f, 119, 122
Fighting mirror, 147f, 148
Fish supply, 199, 226, 229
Fission
  products of, 26
  particulate, 27
  radiation produced by, 22
Flails, for mine clearance, 298
FLAVR SAVR tomato, 247
FLECHT. *See* Full Length Emergency Cooling Heat Transfer program
Fletcher, James C., 147
Flow blockage, and emergency core cooling systems, 32
Flow diverters, for oil well control, 293–294
Flow reduction, and emergency core cooling systems, 42
Food prices, 238
Food production, 226
  bioengineering and, 252–254
Food supply, 202–223
  and agricultural productivity, 206–211
  availability of, 205–206, 206t
  and bioengineering. *See* Bioengineering
  consumption of, 205–206, 206t
  current circumstances of, 237–245
  demands on, 240–241
  distribution of, 205
  effect of population growth on, 213–214
  future of, 214–218, 240–243
    Business-As-Usual scenario, 214, 215–216, 216f, 217t, 227–230
    maintaining, 243–245
    optimistic scenario, 214–215, 217t, 217–218
    pessimistic scenario, 214, 216–217, 217t
    prospects for, 243

Food supply (*continued*)
  history of, 204*f*, 204–205
  improvement of, 211–214
    agricultural technologies and, 212–213
    constraints on, 213–214
    nutrition and, 237–238
    population growth and, 202–206
    prospects for, 229
    quality of, bioengineering and, 252–254
    and soil depletion, 199
    stress on, 180–181
    sustainability of, 230–231
    world yields, 204*f*
Forbes, Ian, 13
Ford, Daniel F., 13, 14, 15–16
Forests, stress on, 199, 225–226
Form factors, 77–78, 79–80
  electric, 83–84
  magnetic, 83–84, 85*f*
  nucleon, 83*f*, 83–84
  proton, 85*f*
Fossil-fuels
  decrease of, 213
  decreasing reliance on, 231
Freedom Forum Environmental Journalism Summit, 6
Friedman, J., 70, 99
Fuel cladding, 24, 40
Fuel-rods
  blockage of, 44
  ruptures of, 43–44
Full Length Emergency Cooling Heat Transfer (FLECHT) program, 44–45, 52, 53
  termination of, 60

G

Galbraith, John Kenneth, 21
Galvin, Robert, 270
Garwin, Richard, 130, 283
Geiger, Hans, 73
Gell-Mann, Murray, 90, 97
Gene markers, 248
Gene transformation, 247–248
Genes, RFLP maps of, 251
Genetic engineering, 212. *See also* Bioengineering
Geography, and bioengineering, 261–263

Geosynchronous defense system, 148
Global food supply. *See* Food supply
Global warming
  speech to Intergovernmental Negotiating Committee on, 189–191
  UCS statement on, 183–188
Gottfried, Kurt, 2, 3, 129–130
Government
  environmental responsibilities of, 234
  technological revolution and, 9–11
Grain production. *See* Food supply
Great Nobel Debate, 180–181
Green Revolution, 204–205, 214, 231, 236, 263
Greenhouse effect, 2, 210
Greenhouse gas, 179–180
Gross, David, 122
Ground penetrating radar, for mine clearance, 299
Gulf War, environmental damages and, 283–295

H

Hadrons, 95
  structure of, 96–99
Haire, J. Curtis, 61
Hanauer, Stephen, 53, 54, 58
Harris, John A., 57
Heisenberg, Werner, 85–86
HEPAP. *See* High Energy Physics Advisory Panel
Hepatitis B vaccine, bioengineering and, 255
Herbicide resistance, bioengineering and, 253
High Energy Physics Advisory Panel (HEPAP), 280
High level waste (HLW), 273
Hofstadter, Robert, 69
Homing Overlay test series, 153, 154*f*
Homing weapons, 153
Human health issues, bioengineering and, 250, 255
Human weakness, and accidental nuclear war, 174
Hybrid seed production, gene transformation and, 248
Hybridization, dangers of, 257

Hydrogen fluoride lasers, 152
Hydrogen, in electron scattering, 81

I

ICBM. *See* Intercontinental ballistic missile
Idaho Nuclear Corporation, 44–45
  official reports of, 31, 41
IIFS. *See* Integrated Intensive Farming Systems
Illness, from radiation cloud, 26
ILTAB. *See* International Laboratory of Tropical Agricultural Biotechnology
India
  agriculture in, 238–239
  food demand in, 241
Industrial nations. *See* Developed nations
Inelastic scattering, 75f. *See also* Scattering
  continuum of, 87
  cross section of, 87, 88f, 89f, 115f
  deep. *See* Deep inelastic scattering
  definition of, 77
  Feynman diagram of, 104f
  neutrino beams and, 93
  from neutron, 90, 91f
  pion production and, 81
  from proton, 90, 91f
  radiative corrections and, 108, 109f
  resonance excitations and, 87
  structure function of, 116f
  theoretical implications of, 119–122
  vs. elastic scattering, 87
INS. *See* Integrated Nutrient Supply
Insecticides, plant-produced, 260
Institute for Defense Analyses, Jason Panel of, 2
Integrated Intensive Farming Systems (IIFS), 266–268
Integrated Nutrient Supply (INS), 267
Integrated Pest Management (IPM), 267
Intercontinental ballistic missiles (ICBM)
  flight of, 142–144, 143f, 143t
  full-scale attack with, 144
  launch-ready status of, 171
    Soviet Union, 172f
    United States, 173f
  pop-up interceptors of, 145–148, 146f, 147f, 151–153, 152f

Intergovernmental Negotiating Committee, of United Nations, speech to, 189–191
Intergovernmental Panel on Climate Change (IPCC), 2, 190
  Second Assessment Report, 242
Interim Acceptance Criteria, of Atomic Energy Commission, 54
  criticisms of, 54–55
International Crop Research Centers, 205
International Laboratory of Tropical Agricultural Biotechnology (ILTAB), 250–251
International Wheat and Maize Improvement Center (CIMMYT), 250
IPCC. *See* Intergovernmental Panel on Climate Change
IPM. *See* Integrated Pest Management
Irradiation, for mine clearance, 300
Irrigation, 209, 213, 239–240, 242

J

Jason Panel of the Institute for Defense Analyses, 2
Journalists, working with, 5–7
Junk science, 4

K

Kavanagh, George, 33, 37
Kendall, Henry, 193–194
Kill radius, 152, 152f
Kinematics, of scattering, 117–118, 118f, 121f
Kuwait, burning oil wells in, 283–295
Kwajalein Atoll, 153, 154f

L

Land mines. *See* Mine clearance
Land resources
  and agricultural productivity, 106–108
  degradation of, 207–208
  quality of, 207–208
  stress on, 225–226
  supply of, 206–207, 207t
Land restrictions, distances for, 27
Lasers
  excimer, 148–150, 149f–150f
  hydrogen fluoride, 152
  x-ray, 151–153, 152f

Latin America, food demand in, 241
Lauben, Norman, 58–59
Launch detection systems, 175
Launch-on-warning strategy, 157, 167–169
Laws of nature, 162
Lawson, C.G., 23, 33, 58
Least-squares, 117, 119f
Lebow, Richard Ned, 140
Lee, John M., 130
Leidenfrost migration, and emergency core cooling systems, 32
Lidsky, L.L., 65
Light water reactors, 22
Lipschutz, Ronnie D., 269
Liquid nitrogen, for oil well control, 294
Livestock management, 212
LLW. See Low level waste
Local observables, 98
LOFT. See Loss-of-Fluid Test
Loss-of-coolant accident, 21–22, 39–40
  computer predictions for, 46–47
  consequences of, 23–28
  core geometry during, 42–43, 45
  emergency core cooling systems in, 28–34
  maximum accident scenario, 61–63
  reasons for, 32
  time line of, 24–25, 40
Loss-of-Fluid Test (LOFT), 30, 41
Low, Francis, 2
Low level waste (LLW), 273

M

Magnetic form factor, 83–84, 85f
Maize, bioengineering of, 250, 252–253
Malnutrition, 237–238
Malthus, Thomas R., 204
Manhattan Project, 276
Marsden, Ernest, 73
Massachusetts Institute of Technology (MIT), 2, 69, 70, 94
Mathematical modeling, 46–47
Matter, wave description of, 85–86
McKnight Foundation, 251
Media, working with, 5–7
Meeting Population, Environment and Resource Challenges: The Costs of Inaction, 224–235

Mesons
  neutral vector, 86–87
  rho, 86
Metallurgical reactions, during reactor accidents, 44–45
Midcourse ballistic missile defense, 164
Middle East, food demand in, 241
Military chain of command, for nuclear weapons, 167
Mine clearance, 296–301
  in peacetime, 297
  techniques for, 291–292
    future, 298–300
    present, 297–298
  vehicles used for, 300
  in wartime, 296
Mineral deficiencies, 238
Mirrors, in space-based missile defense system, 151
MIRV. See Multiple independently targetable reentry vehicles
Missile event, 169
Mission statement, of Union of Concerned Scientists, 3, 9–11
MIT. See Massachusetts Institute of Technology
Momentum
  definition of, 77
  measurement of, 77
Momentum transfer
  and particle location, 76f, 76–77
  in scattering process, 74–76, 75f
Mott cross section, 77, 78f, 79f, 80f, 88f, 103
M.S. Swaminathan Research Foundation, 268
Multiple independently targetable reentry vehicles (MIRV), 136, 143f, 144
Muntzing, L.M., 56, 59, 61
Mutually Assured Destruction, 135, 139, 159–160

N

Nahavandi, Amir N., 59
Naive parton theory, 119–122
National Convocation on the Threat of Nuclear War, 129
National Reactor Testing Station, 29, 45, 52

National security, 127–131
Negative control, over nuclear weapons, 166
Netherlands Directorate General for International Cooperation, 250
Neutral vector mesons, 86–87
Neutrino beams, inelastic scattering and, 93
Neutrino scattering, 93
Neutron
   form factors of, 83f, 83–84
   inelastic scattering from, 90, 91f
   structure of, 71–93
   as target particle, 78–79
Newton's laws, 73, 75f, 77
Nitrogen fixation, bioengineering and, 251
Nixon administration, on national security, 141
Nixon, Richard M., 174
NORAD. *See* North American Air Defense Command
North American Air Defense Command (NORAD), 169
Novide, Sheldon, 39
NRC. *See* Nuclear Regulatory Commission
NSAC. *See* Nuclear Science Advisory Committee
Nuclear arms race, declaration on, 135–138
Nuclear ashes, 22, 37
Nuclear command system
   of Soviet Union, 167, 170f
   of United States, 167, 168f
Nuclear democracy, 96–97, 122
Nuclear hot line, 175
Nuclear power
   control of, 66
   economics of, 16–17, 66
   failure of, 64–68
      factors contributing to, 65–66
   future of, 66–67
   history of, 64–65
   reasons for use of, 21
   safety concerns with, 65–66
   UCS statement on, 16, 19–20
Nuclear Power (Press release), 19–20
Nuclear power reactor(s)
   heat output of, 52
   problems associated with, 22
   safety of, 13–17
      press release on, 19–20
   safety systems of, 14–17
   schematic of, 38f, 51f
   shielding of, 22
   size of, 28
   temperature of, 24, 40
Nuclear power reactor accident
   consequences of, 23–28, 51
   distance effected by, 50–51
   maximum accident scenario, 61–63
   metallurgical reactions during, 44–45
   possibility of, 50–53
   prior to refueling, example of, 25–26
Nuclear power reactor fallout, decay time for, 26
Nuclear Reactor Safety: An Evaluation of New Evidence, 21–36
Nuclear Reactor Test Station, 40–42
Nuclear readiness, lowering levels of, 171, 172f, 173f
Nuclear Regulatory Commission (NRC), 15
Nuclear safety, 49–63
Nuclear Science Advisory Committee (NSAC), 280
Nuclear war
   accidental. *See* Accidental nuclear war
   false alarms of, 169, 170f
   by unauthorized firings, 177–178
Nuclear waste disposal, 22, 269–281
   costs of, 273–274
   future of, 276–279
      within Department of Energy, 276
      by nation, 278–279
      by national laboratories, 277–278, 280
   goals of, 276
   history of, 271–272
   organizational problems with, 274–276
   program for, assessment of, 273–274
   recommendations for, 279–281
   technical challenges of, 272–273
Nuclear waste, volumes of, 273
Nuclear weapons
   coded locks on, 165
   control over, 166
   custody of, 171
   reducing number of, 174–175
   self-destruct devices on, 175
   with separable warheads, 174

Nucleon, 122–123
  form factors of, 83f, 83–84
  structure of, 96–99
  as target particle, 78–79
Nucleus, strong force of, 71
Nutrition, 237–238

O

Oak Ridge National Laboratory, 33, 43–44, 52
  criticisms of Atomic Energy Commission, 56, 59–60
  Nuclear Safety Program of, 56
  testing at, termination of, 60
OCD. *See* Quantum Chromodynamics
Oceans, stress on, 199, 226
OECD. *See* Organization for Economic Cooperation and Development
Office of Environmental Management (EM), 272
Oil wells
  Christmas tree tops of, 288, 289f
  control of
    classical approach to, 286–287
    techniques for, 292–295
  flame control of, 290–291, 295
  flow control of, 289–290
  flow diverters for, 293–294
  in Kuwait, 283–295
  pollution from, 294
  schematic of, 287f, 287–289
  shallow relief, 293
  tunneling into, 293
  typical, 287f
O'Leary, Hazel, 269
Optimistic scenario (OS), for food supply, 214–215, 217t, 217–218
Organization for Economic Cooperation and Development (OECD), 190
ORSTOM, 251
OS. *See* Optimistic scenario
Ozone depletion, 198, 211, 225

P

PAL. *See* Permissive action links
Panofsky, W.K.H., 119

Particle(s)
  location of, 76, 76f
  unstable, 86
  virtual, 84–87
Particle beam weapons, 153
Parton, 71, 92f, 119–122
Parton cloud, 90–91
Parton Model, 89–93
Peacetime
  mine clearance during, 297
  nuclear readiness during, 171
Penetrating bomb, for oil well control, 293
Permissive action links (PAL), 170–171
Pessimistic scenario (PS), for food supply, 214, 216–217, 217t
Pest control, 267
  bioengineering and, 253
  gene transformation and, 247–248
  improving, 212–213
Pesticides, effects on unintended targets, 260
Photons, 86
  scattering of, 93
Physics, laws of, 162
Pinstrup-Andersen, Per, 181
Pion production, and inelastic scattering, 81
Planck's constant, 76f, 76–77, 86
Plant-produced drugs, 260
Plastic, biodegradable, 254
Point quarks, 98
Political stability, and accidental nuclear war, 165
Pollard, Robert, 15, 16
Pollution
  air, 198, 225
  from burning oil, 294
  in developed nations, 201
  water, 198
Population
  current circumstances of, 237
  growth of, 199–200, 203f, 203–204, 227, 228
  effect on food supply, 202–206, 213–214
  stabilization of, 232
Pop-up interceptors, 145–146, 146f
  based on X-ray lasers, 151–153, 152f
  time constraint on, 147–148, 148f

Positive control, over nuclear weapons, 166
Post-Conflict Mine Clearing, 296–301
Post-harvest management, 267
Press releases
  Announcing Declaration on the Nuclear Arms Race, 133–134
  Nuclear Power, 19–20
  World's Leading Scientists Issue Urgent Warning to Humanity, 193–194
Price-Anderson Act, 66
Primordial quarks, 97
Proteins, plant production of, 255
Protons
  form factors of, 83f, 83–84, 85f
  inelastic scattering from, 90, 91f
  spin of, 79–80
  structure of, 71–93
  as target particle, 78–79
PS. *See* Pessimistic scenario
Pugwash, 2

Q

QCD model, 123
Quantum Chromodynamics (QCD), 95, 123
Quark-parton model, 123
Quarks, 71, 90, 95, 122
  1964 theory of, 97
  constituent picture of, 97–98
  and current algebra, 98–99
Quenching the Wild Wells of Kuwait, 285–295

R

Radar, for mine clearance, 299
Radiation
  gaseous, 26
  particulate, 26
  sources of, 22
Radiation cloud, 24–25
  distances covered by, 26–27
  ground-level, 26
  injury from, 26
  and temperature inversion, 26–27
Radiation sickness, 26
Radiative corrections, scattering and, 106–110, 107f, 111f, 112f, 113f–114f

Radioactive waste. *See also* Nuclear waste
  accumulation of, 51–52
  cleanup of, 269–281
  disposal of, 22
Rain, effects on radiation cloud, 27
Rain forests, depletion of, 199
Reagan Administration, 129, 130, 139–140, 156, 160
Refueling, nuclear accident prior to, example of, 25–26
Regge theory, 96
Regulatory Staff, of Atomic Energy Commission, 53–54
Relativity theory, 86
Resonance excitations, scattering and, 87
RFLP maps, 251
Rho mesons, 86
Rice
  bioengineering of, 251, 252
  wild, 258
Rice, C.M., 59
Rio Earth Summit, 180
Ris, Howard, 286
Rittenhouse, Phillip, 58
Road maps, of genes, 251
Robust harrows, for mine clearance, 291, 298–299
Rockefeller Foundation, 251, 252
Rollers, for mine clearance, 298
Rose, David, 61–62
Rosen, Morris, 54, 56–58
Rosenbluth cross section, 80f, 88f, 103
Rutherford, Ernest, 71
  experiments on atoms, 73–77, 75f, 76f
  equipment used for, 74f
Rutherford formula, 77, 78f, 80f

S

Safeguard system, 141
Safety Program Plan, of Atomic Energy Commission, 42
Safety standards, 42–46
Salinization, of agricultural land, 239–240
SALT. *See* Strategic Arms Limitation Treaty
Scale invariance, 106

Scaling, and scattering, 92*f*, 110–117, 116*f*, 117*f*, 120*f*
Scaling variables, 106
Scattering, 71
  antineutrino, 93
  deep inelastic, 94–126
  deuterium in, 81
  elastic. *See* Elastic scattering
  form factors and, 77–78, 79–80
  formalism, 102–110
  fundamental processes of, 102–106
  hydrogen in, 81
  inelastic. *See* Inelastic scattering
  kinematics of, 102*f*, 102–103, 117–118, 118*f*, 121*f*
  momentum transfer and, 74–76, 75*f*
  neutrino, 93
  by object of finite dimensions, 77, 79*f*
  of photons, 93
  probability of, 73
  radiative corrections, 106–110, 107*f*, 111*f*, 112*f*, 113*f*–114*f*
  reasons for, 84
  resonance excitations and, 87
  results of, 110–118
  and scaling, 92*f*, 106, 110–117, 116*f*, 117*f*, 120*f*
  spectrometers and, 81–82, 82*f*, 100, 101*f*
  spectrum of, 83*f*
  Stanford program for, 99–102
  theoretical situation of, 99
  triangle in, 107*f*, 107–108, 108*f*
  types of, 77–81
  typical experiment on, 81
  universal curve for, 90, 92*f*
Scattering cross section, 73–74, 75*f*
Science
  and environmental responsibilities, 233
  limitations of, 163
  misuse of, 4–5
Scientific community
  involvement in public problems, 1–7
  responsibilities of, 10–11
Scientific principles, 162
Scientific truth, 6
Scientists' Declaration on the Threat of Nuclear War, 129

The Scripps Institute, 250–251
Seaborg, Glenn T., 35
Sea-launched cruise missiles, accidental launch of, 171
Second Assessment Report, 242
Secrecy, regarding Star Wars, 161–164
Serageldin, Ismail, 181–182, 237
Shallow relief wells, 293
Shaw, Milton, 47, 55, 59
Shielding, of nuclear power reactors, 22
Shock waves, 45
Shurcliff, William, 5
Simmons, Adele, 283
60 Minutes, 15–16
SLAC. *See* Stanford Linear Accelerator Center
SLBM. *See* Submarine-launched ballistic missile
Slow neutron irradiation, for mine clearance, 300
S-matrix theory, 96, 97
Sniffing, for mine clearance, 299
Soil
  care of, 266
  stress on, 199
Soil conservation, 212
Soil erosion, 207–208, 208*t*, 228, 239
Sound bites, 6
Soviet Union
  and accidental nuclear war, 165–166
  and evolution of space weapons, 159–161
  intercontinental ballistic missiles of, 172*f*
  invasion of Czechoslovakia, 166–167
  and nuclear arms race, 135–138
  nuclear command system of, 167, 169, 170*f*
  nuclear hot line with U.S., 175
  response to space-based ballistic missile defense system, 157–158
Space weaponry, evolution of, 159–161
Space-based ballistic missile defense, 139–158. *See also* Star Wars
  cost of, 148–151
  effectiveness of, 142
  energy needed for, 148
  midcourse, 164
  number of mirrors in, 151

outcomes of, 142
political implications of, 164
and pop-up interceptors, 145–148, 146f, 148f
proposal of, 130
rationales for, 157–158
Soviet response to, 157–158
successful test of, 153, 154f
terminal phase, 164
Spectrometers, electron scattering and, 81–82, 82f, 100, 101f
Spin
of electrons, 77
of proton, 79–80
Spin-one partons, 122
Spin-zero partons, 122
Stanford Linear Accelerator Center (SLAC), 69, 71, 72f, 81–82, 82f, 94–95, 99–102, 100f, 101f
Star Wars. *See also* Space-based ballistic missile defense
assessment of, 164
fallacy of, 159–164
misconceptions about, 161–164
proposition of, 160
systemic problems with, 164
Starch content, increase of, bioengineering and, 254
Steam expansion, and emergency core cooling systems, 32
Stove pipes, for oil well control, 294
Strategic Arms Limitation Treaty (SALT), 129
Strategic Defense Initiative, 139–140. *See also* Space-based ballistic missile defense; Star Wars
Stress, and accidental nuclear war, 174
Strong force, 71
Strontium-90, 27
Submarine-launched ballistic missile (SLBM), 143f, 144
Sub-Saharan Africa, food demand in, 241
Sustainable food supply, 230–231
Synergistic effects, 229–230

T

Target particle, 78–79
Taylor, R.E., 70, 99

Technology
control of, 9–11
and environmental responsibilities, 233
limitations of, 163
Temperature inversion, radiation cloud and, 26–27
Temperature, of nuclear power reactor, 24, 40
Terminal phase ballistic missile defense, 164
Theoretical Possibilities and Consequences of Major Accidents in Large Nuclear Power Plants, 25–26
Third Annual World Bank Conference on Effective Financing of Environmentally Sustainable Development (1995), 224, 236
Threat cloud, 145
Three Mile Island, 16
Tissue culture, bioengineering and, 249
Transuranic waste (TRU), 273
TREAT (experimental reactor), 43–44
Triangle, in scattering, 107, 107f, 108f
Tri-Party Agreement, 272, 274, 276
Tropical rain forests, depletion of, 199
TRU. *See* Transuranic waste

U

UCS. *See* Union of Concerned Scientists
Ultra-violet radiation, 198
Uncertainty Principle, 85–86
Union of Concerned Scientists (UCS)
ASAT program of, 130
on global warming, 183–188
history of, 2–3
mission statement of, 3, 9–11
on national security, 127–131
in 1970s, 13–17
statement on nuclear power, 16, 19–20
World Scientists' Warning to Humanity, 195–201
United Nations Intergovernmental Negotiating Committee, speech to, 189–191
United States
intercontinental ballistic missiles of, launch-ready status of, 173f
nuclear command system of, 167, 168f
nuclear hot line with Soviet Union, 175

Universal curve, for scattering, 90, 92f
Unstable particles, 86
USAID, 250

V

Vaccines, and bioengineering, 255
Valence quarks, 97, 122
"Valley of Death," 275
Vandenberg, Arthur, 164
Vector-Meson-Dominance (VMD) theory, 97, 122
Vegetarian diet, 211
Vietnam War, 2
Violence, reduction of, 201, 231
Viral diseases
   bioengineering and, 258–260
   concerns regarding, 259–260
   pathogen derived resistance, 259
Virtual particles, 84–87
Virtual photons, 86
Virtual protons, 105
Vitamin deficiencies, 238
VMD. *See* Vector-Meson-Dominance theory

W

Waddell, R.D., Jr., 43
War, reduction of, 201, 231
Wartime, mine clearance during, 296
WASH-740, 25–26
WASH1400, 15
Waste accumulation, 51–52
Waste disposal. *See* Nuclear waste disposal
Water conservation, 212, 266–267
Water harvesting, 266–267
Water pollution, 198
Water Reactor Safety Program Plan, 34, 52–53, 55, 61
   computer coding in, 59
Water resources
   and agricultural productivity, 209–210
   irrigation and, 209
   stress on, 198, 226
   use of, 209
Water shortage, 209–210, 229, 240
Waterhammer effects, 45
Wattenburg, W.H., 291, 298–299
Wave description of matter, 85–86
Weak force, 80
Weapons fallout, decay time for, 26
Weather conditions, effect on accident spread, 26, 27
Weeds, bioengineering and, 256–257
Weinberg, Alvin, 56
Weisskopf, Victor F., 2, 3
Wellheads, 288, 289f
Wild plants, bioengineering and, 257–258
Wild rice, 258
Wilson, E.O., 302
World Bank, 224
World Scientists' Warning to Humanity, 181, 198–201
   summary information, 195–197
World War II, scientific involvement in, 1–2
World's Leading Scientists Issue Urgent Warning to Humanity (press release), 193–194

X

X-ray lasers, as pop-up interceptors, 151–153, 152f

Z

Zr-3, 44–45
Zr-4, 44–45
Zr-5, 44–45
Zweig, George, 90, 97

GPSR Compliance

The European Union's (EU) General Product Safety Regulation (GPSR) is a set of rules that requires consumer products to be safe and our obligations to ensure this.

If you have any concerns about our products, you can contact us on

ProductSafety@springernature.com

In case Publisher is established outside the EU, the EU authorized representative is:

Springer Nature Customer Service Center GmbH
Europaplatz 3
69115 Heidelberg, Germany

www.ingramcontent.com/pod-product-compliance
Lightning Source LLC
LaVergne TN
LVHW010336260326
834688LV00036B/738